elements of
INFRARED
TECHNOLOGY:
generation, transmission, and detection

PAUL W. KRUSE

Staff Scientist
Honeywell Research Center

LAURENCE D. McGLAUCHLIN

Research Section Head
Honeywell Research Center

RICHMOND B. McQUISTAN

Associate Professor of Electrical Engineering
University of Minnesota
Consultant
Honeywell Research Center

JOHN WILEY & SONS, INC., NEW YORK · LONDON

To our wives and parents

Preface

The infrared region of the electromagnetic spectrum is generally defined as those wavelengths lying between the visible and microwave regions. Its importance arises from the fact that every material object emits, absorbs, transmits, and reflects infrared radiation in a characteristic manner. This has a twofold implication. First, from a study of the intensity and wavelength distribution of the radiation which has arisen from or interacted with an object, information concerning the object may be obtained. This information may be used, for example, to distinguish a body from its surroundings or to identify an unknown material. Second, the radiation is of itself important. The manner in which it interacts with a material may be used to change the characteristics of that material. Or again, it may be employed as a medium for conveying information from point to point.

Although the existence of infrared radiation was realized well over a century ago, only in the last two decades have applications been widespread. The field of infrared technology, the use of infrared radiation for military and commercial purposes, may be considered to have originated in World War II. Since then interest has increased greatly to the point where infrared technology has become of major military importance. Although the commercial applications have not seen such explosive growth, they have nevertheless increased steadily in recent years.

At the time the manuscript for this book was prepared, little organized information related to infrared technology was available. The existing literature treated selected topics in this area, but did not provide an integrated presentation of the entire subject. This work is an endeavor to fulfill this need.

Broadly speaking, the field of infrared technology can be subdivided into four main categories. The first deals with the nature of infrared radiation. Examples of topics in this area are the spectral and angular distribution of radiation emitted from heated bodies; the reflection, refraction, absorption, diffraction, and scattering of radiation by media; and the several photoelectric effects. The second major category is comprised of infrared components and materials including sources, window materials, and detectors.

The integration of infrared components into systems constitutes the third major category. These include optical, electronic, and cooling systems. Applications, both commercial and military, make up the fourth and final category. This book treats the first two categories. The last two are left to a companion volume now in preparation.*

It is our intent in this volume to provide a discussion useful to two different groups of readers. One of these groups is composed of senior or first-year graduate students having an academic interest in the subject. The other group consists of scientists and engineers in both commercial and military laboratories who are engaged in the development of infrared components and the evaluation of materials and components for use in infrared applications. Accordingly, for the sake of logical completeness, we have included some material which can be found in books treating subjects other than infrared technology. Inclusion of this material will allow the reader to understand the principles underlying the behavior of infrared components while relying on a minimum number of assumptions. As examples of this, we have included a mathematical discussion showing the relationships between the optical constants and more basic parameters; we have developed expressions describing the attenuation of radiation by the atmosphere; and we have treated the interactions between radiation and charged particles leading to such phenomena as dispersion and free carrier absorption. In addition, we have included some material as a matter of convenience for those working in the field of infrared technology. Thus we have included a chapter on the physics of semiconductors to enable the designer of infrared detectors and optical components to make predictions concerning the behavior of new materials; we have treated fluctuation phenomena to clarify the nature of those effects which set a fundamental limit to the performance of detectors and the electronic components of electrical systems; and we have included sections describing the measured characteristics of detectors and optical materials.

It is a pleasure to acknowledge the encouragement and cooperation given us by Minneapolis-Honeywell. We were greatly aided by the diligence and care of Mrs. L. Lehr, who often deciphered illegible script and

* Elements of Infrared Technology: Systems and Applications.

transformed it into well-organized manuscript. We are also indebted to John F. Ready for his painstaking review of the entire work and his numerous helpful comments. Mrs. E. L. Swedberg and Mrs. R. T. Squier provided valuable assistance in mathematical computations and the preparation of figures. We should like to acknowledge the helpful comments of Dr. W. N. Arnquist, Dr. S. S. Ballard, Dr. C. Hilsum, and Dr. J. N. Howard, and the cooperation of Dr. D. E. Bode, Dr. P. H. Cholet, Dr. M. Garbuny, Dr. R. M. Langer, Dr. H. Levinstein, Mr. G. McDaniel, and Dr. R. H. McFee. Finally any acknowledgment would be incomplete without mention of our wives, whose constant encouragement and long hours of proofreading contributed greatly to the manuscript.

PAUL W. KRUSE
LAURENCE D. MCGLAUCHLIN
RICHMOND B. MCQUISTAN

Honeywell Research Centre
October 1961

Contents

List of most important symbols

Symbol	Definition	Page
$\mathbf{E}$	electric field vector	89
E	radiant intensity	12
E_λ	spectral radiant intensity	12
$\mathscr{E}$	energy	24
$\mathscr{E}_0, \mathscr{E}^*$	Fermi level energy	24, 205
f	electrical frequency	236
Δf	electrical bandwidth	236
g	conductance	246
g_m	transconductance	248
h	Planck's constant, 6.6252×10^{-34} joule sec	25
$\mathbf{H}$	magnetic field intensity vector	89
$\mathscr{H}$	irradiance	13
$i_N, (\overline{i_N^2})^{1/2}$	rms noise current	239
$i_{S,\mathrm{PC}}$	photoconductive short circuit current per unit detector width	328
$i_{S,\mathrm{PEM}}$	photoelectromagnetic short circuit current per unit detector width	333
I_p	plate current	244
j	imaginary number, $\sqrt{-1}$	92
$\mathbf{J}$	current density vector	89, 211
$\mathbf{J}_e$	electron current density vector	225
$\mathbf{J}_h$	hole current density vector	225
J_s	short circuit current density	341
k	Boltzmann's constant, 1.3805×10^{-23} joules/deg K	23
k	absorption constant	94
k_0	thermal conductivity	349
K	thermal conductance	346
K_e	relative capacitivity	93
K_e	effective thermal conductance	347
K_e^*	complex dielectric constant	94
K_m	relative permeability	93
K_0	thermal conductance at ambient temperature	346
L_D^*	ambipolar diffusion length in a magnetic field	328
L_e	electron diffusion length	227
L_h	hole diffusion length	227
m	reduced dimensionless variable	330
m_e	effective mass of a free electron	205
m_h	effective mass of a free hole	207
M	gram molecular weight	70
M	electric moment	108
$M(b)$	density of particles whose radii are between b and $b + db$	182
$M(\nu, T)$	Planck distribution function	58
n	index of refraction	45
n	electron concentration in conduction band	206
$n(\mathscr{E})$	fraction of particles having energy in the interval $\mathscr{E}$ to $\mathscr{E} + d\mathscr{E}$	26
n_a	concentration of absorbing elements	102
$\bar{n}_a$	average absorber concentration	168
n_e	carrier concentration	126
n_i	intrinsic concentration	207
n_n	electron concentration in n-region	232

Symbol	Definition	Page
n_p	electron concentration in p-region	232
n_s	concentration of scattering elements	107
n_0	electron concentration at thermal equilibrium	326
n_{p0}	electron concentration in p-region at thermal equilibrium	338
N	number of photons per unit front surface area per second absorbed by detector	327
$\overline{N}$	average number of photons per second per unit area	58, 357
$\overline{N^2}$	mean square deviation in number of photons per second per unit area	357
N_a	concentration of acceptors	209
N_d	concentration of donors	208
N_s	total number of scattering particles	107
p	hole concentration in valence band	207
p_n	hole concentration in n-region	232
$p_N(f)$	mean square deviation in number of emitted photons per unit bandwidth	60
p_P	mean square deviation in emitted power	59
$p_P(f)$	mean square deviation in emitted power per unit bandwidth	59, 351
$\overline{p_P^2}$	mean square deviation in emitted power in a given electrical bandwidth	59, 351
p_0	hole concentration at thermal equilibrium	325
p_{n0}	hole concentration in n-region at thermal equilibrium	340
P	radiant power emitted by or incident upon a surface	12
P	electrical polarization	105
P_N	noise power	236
P_N	noise equivalent power	270
P_λ	spectral radiant power emitted or absorbed	12, 329
q	electronic charge	104
Q	quality factor of narrow band quantum counter	372
r	amplitude reflection coefficient	100
$r_{\parallel}$	amplitude reflection coefficient for parallel polarized radiation	183
$r_\perp$	amplitude reflection coefficient for perpendicularly polarized radiation	183
R	electrical resistance	236
R	radiant emittance	12
R	universal gas constant	70
$\mathscr{R}$	responsivity	272
R_B	bolometer resistance	346
R_L	load resistance	291
R_N	equivalent noise resistance	246
R_λ	spectral radiant emittance	13
$\mathscr{R}_\lambda$	spectral responsivity	272
R_ω	radiance	13
R_{bb}	black body radiant emittance	29
$R_{\omega\lambda}$	spectral radiance	13
s	surface recombination velocity	222
s_1	front surface recombination velocity	326
s_2	back surface recombination velocity	326
S	scattering area ratio	182
S	intensity of an absorption line	169

Symbol	Definition	Page
t	amplitude transmission coefficient	101
t_r	radiative lifetime	69
T	absolute temperature	17
T_c	cathode temperature	245
T_n	equivalent noise temperature	239
T_1	detector temperature	346
T_2	background temperature	357
T_3	black body source temperature	362
$\overline{\Delta T^2}$	mean square deviation in temperature	355
$\overline{\Delta T_f^2}$	mean square deviation in temperature per unit bandwidth	355
$v_N, \left(\overline{v_N^2}\right)^{1/2}$	rms noise voltage	237
v_{R_L}	voltage developed across load resistor	348
v_0	open circuit voltage per unit detector width	329
V	applied voltage	346
V	visual range	190
V_p	plate voltage	245
V_0	open circuit voltage	341
w	water vapor concentration	177
w_i	value of w at which the absorption in window i undergoes a transition from weak band to strong band absorption	177
Z	atomic number	61
α	absorptivity	13
α	generalized temperature coefficient of resistance	346
α	polarizability	105
α_t	absorptance	139
α_1, α_2	reduced dimensionless variables	330
β	scattering coefficient	102
β	temperature coefficient of resistance of a semiconductor	346
β'	scattering coefficient per unit concentration (scattering cross section)	102
γ	temperature coefficient of resistance of a metal	310
$\gamma_\lambda(\phi)$	differential scattering cross section	107
Γ	propagation coefficient	93
Γ'	complex index of refraction	94
δ	Dirac delta function	58
δ	half-width of an absorption line	169
ε	emissivity	13
ε	absolute capacitivity or absolute permittivity	90
ε_0	capacitivity of free space	91
ε_{eff}	effective hemispherical emissivity	40
$\varepsilon_{\lambda\text{hem}}$	spectral hemispherical emissivity	40
$\varepsilon_{\omega\lambda}$	spectral goniometric emissivity	40
$\varepsilon_{n\omega\lambda}$	normal spectral emissivity	46
η	reduced energy	206
$\eta(\nu)$	quantum efficiency	357
η^*	reduced Fermi energy	206
η_a	reduced acceptor ionization energy	210
η_d	reduced donor ionization energy	209
η_i	reduced intrinsic excitation energy	207

Symbol	*Definition*	*Page*
κ	extinction coefficient	103
λ	electromagnetic wavelength	2
λ_0	semiconductor absorption edge	124
λ_0	long wavelength limit of detector	273
μ	absolute magnetic permeability	47
μ_e	electron mobility	213
μ_h	hole mobility	213
μ_0	magnetic permeability of free space	91
ν	electromagnetic frequency	22
ν_0	electromagnetic frequency corresponding to long wavelength limit	289
ρ	electrical resistivity	215
ρ	reflectivity	13
ρ_t	reflectance	139
$\rho_{\parallel}$	reflectivity for radiation polarized parallel to the plane of incidence	46
$\rho_{\perp}$	reflectivity for radiation polarized perpendicular to the plane of incidence	46
σ	electrical conductivity	47, 211
σ	Stefan-Boltzmann constant, 5.6687×10^{-8} watts meter^{-2} (deg K)$^{-4}$	19
$\sigma(\lambda)$	attenuation coefficient	101
σ_0	zero frequency conductivity	126
τ	carrier lifetime	221
τ	detector time constant	274
τ	transmissivity	13
$\bar{\tau}$	average transmittance due to the combined effects of absorption and scattering	174
τ_a	transmissivity as affected by absorption processes only	103
$\bar{\tau}_a$	average transmissivity, as affected by absorption only, within the spectral interval $\Delta\lambda$	168
τ_e	electron lifetime	227
τ_h	hole lifetime	227
τ_r	carrier relaxation time	139
τ_t	transmittance	139
τ_{ai}	transmission of the ith window as affected by absorption	173
τ_{si}	transmission of the ith window as affected by scattering	173
χ_e	electric susceptibility	105
Ψ	radiant energy density	15
Ψ_λ	spectral radiant energy density	23

1
Definition and history

1.1 DEFINITION

The electromagnetic spectrum is the continuum consisting of the ordered arrangement of radiation according to wavelength, frequency, or photon energy. It has been established experimentally that the electromagnetic spectrum includes waves of every length from an extremely small fraction of a millimeter to many kilometers. There is no single source or detection mechanism that is useful over the entire electromagnetic spectrum, and as a result the spectrum has been separated into rather loosely defined spectral regions. The bases for these broad subdivisions have generally been in accordance with the various means of generating, isolating, and detecting the radiations involved.

Although all electromagnetic radiation when absorbed by matter produces heat, radiation in a certain spectral region, namely, the infrared region, can be more readily detected by the heat it produces. In addition, it turns out that heated bodies provide excellent sources of this type of radiation; thus infrared radiation is sometimes referred to as thermal radiation. The further subdivision of the infrared into near, intermediate and far (depending upon the "distance" from the visible region) has also met with some acceptance.

Another definition of infrared radiation is: it is that portion of the electromagnetic spectrum lying between the visible and microwave regions, that is, the wavelengths between 7.5×10^{-4} mm and approximately 1 mm. These limits can be expressed in various forms involving wavelength, frequency, photon energy, or wave number. In terms of commonly used units, the limits are given in Table 1.1.

1

TABLE 1.1. Limits of the infrared region

	Lower Limit	Upper Limit
Wavelength	7.5×10^{-4} mm	1 mm
	$0.75\ \mu$ (microns)	$10^3\ \mu$
	750 mμ (millimicrons)	10^6 mμ
	7500 Å (angstroms)	10^7 Å
Frequency	3×10^{11} cps	4×10^{14} cps
Photon energy	1.23×10^{-3} ev	1.72 ev
Wave number	10 cm^{-1}	1.3×10^4 cm^{-1}

1.2 HISTORY

In the following short history of the subject, only the briefest mention can be made of the great men and inspiring events that have made possible this or any book concerned with infrared radiation. The reasons for including this chapter are twofold: to provide a sense of continuity regarding subsequent expositions, and to pay tribute to those who have given us our subject.

When studying the fascinating story concerning the advances and triumphs of science, one is struck by the degree of interdependence of the various disciplines. An advance or discovery in one field frequently results in progress in one or more other areas of scientific endeavor. It is, therefore, difficult, if not impossible, to discuss any particular aspect of science without occasional reference to the entire subject. In light of this consideration, we have attempted to present a discussion of the origin, development, and present status of the subject in relation to other associated fields of science.

Although it may be said that Galileo Galilei (1564–1642) and Sir Isaac Newton (1642–1727) are the founders of the sciences of optics and spectroscopy, respectively, they were not without antecessors in these related fields. Indeed, prehistoric man when observing the ethereal, transient beauty of a rainbow probably gave considerable thought to his vision, and in his ignorance was likely to have ascribed some mystical, supernatural significance to it. Ancient civilizations, cognizant of the simple laws of reflection, aware too of the more elementary aspects of refraction, did not associate them with dispersion and the rainbow. Seneca (4 B.C.–64 A.D.) and the Arabian scientist Alhazen (965–1038 A.D.) discussed various topics related to optics, particularly to refraction. Even Willebrord Snell (1591–1626), the discoverer of the true law of refraction, and Sir Isaac Barrow, Newton's mentor, were unaware of the composite nature of white light. Until Newton's fundamental experiments in 1666, it was thought that the visible spectrum resulted from "the mixing, in various proportions, of light with darkness." Newton showed in his experiments, which are reported in his *Optiks* (1704), that "the prism parts or sorts the heterogeneous mixture of rays of which the sun's light is composed." By showing that a second prism could be used to recombine the spectrum, he was able to reconstitute white light from the various colors. With this simple, definitive experiment Newton erected the whole foundation of the science of spectroscopy.

Unfortunately, Newton's explanation of these phenomena was incorrect. He felt that all colors have the same origin as the colors arising from interference effects in thin films. Moreover, his explanations included the concept that light particles travel faster in a dense medium because they are attracted to its surface. As a result of these ideas, many subsequent investigators gave incorrect interpretation to their work. For example, Young, in 1802, attributed the yellow lines in a candle's spectrum to an interference effect within the flame. So great was the prestige of Newton's name that his theories were accepted for 160 years.

It is obvious from Newton's writings that he did not regard his corpuscular theory or his theory of refraction as definitive and that he realized their validity was still to be proved experimentally. This illustrates the thesis that even the greatest intellects must work within the framework of knowledge existing in their age. Consequently their conjecturing should be regarded as such—the basis for further experimental investigation.

It was not until 1850 that Foucault showed that the velocity of light was *less* in a denser medium; and therefore that the explanation of refraction, based on the wave theory of light advanced by Huygens (1629–1695), a contemporary of Newton, was correct.

The eighteenth century was a barren period for the sciences of optics and spectroscopy, the notable exceptions being Bradley's discovery of aberration in 1728 and the work of Thomas Melville (1721–1794) who studied the spectrum of the sodium flame.

In contrast, the dawn of the new century brought with it many new and fascinating discoveries. At this time it had occurred to no one that the spectrum of the sun extended beyond that narrow range which could be detected by the human eye. However, in 1800 Sir William Herschel, British Astronomer Royal, while studying the heating effect produced by various portions of the solar spectrum, established that it contained some form of radiant energy which could not be seen. He arrived at this fact by placing sensitive thermometers at different places in the spectrum arising from the refraction of sunlight by a prism. He found that the greatest heating power lay outside of the visible, just beyond the red end of the spectrum. Although Herschel was able to show that this energy obeyed some of the same optical laws as visible light, it is difficult to determine from his writings whether he felt that this new form of energy was similar to light or whether he regarded it as possessing characteristics intrinsically different. Another important event in the early years of the nineteenth century was the first wavelength determination by Young in 1802. Young's correct interpretation of interference provided the basis for these measurements.

Sir John Herschel, son of the discoverer of infrared radiation, was able to demonstrate in 1840 the existence of infrared absorption and transmission bands by noting variations in the rate of evaporation of alcohol from blackened paper upon which the solar spectrum was projected. Just before this, in 1833, L. Richie experimentally established the validity of a fundamental law of thermal radiation, namely, that a material which is a good emitter of infrared radiation is an equally good absorber. A more cogent statement of this law was given by G. Kirchhoff in 1859 (and independently by B. Steward); as a result this law bears Kirchhoff's name.

For thirty years following the pioneering experiments of the elder Herschel, progress in the study of infrared was rather slow because of, among other reasons, the lack of sensitive and accurate detectors. But by the 1830's detectors were made available that depended on some of the numerous physical characteristics of matter which are functions of temperature, the thermoelectric effect (discovered by Seebeck in 1826) and electrical resistivity being two of the more sensitive. In addition, at that time instrumentation and techniques for measuring electrical quantities had been developed to such a degree that they could be used in conjunction with these effects to make possible the accurate detection of infrared radiation.

The first step in this direction was the invention in 1830 of the radiation thermocouple by L. Nobili. The logical extension of this idea to a number of thermocouples in series, that is, a thermopile, was accomplished by M. Melloni three years later. He was to use this device to extend his experimentation farther into the infrared region. That approximately 1μ radiation produced phosphorescent and photographic effects was shown by E. Becquerel in 1843. Although physical effects such as heating were involved in most of the advances in infrared detection, W. Abney in 1880 was able to produce photographic plates which were sensitive to 2μ. Beginning in 1881, S. Langley was able to make bolometers that were more sensitive than the thermocouples available at that time.

By utilizing these detectors when they became available, the science of infrared moved steadily ahead and the idea that infrared radiation was quite similar to light was beginning to be accepted.

Starting in 1814, J. Fraunhofer contributed greatly to spectroscopy and was responsible for the application of the diffraction grating to spectroscopic studies. Despite the fact that most of his work was done in the visible portion of the spectrum, the techniques and instrumentation he developed also implemented the study of infrared. The development of the first practical spectroscope by G. Kirchhoff and R. Bunsen in 1859 was a major step forward.

Wavelengths up to 1.5 μ were measured by A. Fizeau and J. Foucault in 1847. Subsequent investigators extended measurements of the infrared range to even longer wavelengths: J. Müller to 1.9 μ in 1859; M. Moulton to 2.14 μ in 1879; Curie and Desains to 7 μ in 1880. Following his invention of the bolometer, Langley extended measurements of the solar spectrum to 18 μ. In the last decade of the nineteenth century Paschen, Rubens, and co-workers, active in many phases of infrared, were able to extend the range to 20 μ by utilizing residual ray techniques, which depend on the ability of some materials to selectively reflect radiation. Other groups, primarily in Germany, were able to make measurements to 300 μ.

All this experimental activity in the field of infrared had far-reaching theoretical consequences. For example, on the basis of measurements made by J. Tyndall of the power transferred by radiation from one heated body to another, Stefan in 1879 concluded that the rate at which heat was transferred from one body to another was proportional to the difference of the fourth power of their absolute temperatures. This strictly experimental result was then derived theoretically by Boltzmann in 1884 from thermodynamic considerations. This law is now called the Stefan-Boltzmann law.

Probably the outstanding experimental achievement relating to the study of thermal radiation was the verification of Maxwell's classical electromagnetic theory of radiation (1862). Because longer wavelengths are involved, substantiation of Maxwell's theory is more readily evident from experiments involving infrared rather than ultraviolet or visible radiation. In the course of just such experiments concerned with the examination of the electromagnetic theory, H. Hertz, in 1887, was able to produce by electrical means very long wavelength infrared radiation. It became increasingly apparent that there was no essential difference between thermally and electrically produced electromagnetic waves.

Strangely enough, these same experiments of Hertz, which did so much to establish the existence of electromagnetic waves, were also instrumental in the demise of the concept that the wave picture is the only manner by which radiation may be described, and led to the reassertion of the particle or photon theory of radiation. Hertz noticed during his work that when radiation fell on an air gap, the gap could conduct electricity more easily.

This observation initiated an extended investigation of the photoelectric effect by which radiation of sufficiently short wavelength is able to ionize atoms.

One of the outstanding problems facing scientists of the last decades of the nineteenth century was that of explaining the wavelength distribution of thermal radiation emitted from a small hole in the side of a heated enclosure. Attempts to describe the observed wavelength distribution of this thermal radiation in terms of the classical concepts of the electromagnetic and kinetic theories were fruitless, although Wien had been able to discover the form of the functional relationship. In addition, he was able to relate the temperature of the enclosure to the wavelength interval at which the maximum power was emitted (Wien's displacement law). In analyzing this phenomenon, Planck in 1900 was led to the concept of energy quantization, asserting that the emission of radiation took place not continuously but in discrete steps. Einstein's successful explanation of the photoelectric effect, based on Planck's hypothesis, also contributed to the establishment of the validity of the quantum theory.

Thus we have two different theories, the corpuscular and the wave theory, each of which describes some, but not all, of the observed facts. It is disturbing to have two different concepts of radiation; either light is a shower of particles or it is a wave, but we are tempted to say it cannot be both.

It must be realized that this apparent dichotomy concerning the nature of light arises because of the limitations in our own means of perception. The following may illustrate this point. Consider a man who on alternate days is either deaf or blind but never both at the same time. He goes to see plays and on the days he can see but not hear, these plays are pantomimes; and on the days he can hear but not see, the plays are similar to a radio presentation. By two completely different mechanisms he can consistently explain this world of sense impressions. In time he may be able to combine these separate impressions into one coherent experience; so it may be with the scientist who attempts to reconcile conflicting theories.

We have seen how the study of thermal radiation played an extremely important role in the establishment of the dual nature of light and of the quantum theory in the early part of the twentieth century. By this time the experimental aspects of infrared had moved ahead. Bolometers, thermopiles, and associated electrical apparatus combined with interferometric and wavelength measuring techniques permitted a vigorous approach to infrared experiments. W. Coblentz, by obtaining infrared absorption, emission, and reflection spectra of a large number of organic and inorganic materials, provided evidence of the analytical value of infrared. Intensive interest in the study of molecular quantum mechanics was stimulated by the observation in 1913 of rotational spectra and the concept of molecular vibrational modes of motion. The use of infrared absorption spectra remains a powerful tool in fundamental studies of chemical bonding and molecular structure.

This renewed interest in infrared, coupled with its obvious military applications, was responsible for subsequent development of more sensitive and rapid detectors. Although the detection capabilities of bolometers and thermopiles had been increased tremendously, these devices remained rather slow. More rapid and sensitive detection mechanisms were to be found in sensors which depended on the photon or particle characteristics of thermal radiation. Photon detectors do not rely on the heating effect of infrared; therefore they do not depend on the thermal diffusivity of the detector material and as a result are much faster.

Photon effects, such as the photoemissive, photoconductive, and photo-voltaic effects, had been observed with visible and ultraviolet radiation in the nineteenth century; but infrared detection mechanisms utilizing the photon characteristics of thermal radiation were not exploited until much later.

One of the earliest sensitive photon detectors utilized the photoemissive effect in which photons, striking a cathode made of certain materials, eject electrons which are collected by an anode. Unfortunately, although much work has been done, it has not been possible to extend the useful range of this effect very far into the infrared beyond about 1.3 μ. This was the photoeffect employed in the sniperscope and snooperscope used in World War II.

The phenomenon by which absorbed infrared radiation lowers the electrical resistance of certain materials without a change in temperature is known as the photoconductive effect. It was first observed by W. Smith in 1873, who noted that the resistance of selenium decreased when exposed to light. In 1917, Case in the United States was able to produce thallous sulfide photoconductive detectors which were sensitive to 1.2 μ. However, the mechanisms responsible for the operation of these cells were not understood until the work of Gudden and Pohl. These scientists began their work in Germany in the early 1920's. Fournier of France, Todesco of Italy, Sewig of Germany, and Asao of Japan also conducted research on thallous sulfide photoconductive detectors. Lead sulfide cells were developed and studied intensively by the Germans beginning before World War II. They were able to extend the PbS sensitive range to approximately 4 μ. The Germans also carried on extensive infrared system developments; the most notable of these was the Kiel IV, an infrared airborne system with outstanding range capability. This equipment was developed at the Carl Zeiss works in Jena under the direction of Werner K. Weihe. In 1941, Cashman at Northwestern University conducted research efforts which were successful in solving the problems in fabricating stable thallous sulfide cells. Since that period, much time, effort, and money have been expended to study the photoconductive effect. Among other prominent materials investigated have been lead telluride, lead selenide, silicon, tellurium, indium antimonide, indium arsenide, and doped germanium.

Infrared photovoltaic detectors which yield a voltage when irradiated have also been of great interest. With the advent of the junction transistor, the technology associated with the fabrication and study of *p-n* junctions has been employed to develop and refine detectors utilizing this effect. In addition to these three photoeffects, the photoelectromagnetic (PEM) effect holds much promise of yielding a very useful type of detector.

It must not be inferred from the preceding outline that work on thermal detectors has not also been actively pursued. Indeed, some extremely interesting and important developments have taken place along these lines.

In 1947, for example, Golay constructed an improved pneumatic infrared detector in which the radiation heats a small amount of occluded gas, causing a flexible mirror which forms part of the gas chamber to be distended. This movement is detected by the change in intensity of light reflected from the mirror to a photocell. Another thermal detector has been developed which utilizes the temperature shift of the absorption edge in certain materials to modulate radiation passing through that material. The thermistor bolometer (temperature sensitive resistor) has found widespread use in detecting radiation from low temperature sources. The superconducting effect has been used to make extremely sensitive bolometers.

Another important area of infrared technology pertains to the development of radiation sources. The nature of emission from thermal sources is such that they have two serious deficiencies; they cannot be modulated at the frequencies at which fast detectors operate, and they are not coherent. Progress in this area has been slow, although developments in laser techniques show promise of overcoming these deficiencies.

Investigations of materials useful as windows, lenses, and prisms in the infrared were probably begun by M. Melloni who discovered that sodium chloride was transparent to a large part of the infrared spectrum. Later P. Desains found that calcium fluoride (fluorite) was also a good infrared optical material with the added advantage that it was not deliquescent. Following the experiments of Desains, other investigators reported still more useful transmissive materials. Among these silver chloride and silver bromide were found to be good windows at long wavelengths.

During World War II the Germans introduced two types of mixed crystals, KRS-5 and KRS-6, which could be grown from the molten state. Recently, semiconductor materials such as germanium and silicon have been adopted for use as lenses and windows. The last decade has seen the development of a wide variety of infrared transmitting glasses.

Optical filters are another class of components which has recently occupied the attention of workers in the field of infrared technology. The art of making multilayer interference filters has progressed to such a point that these items are now readily available. Semiconducting materials have also been used for infrared filtering.

In addition to this work, much has been done to develop means of detecting entire infrared images. These efforts have been along two lines. The first technique utilizes a single element or a multielement detector. Either the detector or the associated optics may be moved to scan the infrared scene. Electronic scanning of an extended-area detector upon which the entire scene is imaged is the second and more sophisticated manner by which an entire infrared picture can be obtained. Each of these methods can be used with both thermal and photon detectors, although because of its faster response the latter type is preferred.

The commercial uses of infrared are expanding rapidly. The utility of infrared as a tool for the identification of molecules and functional groups was realized by chemists in the late 1920's. Today this nondestructive means of material identification is used by industrial, medical, and other research scientists. Infrared is also being used to measure temperature whenever it is not practical or convenient to touch the heated object with a sensor. Among the many possible examples of this are the monitoring of metal rolling or extruding processes, railroad hotbox detection, and temperature determination within an explosion chamber.

The military uses of infrared are legion. There are several reasons for this. First, most targets of military interest, vehicles, troops, airfields, factories, and the like differ from the general terrain either in temperature or emissivity or both. They therefore have different radiating characteristics which are difficult to camouflage and which can readily be seen by infrared equipments. Second, infrared systems can perform many tasks passively, that is, by utilizing the radiation which is emitted by the targets they seek. This has the advantage that the detection system does not disclose its presence in the way that a radar system does. Third, infrared, when compared to radio or radar, is capable of revealing considerably greater detail because of the shorter wavelengths being used.

Some of these advantages of infrared are utilized in each of the following types of equipment. Very compact and lightweight infrared equipments have been developed to guide missiles. Infrared can be used to acquire, track, and direct fire toward enemy targets. This field of technology makes it possible also to keep enemy movements under surveillance even in complete darkness. Infrared equipment can be used by the military to form a thermal image of terrain disclosing such things as streets, runways, buildings, and missile launching sites. Although active infrared systems forego one of the previously mentioned advantages, this type of system has been effectively used, for example, in the sniperscope, snooperscope, and communications devices.

Both commercial and military applications are discussed in detail in a companion volume.*

* *Elements of Infrared Technology: Systems and Applications;* in preparation.

2
Infrared sources

A study of source characteristics is important because of the help such knowledge provides in the choice of detectors, the design of optical elements, and in certain other aspects of system planning. In many applications the radiation source is not subject to control. This is most commonly true, of course, for "passive" systems where the objects are detected by their natural radiation. It may be partially true even in "active" systems, which utilize a source for the illumination of a scene so that objects may be detected by their reflected radiation. In active systems it may be necessary to adapt the system to a source which was designed elsewhere. Then the source can sometimes be modified in an indirect way, as, for example, by the use of spectral or spatial filtering. An understanding of these techniques depends upon knowledge of the properties of infrared radiators.

There is a natural division or grouping of infrared sources which depends upon the nature of the wavelength distribution of the emitted energy. One type of source emits radiation over a very broad and continuous band of wavelengths. A plot of its emission versus wavelength is a smooth curve which usually passes through only one maximum. This type is called a continuous spectrum source, or more briefly, a continuous source. Another type of source is one which radiates strongly in some relatively narrow spectral intervals, but, in other wavelength intervals, the source does not radiate at all. A plot of emission versus wavelength reveals a series of emission bands or lines. The curve is discontinuous and the source is called a discontinuous spectrum source, a line source, or a band source.

This chapter first describes continuous spectrum sources and then those having predominantly line spectra. It is convenient with continuous sources to divide the discussion into one part dealing with the amount of power

11

radiated and the other describing the distribution of that power in wavelength, in space, and in time.

It is interesting to note that each type of infrared source has been subjected to extensive basic research for reasons far removed from consideration of its utility in infrared systems. The study of thermal radiators by Nernst, Planck, and others provided the starting point for the development of the quantum theory. The study of line sources has contributed greatly to our knowledge of the structure of atoms and molecules.

2.1 DEFINITION OF TERMS

Certain parameters of radiation sources need to be considered frequently. Therefore, some of the terms are defined at the start of the discussion. Actually, the list which can be used to describe source characteristics is much longer than the one given here. An effort has been made, however, to restrict the number of terms introduced, choosing only those of the greatest utility in system design calculations.

TABLE 2.1. Definition of radiation terms

Term	Units	Definition
Radiant power, P	Watts	The radiant energy emitted by or incident upon a surface per unit time.
Spectral radiant power, P_λ	Watts micron^{-1}	The amount of radiant energy emitted by or incident upon a surface per unit time within a specified small spectral interval centered at the wavelength λ.
Radiant intensity, E	Watts steradian^{-1}	The radiant power emitted by a source into a unit solid angle.
Spectral radiant intensity, E_λ	Watts steradian^{-1} micron^{-1}	The radiant power emitted by a source into a solid angle within a specified small spectral interval centered at the wavelength λ.
Radiant emittance, R	Watts meter^{-2}	The radiation power emitted by unit area of the source (into an entire hemisphere).

TABLE 2.1. (*continued*)

Term	Units	Definition
Spectral radiant emittance, R_λ	Watts meter^{-2} micron^{-1}	The radiation power emitted by unit area of the source within a specified small spectral interval centered at the wavelength λ.
Radiance, R_ω	Watts meter^{-2} steradian^{-1}	The radiation power emitted by unit area of the source into a unit solid angle.
Spectral radiance, $R_{\omega\lambda}$	Watts meter^{-2} steradian^{-1} micron^{-1}	The radiation power emitted by unit area of the source per unit solid angle within a specified small spectral interval centered at the wavelength λ.
Irradiance, $\mathscr{H}$	Watts meter^{-2}	The radiation power incident on unit area of a surface.
Reflectivity, ρ	Numeric (usually expressed as a decimal but may be expressed in per cent).	The fraction of the incident radiation power which is reflected by a surface.
Absorptivity, α	Numeric	The fraction of the incident radiation power which is absorbed by a surface.
Transmissivity, τ	Numeric	The fraction of the incident radiation power which is transmitted by the medium.
Emissivity, ε	Numeric	A ratio which compares the radiating capability of a surface to that of an ideal radiator or "black body." The value of ε for a black body is unity.

2.2 EMISSION FROM HEATED BODIES

All objects are continuously emitting and absorbing radiation. The emitted radiation results from the acceleration of electrical charges within the material. The interactions of these charges in solid bodies are very complex. Thus, it is very difficult to apply electromagnetic theory to the electrons in a solid to explain such things as the observed variations in the spectral emissivity of real solids. Instead, thermodynamics has been utilized to predict and explain the radiative behavior of such bodies. This mode of attack has led to the very useful concept of the ideal radiator. Certain radiating surfaces exist whose radiation characteristics are completely specified if their temperature is known. These surfaces, which are continuous spectrum sources, are known as ideal thermal radiators, or "black bodies." The following section shows why a so-called black body is an ideal radiator.

2.2.1 Kirchhoff's Law

In Fig. 2.1 a small solid object S is located within the walls of an evacuated cavity which is maintained at a uniform temperature, T_0. No matter what material S is made of and whatever its initial temperature, it will finally come to and remain at temperature T_0. In a later part of this chapter it will be shown that the hotter an object is, the greater the power radiated by it. Thus, as long as S and the walls of the cavity differ in temperature, there will be a net flow of radiant energy toward the cooler of the two. Eventually the stream of radiation will be the same in all directions and it will be the same in all enclosures which are at the same temperature no matter what materials are used to make them.

At thermal equilibrium, then, the power radiated by S must equal the power absorbed by it.

$$R = \alpha \mathscr{H}, \qquad (2.1)$$

where $R \equiv$ the radiant emittance, or power per unit area emitted by S,

$\quad \alpha \equiv$ the absorptivity of S,

$\quad \mathscr{H} \equiv$ the power per unit area striking S.

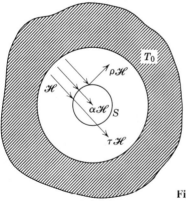

Figure 2.1. Object within cavity.

Since α cannot be greater than unity, the emitted power is a maximum for a perfect absorber, a so-called black body. For such ideal radiators $\alpha = \varepsilon = 1$. For other less efficient absorbers, the emissivity is correspondingly lower so that in all cases

$$\varepsilon = \alpha. \tag{2.2}$$

It turns out that this is also true for each spectral component of the radiation, that is, $\varepsilon_\lambda = \alpha_\lambda$. This can be proven quite rigorously using the second law of thermodynamics. If it were not true, it would be possible to set up a Carnot engine in the enclosure transferring heat continuously into work without causing other changes in the system. Equation 2.1 is a statement of Kirchhoff's law. The relationship

$$P_i = P_\alpha + P_\rho + P_\tau,$$

where P_i is the incident radiant power, P_α, the absorbed power, P_ρ, the reflected power, and P_τ, the transmitted power, is another statement of the conservation of energy. By dividing both sides by P_i, this relationship may be rewritten as

$$\alpha + \rho + \tau = 1, \tag{2.3}$$

where α, ρ, and τ have the meaning stated in Table 2.1.

For an opaque body $\tau = 0$, so that Eq. 2.3 becomes

$$\alpha = 1 - \rho,$$

indicating that surfaces of high reflectivity are poor emitters!

This can be tested by a very straightforward experiment if a disappearing filament pyrometer is available. If we examine the incandescent surface of a coiled-coil filament in a light bulb, the convoluted surface appears brightest just where the unheated filament has dark, highly absorbing recesses. Similarly, if a china plate having a highly reflecting design printed on it is heated to incandescence in an oven, the design has a darker appearance because it is a poorer emitter.

2.2.2 The Stefan-Boltzmann Law

In 1884, Boltzmann deduced a law relating the total power emitted by a black body to its absolute temperature. His reasoning can be illustrated as follows: Consider an ideal Carnot engine which uses radiation as the working substance. Its walls including those of the piston are impervious to heat. Opposite the piston in the base there is a small opening which can be covered at any time by a cover. Both the cover and the rest of the interior of the cylinder are perfectly reflecting.

We will need to use the relationship between the energy density, Ψ, of isotropic radiation and its pressure $\mathscr{P}$. Isotropic radiation is radiation which delivers the same power density and wavelength distribution to a sampling surface no matter how that surface is oriented within the volume which contains the isotropic radiation. Energy density is the term used to specify the amount of radiant energy contained within unit volume of the region under consideration.

If radiation is incident on a surface at an angle θ measured from the normal, the energy incident on a unit area is $\cos \theta$ times as great as that received on unit area normal to the beam. Furthermore, the component of momentum normal to the surface is proportional to $\cos \theta$ so that

$$p = w \cos^2 \theta, \tag{2.4}$$

where $p \equiv$ the pressure due to radiation which is incident at the angle θ (we assume complete absorption),

$w \equiv$ the energy density due to the radiation which is incident at the angle θ.

Equation 2.4 may appear to be more plausible if we recall that pressure has the dimensions of force/length^2 and energy density has the dimensions force $\times$ length/length^3.

In order to determine the pressure on a surface due to radiation approaching from all angles, we sum over N beams which are uniformly distributed in angle of incidence. (See Fig. 2.2). The fraction of the total number of beams incident from within an annular ring of width $d\theta$ is

$$\frac{dN}{N} = \frac{d\omega}{2\pi} = \frac{2\pi \sin \theta \, d\theta}{2\pi} = \sin \theta \, d\theta, \tag{2.5}$$

where $d\omega$ is the elemental solid angle subtended by the annular ring. So that

$$\mathscr{P} = \int_0^{\pi/2} w \cos^2 \theta \, dN = wN \int_0^{\pi/2} \cos^2 \theta \sin \theta \, d\theta = w \frac{N}{3}, \tag{2.6}$$

where $\mathscr{P} \equiv$ the radiation pressure due to beams incident at all angles.

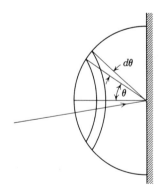

Figure 2.2. Geometry used for derivation of radiation pressure expression.

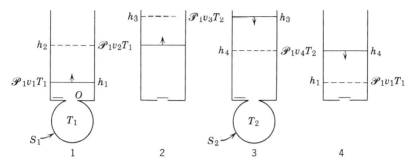

Figure 2.3. Boltzmann's radiation engine. [After Richtmyer, Kennard and Lauritsen, *Introduction to Modern Physics*, McGraw-Hill Book Co., New York 1955, p. 111.]

Since by our previous definitions the total energy density Ψ is equal to the sum of the density in all N beams, we have

$$\Psi = wN. \tag{2.7}$$

Therefore we see that

$$\mathscr{P} = \tfrac{1}{3}\Psi. \tag{2.8}$$

Let us return to our hypothetical experiment with the Carnot engine. Referring to Fig. 2.3, in Step 1, the opening O is uncovered allowing the cylinder to fill with radiation and come to equilibrium at temperature T_1. The piston is then allowed to move to the height h_2 so that the volume increases by an amount $(v_2 - v_1)$ while the radiation pressure is kept at $\mathscr{P}_1 = \tfrac{1}{3}\Psi_1$, the radiation pressure at the beginning of the expansion. To maintain this density, additional radiation must enter the cylinder through O for two reasons: First, the radiation has done work amounting to

$$\mathscr{P}_1\, \Delta v = \tfrac{1}{3}\Psi_1 (v_2 - v_1).$$

Second, the cylinder contains more energy because the same energy density is now present in a larger volume.

The increase in energy is equal to the energy density multiplied by the increase in volume, that is, $\Psi_1(v_2 - v_1)$.

The total heat flow into the cylinder during Step 1 is then

$$\mathscr{E}_{\text{in}} = \mathscr{P}_1\, \Delta v + \Psi_1\, \Delta v,$$

or

$$\mathscr{E}_{\text{in}} = \tfrac{4}{3}\Psi_1 (v_2 - v_1),$$

and this must be supplied to the heat source S_1 in order to keep it at T_1.

In Step 2 the perfectly reflecting cover is placed over the opening and the piston is allowed to expand very slowly. In this adiabatic process the

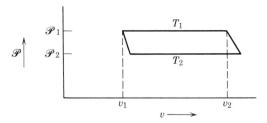

Figure 2.4. Carnot cycle.

radiation again does work on the piston. Partly because of this work and partly because of the increase in volume, the energy density is decreased; this is accompanied, of course, by a change in radiation pressure so that the radiation in the cylinder becomes characteristic of that to be expected at a new temperature T_2. If the change in $\mathscr{P}$ during this step is small, $\mathscr{P}_1 - \mathscr{P}_2$ may be taken as $d\mathscr{P}$ and

$$d\mathscr{P} = \tfrac{1}{3}\, d\Psi. \tag{2.9}$$

In Step 3 the cylinder is placed over the second radiation source S_2, the opening O uncovered, and the piston moved by means of a very slight compressive force so that the radiation density rises infinitesimally above the value Ψ_2, forcing radiation through O into S_2. This process is stopped at a point such that in Step 4, with O again closed, an adiabatic compression will have increased the pressure and hence Ψ to the value Ψ_1, at the same time as the volume has been reduced to the original value v_1.

Figure 2.4 shows the Carnot cycle for this sequence of operations. The area within the parallelogram is a measure of the external work done by the engine during this cycle. If $d\mathscr{P}$ approaches zero, the difference between the areas of the two triangles at the ends of the rectangle becomes negligible and the work done is

$$dW = (v_2 - v_1)\, d\mathscr{P} = \tfrac{1}{3}(v_2 - v_1)\, d\Psi.$$

Since the efficiency of a Carnot engine is given by

$$\frac{dW}{\mathscr{E}_{\text{in}}} = \frac{dT}{T_1},$$

then

$$\frac{dW}{\mathscr{E}_{\text{in}}} = \frac{\tfrac{1}{3}(v_2 - v_1)\, d\Psi}{\tfrac{4}{3}(v_2 - v_1)\, \Psi_1},$$

and in general

$$\frac{d\Psi}{\Psi} = 4\,\frac{dT}{T},$$

which integrates to

$$\ln \Psi = 4 \ln T + C,$$

or

$$\Psi = aT^4. \tag{2.10}$$

The relationship between Ψ and R can be shown to be

$$R = \frac{c_0 \Psi}{4},$$

where c_0 is the speed of light. This appears plausible when viewed in the following way. Imagine a cavity with a tiny hole in it. The cavity is filled with black body radiation whose energy density is Ψ. The amount of this energy which issues forth, or is extruded from the hole per unit of time, certainly depends upon c_0, the velocity with which it is propagated. Only half of the energy within a given volume will be moving out of the hole since half of the radiation, at any instant, is moving so as to cross a hemisphere oriented rearward in the cavity. The remaining factor of two arises because of the Lambertian energy distribution in the forward hemisphere. This last statement is amplified later in this chapter.

Since R, the radiant emittance, is proportional to Ψ, the energy density, then

$$R_{bb} = \sigma T^4, \tag{2.11}$$

where R_{bb} is the power per unit area radiated into a hemisphere by a perfect radiator, or black body; σ is known as the Stefan-Boltzmann constant and T is the absolute temperature. The value of σ has been found to be 5.6687×10^{-8} watts m^{-2} (deg K)$^{-4}$.

Equation 2.11 provides a simple means for computing the radiant emittance or radiance of any black body source for which the temperature can be determined. This is not the limit of its usefulness. It turns out that many common surfaces are "gray" or nearly so. In other words, their emissivity (not their radiant emittance) is independent of wavelength. Thus a relationship very useful in computing the radiant emittance of a real body having an emissivity ε is

$$R = \varepsilon \sigma T^4. \tag{2.12}$$

The relationship 2.12 requires the use of true temperature. True temperature can be determined in some cases with a thermometer or thermocouple directly. When this is impractical, true temperature may be obtained by measuring brightness temperature or color temperature using a radiation pyrometer. These terms are defined as follows:

True temperature: the temperature which would be indicated by a thermocouple or thermometer in equilibrium with the radiating body. It is this temperature which must be used in Eq. 2.12 or in the Planck formula developed later in this chapter (Eq. 2.38). True temperature may be computed from the brightness temperature using

$$T_{tr} \cong \frac{T_{br}}{T_{br} \dfrac{k\lambda}{hc_0} \ln \varepsilon + 1} ,$$

where $k \equiv$ Boltzmann's constant $= 1.3805 \times 10^{-23}$ joules/deg K and $h \equiv$ Planck's constant $= 6.6252 \times 10^{-34}$ joule sec.

Brightness temperature: the temperature at which a black body would need to be operated in order to emit the same amount of radiation as the gray body being observed. The foregoing comparison is usually made in some narrow spectral interval. This is often done by the use of a filter centered at 650 mμ.

Color temperature: the temperature at which a black body would need to be operated so that the ratio of the spectral radiant emittance emitted at two specified wavelengths is the same as the ratio of the spectral radiant emittance emitted by the body being described (at the same wavelengths). Two wavelengths often used for the color matching are 467 mμ and 650 mμ.

2.3   WAVELENGTH DISTRIBUTION

The manner in which radiation power varies with wavelength is termed the wavelength distribution. It is customary to determine the wavelength distribution of a source by a spectrometer which separates the radiation into wavelength increments and measures the relative power in each interval. The measurements are usually reported for intervals of equal $\Delta\lambda$. However, good arguments can be constructed to urge the usage of constant spectral purity, that is, equal values of $\Delta\lambda/\lambda$.

A number of phenomena dependent upon wavelength distribution will be discussed throughout this book. The transmission of the atmosphere, reflectivity of optical materials, the response of detectors, all of these are dependent upon the wavelength of the radiation employed. For this reason it will be important to know, or to be able to compute, the wavelength distribution of the particular source which is to be utilized in a given application.

2.3.1   The Rayleigh-Jeans Expression

The Rayleigh-Jeans expression, an approximate relationship between wavelength and spectral radiant emittance, is discussed for two reasons. The derivation illustrates the type of reasoning which had been applied in order to explain the spectral distribution of sources. In addition, a large amount of the background material needed for a presentation of the Rayleigh-Jeans expression can be effectively utilized in the following section which treats the Planck radiation law.

In the attempt to account for the radiation characteristics of a cavity, Rayleigh and Jeans did not specify the nature of the radiating and absorbing mechanisms at the walls. They simply computed the number of "degrees of freedom" per unit volume and per unit wavelength of the electromagnetic field within the cavity, and multiplied by the average energy associated with each to compute the total energy density per unit wavelength.

The concept of a "degree of freedom" deserves some amplification. A block sliding in a groove is free to move back and forth. It cannot move from side to side, up and down, nor can it rotate. The block is, therefore, said to have one degree of freedom of motion. A billiard ball rolling on a table has five degrees of freedom, two translational, plus the freedom to rotate about each of three mutually perpendicular axes. A nonrigid diatomic molecule possesses six degrees of freedom, three translational, freedom to rotate about two axes, plus one vibrational mode of motion.

With a vibrating string, a degree of freedom is assigned to each of the vibrational modes which it may adopt. The waves on the string may have a variety of wavelengths (a fundamental plus many overtones), subject only to the restriction that

$$n_L \frac{\lambda}{2} = L, \quad \text{or} \quad \frac{n_L}{L} = \frac{2}{\lambda}, \tag{2.13}$$

where L is the length of the string between rigid supports, n_L is an integer, the number of modes for vibration in one plane, and λ is the wavelength. Then N, the number of degrees of freedom, is $4L/\lambda$ where the additional multiplier 2 arises from the fact that the string is free to vibrate in each of two mutually perpendicular planes.

If this same reasoning is applied to sound waves in a box whose sides have lengths L_x, L_y, L_z, then standing waves will form due to reflection by the sides of the box if

$$\left(\frac{n_x}{L_x}\right)^2 + \left(\frac{n_y}{L_y}\right)^2 + \left(\frac{n_z}{L_z}\right)^2 = \left(\frac{2}{\lambda}\right)^2. \tag{2.14}$$

A complete derivation of Eq. 2.14 is quite lengthy and we will therefore sketch only the method for obtaining this relationship.

The equation for an acoustical wave in a rectangular parallelepiped with smooth rigid walls[1] is:

$$\frac{\partial^2 p}{\partial x^2} + \frac{\partial^2 p}{\partial y^2} + \frac{\partial^2 p}{\partial z^2} = \frac{1}{v^2} \frac{\partial^2 p}{\partial t^2}, \tag{2.15}$$

where p is the pressure at any point in the enclosure, and v is the velocity of the wave. A solution of this equation is

$$p = \frac{\cos}{\sin}\left(\omega_x\frac{x}{v}\right)\frac{\cos}{\sin}\left(\omega_y\frac{y}{v}\right)\frac{\cos}{\sin}\left(\omega_z\frac{z}{v}\right)\exp(-2\pi jvt).$$

Either sine or cosine functions can be used, and

$$v = \frac{1}{2\pi}(\omega_x{}^2 + \omega_y{}^2 + \omega_z{}^2)^{\frac{1}{2}}. \tag{2.16}$$

From the solutions given above, we can determine the velocity of the particles of the medium in any direction. The particle velocity u in the x direction is given by

$$u = -\frac{1}{\rho}\int\frac{\partial p}{\partial x}\,dt$$

$$= \frac{\omega_x}{2\pi jv\rho v}\frac{-\sin}{\cos}\left(\omega_x\frac{x}{v}\right)\frac{\cos}{\sin}\left(\omega_y\frac{y}{v}\right)\frac{\cos}{\sin}\left(\omega_z\frac{z}{v}\right)\exp(-2\pi jvt), \tag{2.17}$$

where ρ is the density of the medium. Any standing wave must surely be symmetrical with respect to the midpoint of the enclosure. Since the particle velocity must be zero at the walls, $x = \pm L_x/2$; then ω_x is restricted to the values: $\omega_x = v\pi n_x/L_x$. When n_x is 0 or an even integer, $-\sin(\omega_x x/v)$ is used in the expression for u. When n_x is an odd integer $\cos(\omega_x x/v)$ is used.

In the same way characteristic values can be determined for ω_y and ω_z. So that using expressions like 2.17 in 2.16 we have

$$v = \frac{v}{\lambda} = \frac{v}{2}\left[\left(\frac{n_x}{L_x}\right)^2 + \left(\frac{n_y}{L_y}\right)^2 + \left(\frac{n_z}{L_z}\right)^2\right]^{\frac{1}{2}}, \tag{2.18}$$

and 2.14 follows immediately from this.

The number of modes having wavelengths greater than any given λ_{min} will be equal to the number of possible combinations of positive integers which makes the left side of 2.14 less than $(2/\lambda_{min})^2$. This number of combinations can be found in a systematic way by letting coordinates X, Y, and Z take on positive integral values: $X = n_x/L_x$, $Y = n_y/L_y$, $Z = n_z/L_z$, so that they locate an array of points. These points will lie at the corners of a set of parallelepipeds or cells, the edges of which are $1/L_1$, $1/L_2$, $1/L_3$ long. They therefore generate what is referred to as a reciprocal lattice. Each cell has eight points and each point is shared by eight cells. The volume of a cell is $1/L_1L_2L_3$ so that there are $L_1L_2L_3$ cells or points per unit volume. The points to be counted lie inside one octant

(X, Y, and Z take on only positive integral values) of a sphere whose radius is given by

$$(\text{radius})^2 = X^2 + Y^2 + Z^2 = \frac{4}{\lambda_{\min}^2}. \tag{2.19}$$

The volume of the octant is

$$\frac{1}{8} \cdot \frac{4}{3}\pi \left(\frac{2}{\lambda_{\min}}\right)^3 \cdot L_1 L_2 L_3 = \frac{4}{3}\frac{\pi L_1 L_2 L_3}{\lambda_{\min}^3}.$$

Dividing by the volume of a cell yields the number of modes of vibration for which $\lambda > \lambda_{\min}$.

$$n_v = \frac{4\pi}{3}\frac{1}{\lambda^3}, \tag{2.20}$$

and

$$-\frac{dn_v}{d\lambda} = \frac{4\pi}{\lambda^4}. \tag{2.21}$$

Electromagnetic waves, which are transverse, have two possible planes of polarization for each mode, so that the number of degrees of freedom within a wavelength interval $d\lambda$ is

$$dn_v = \frac{8\pi}{\lambda^4}\, d\lambda, \tag{2.22}$$

where we have disregarded the minus sign. Degrees of freedom are additive, that is, the total number of degrees of freedom associated with a group of particles, or objects, or waves, is the sum of the degrees of freedom possessed by the individual members of the group. Furthermore, the principle of equipartition of energy states that on the average each degree of freedom in one group of waves has the same amount of kinetic energy as a degree of freedom in any other group.

Most college physics textbooks[2] use the kinetic theory of gases to show that the kinetic energy per degree of freedom is

$$\bar{\mathscr{E}}_{\text{kin}} = \tfrac{1}{2}kT, \tag{2.23}$$

where k is Boltzmann's constant, 1.3805×10^{-23} joules/deg K, and T is the absolute temperature. The energy of each degree of freedom in the radiation within a cavity is the sum of the average values of the kinetic and potential energies.

In vibratory motion, the average potential energy is equal to the average kinetic energy, so that the spectral energy density is kT per degree of freedom multiplied by the number of degrees of freedom, or

$$\Psi_\lambda\, d\lambda = \frac{8\pi kT}{\lambda^4}\, d\lambda. \tag{2.24}$$

This is the Rayleigh-Jeans formula. It predicts an infinite spectral radiant emittance as λ approaches zero. However, measurements show that as the wavelength approaches zero the spectral radiant emittance rises, passes through a maximum, then approaches zero. Thus we see that the Rayleigh-Jeans expression, although helpful, does not properly describe the facts. The proper expression, the Planck radiation formula, expresses the correct relationship between radiant emittance and wavelength.

2.3.2 The Planck Radiation Formula

We will not indulge in lengthy conjecture as to Planck's mode of reasoning; he seems to have proceeded in a somewhat empirical way. However, he distrusted one of the consequences of the theorem of the equipartition of energy, namely, that the average energy of molecular oscillators is kT.

The essential features of his theory can be conveniently described in the following way:

Consider a classical harmonic oscillator. Elementary methods can be used to show that its energy $\mathscr{E}$ is given by

$$\mathscr{E} = \frac{k_1}{2} x^2 + \frac{p^2}{2m}, \tag{2.25}$$

where p is the momentum and k_1 the force constant. This is the equation of an ellipse of semiaxes p_{max} and x_{max}. During its motion the particle moves on a curve of constant energy. The motion can be represented as the motion of a point along the ellipse, the energy at times being entirely potential and at other times all kinetic. (See Fig. 2.5.) The energy of the oscillator is proportional to the area of the ellipse. This area is

$$A = \pi x_{max} p_{max}, \tag{2.26}$$

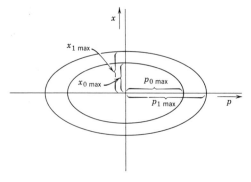

Figure 2.5. Permissible oscillator energies.

and since

$$p_{max} = (2m\mathscr{E})^{1/2},$$

whereas

$$x_{max} = \left(\frac{2\mathscr{E}}{k_1}\right)^{1/2},$$

we see that

$$A = 2\pi\mathscr{E}\left(\frac{m}{k_1}\right)^{1/2}. \tag{2.27}$$

Now $2\pi(m/k_1)^{1/2} = 1/\nu$ where ν is the frequency of the oscillator. Using 2.27 we see that

$$\mathscr{E} = A\nu. \tag{2.28}$$

For an oscillator of higher energy p_{max} and x_{max} are correspondingly greater. Figure 2.5 shows the energy curves of two oscillators having energies $\mathscr{E}_0$ and $\mathscr{E}_1$, where $\mathscr{E}_1 > \mathscr{E}_0$. The difference in area is related to the difference in energy by

$$\Delta A = \Delta\mathscr{E}/\nu. \tag{2.29}$$

Planck's innovation was that he assumed that there was a certain minimum energy which an oscillator had to have. He further assumed that the various oscillators had to differ in energy by that same amount. The amount was assumed to be proportional to ν, the proportionality constant being h, which we now call Planck's constant. The value of h has been determined to be 6.6252×10^{-34} joule sec. In Fig. 2.6 we show this diagrammatically where the solid lines correspond to oscillators which were permitted to exist according to Planck's assumptions and the dotted lines represent the motions of oscillators which were not permitted.

The population of oscillators was assumed to consist, therefore, of those

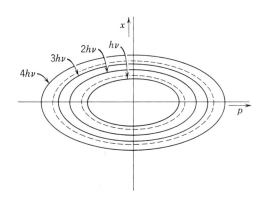

Figure 2.6. Permissible and forbidden oscillator energies.

having energy amounting to $1h\nu$, $2h\nu$, ... $nh\nu$ ... where n was always an integer. The question now arises: What fraction of them have energies amounting to $h\nu$, what fraction have energies amounting to $2h\nu$, and so on; that is, how are these oscillators distributed in energy? This problem occurs frequently in physics; given an assemblage of individual elements, we wish to know the fraction possessing a particular energy, or couched in slightly different terms we wish to determine the fraction of individual elements possessing an energy which lies in the energy range between $\mathscr{E}$ and $\mathscr{E} + d\mathscr{E}$. The fraction, dn, of the assemblage having an energy in this interval is usually written

$$dn = g(\mathscr{E})f(\mathscr{E})\,d\mathscr{E},$$

where $g(\mathscr{E})\,d\mathscr{E}$ is the number of permissible states in this range and
$f(\mathscr{E})$ is the probability that a state of energy $\mathscr{E}$ is occupied.

In many instances the particles may be considered free. Then the number of permissible states between $\mathscr{E}$ and $\mathscr{E} + d\mathscr{E}$ is given by

$$g(\mathscr{E})\,d\mathscr{E} = D\mathscr{E}^{\frac{1}{2}}\,d\mathscr{E},$$

where D is independent of energy.

The probability that a state of energy $\mathscr{E}$ is occupied depends on the type of statistics governing the behavior of the particles: Maxwell-Boltzmann, Fermi-Dirac, or Bose-Einstein. This topic is extensively discussed and these expressions are derived in many well-known books.[3,4,5]

These three distributions are given below as well as the special conditions under which each is valid. There is a general condition which is necessary for any of them to be valid, namely, that thermodynamic equilibrium must exist.

Maxwell-Boltzmann statistics. In this case the probability that a state of energy $\mathscr{E}$ will be occupied is given by

$$f(\mathscr{E}) = A \exp\left[-\mathscr{E}/kT\right], \tag{2.30}$$

where A is a normalizing constant,
$\mathscr{E}$ is the energy,
k is the Boltzmann constant,
T is the absolute temperature.
Maxwell-Boltzmann statistics are valid when:

1. The particles are distinguishable.
2. Theory permits a number of particles in the same energy interval to occupy the same small region in space. This second condition is usually

stated more precisely by saying that when particles obey Maxwell-Boltzmann statistics, a number of particles may occupy the same cell in phase
space.

Fermi-Dirac statistics. In this case the probability that a state of energy
$\mathscr{E}$ is occupied is given by

$$f(\mathscr{E}) = \frac{B}{\exp\,(\mathscr{E} - \mathscr{E}_0/kT) + 1}, \qquad (2.31)$$

where B is a normalizing constant,
$\quad \mathscr{E}_0$ is a reference energy, the so-called Fermi level of energy, and the
$\qquad$ other symbols have their former meaning.
The Fermi-Dirac statistics are valid when:

1. The particles are indistinguishable.
2. Theory permits only one particle of a given energy to occupy a small
region in space at any given time. This second condition is usually stated
more precisely by saying that the Pauli exclusion principle applies and only
one particle can occupy a cell in phase space.

Einstein-Bose statistics. In this case the probability that a state of
energy $\mathscr{E}$ is occupied is given by

$$f(\mathscr{E}) = \frac{C}{\exp\,(\mathscr{E} - \mathscr{E}_0/kT) - 1}, \qquad (2.32)$$

where C is a normalizing constant and the other symbols have their former
meaning.
Einstein-Bose statistics are valid when:

1. The particles are indistinguishable.
2. Theory permits a number of particles in the same energy interval to
occupy the same small region in space at the same time. This condition
is usually more precisely stated by saying that the particles do not obey the
Pauli exclusion principle, that is, that a number of particles may occupy the
same cell in phase space.

The frequency of occurrence of oscillators of energy $\mathscr{E}$ was taken to be
proportional to $\exp\,(-\mathscr{E}/kT)$ in accordance with classical Maxwell-
Boltzmann statistics. The statistics applicable in this case are the
Einstein-Bose statistics but since $\mathscr{E}$, that is, $h\nu$, is so large by comparison
with kT the expression 2.32 simplifies to 2.30. Thus, if N_0 is the number of
oscillators having zero energy, $N_1 = N_0 \exp\,(-h\nu/kT)$, the number of oscillators in the first ellipse, $N_2 = N_0 \exp\,(-2h\nu/kT)$, the number of oscillators

in the second ellipse, etc., then the total number of oscillators N_T is

$$N_T = N_0 + N_1 + N_2 + \dots$$
$$= N_0 + N_0 \exp(-h\nu/kT) + N_0 \exp(-2h\nu/kT) + \dots$$
$$= N_0 [1 - \exp(-h\nu/kT)]^{-1}. \quad (2.33)$$

Similarly, Planck's assumptions gave for the total energy possessed by this number of oscillators the expression

$$\mathscr{E}_T = (N_0 \times Oh\nu) + N_0 h\nu \exp(-h\nu/kT)[1 + 2\exp(-h\nu/kT)$$
$$+ 3\exp(-2h\nu/kT) + \dots]$$
$$= N_0 h\nu \exp(-h\nu/kT)[1 - \exp(-h\nu/kT)]^{-2}. \quad (2.34)$$

From 2.33 and 2.34 the average energy $\bar{\mathscr{E}}$ of an oscillator is

$$\bar{\mathscr{E}} \equiv \frac{\mathscr{E}_T}{N_T} = \frac{h\nu}{[\exp(h\nu/kT) - 1]}. \quad (2.35)$$

As $h\nu$ approaches zero, 2.35 should approach the value given by the equipartition of energy theorem. In deriving that theorem the energies of the oscillators are not restricted to be integral multiples of some value $h\nu$. They may possess any energy, so that the situation assumed by those physicists who came before Planck could be represented by a figure like 2.6 in which the width of the annular regions was infinitesimally small. In 2.35 the denominator can be expanded as

$$\left[\left(1 + \frac{h\nu}{kT} + \frac{1}{2}\left(\frac{h\nu}{kT}\right)^2 + \dots \right) - 1 \right],$$

so that as $h\nu$ approaches zero, $\bar{\mathscr{E}}$ is seen to approach kT.

Let us return now to the problem of determining the spectral radiant emittance function. As before, we multiply the number of degrees of freedom per unit volume per unit wavelength by the energy per degree of freedom. Here the energy per degree of freedom is given by Eq. 2.35. It was shown previously that the number of degrees of freedom per unit volume and per unit wavelength was given by

$$8\pi \frac{d\lambda}{\lambda^4}.$$

Using this and 2.32, we have for the spectral energy density

$$\Psi_\lambda \, d\lambda = \frac{8\pi}{\lambda^4} \frac{h\nu}{[\exp(h\nu/kT) - 1]} \, d\lambda. \quad (2.36)$$

This is one form of Planck's radiation formula. By analogy with the relationship between Ψ and R_{bb}, the black body spectral radiant emittance

$R_{bb\lambda}$ is $c_0/4$ times the spectral energy density.

$$R_{bb\lambda} \, d\lambda = \frac{2\pi c_0 h\nu}{\lambda^4 [\exp (h\nu/kT) - 1]} \, d\lambda. \qquad (2.37)$$

Since $\nu = c_0/\lambda$,

$$R_{bb\lambda} \, d\lambda = \frac{2\pi c_0^2 h}{\lambda^5 [\exp (hc_0/\lambda kT) - 1]} \, d\lambda. \qquad (2.38)$$

Equation 2.38 gives the functional relationship between $R_{bb\lambda}$ and λ. This is the more common form of Planck's radiation formula, an extremely useful relationship. It can be used to compute the spectral radiant emittance in any wavelength interval for an ideal radiator if the temperature is known. It forms the basis for the derivation of other radiation formulas. For example, integration yields the Stefan-Boltzmann law and differentiation gives the Wien displacement law. (See Section 2.3.3.) Notice that $R_{bb\lambda} \, d\lambda$ has the dimensions of power area^{-1}. The reader is cautioned against assuming that the radiant emittance at any single wavelength has a value other than zero. In any real situation we are forced to consider the power radiated within a stated spectral interval of finite extent. As written, 2.38 states the spectral power radiated into an entire hemisphere. R_ω, the radiance, is dependent upon direction. A special case, that of a Lambertian surface, will be discussed in Section 2.4.2. Figure 2.7 shows a plot of $R_{bb\lambda}$ versus wavelength for various values of T.

Since many infrared detectors in common use have a response which depends upon the rate of photon arrival rather than upon the incident radiation power, it is often useful to have an expression for the number of photons radiated per unit time per unit spectral interval by a thermal source into a hemispherical solid angle. This number, $N_{bb\lambda}$, is given by 2.38 divided by the energy per photon. Thus

$$N_{bb\lambda} \, d\lambda = \frac{2\pi c_0 \, d\lambda}{\lambda^4 [\exp (hc_0/\lambda kT) - 1]}. \qquad (2.39)$$

Figure 2.8 shows a plot of $N_{bb\lambda}$ versus wavelength for various values of T.

2.3.3   The Wien Displacement Law

The Planck radiation formula shows that the spectrum of the radiation shifts toward shorter wavelengths as the temperature of the radiator increases. This is a common observation with practical sources. A metal, when heated, radiates first at very long wavelengths in the infrared. As it becomes hotter it begins to radiate more at shorter and shorter wavelengths, first in the red end of the visible spectrum and finally more and more toward the blue. Thus the metal appears successively deep red, red, orange, yellow, and finally (when sufficient blue and violet light are being radiated), white.

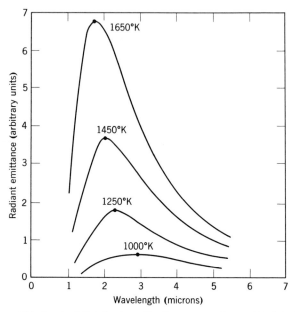

Figure 2.7. Spectral distribution of radiant power from a black body.

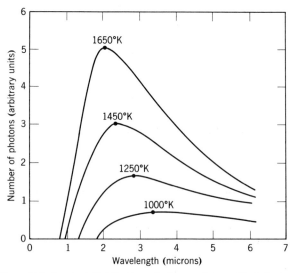

Figure 2.8. Spectral distribution of photons from a black body.

The Wien displacement formula relates the wavelength of maximum emission to the temperature of the body. It can be derived by utilizing thermodynamic reasoning, wherein a hypothetical cylinder filled with radiant energy is acted upon by a piston. We choose to determine the maximum by taking the derivative of Ψ'_λ with respect to λ and setting it equal to zero.

$$\frac{d\Psi'_\lambda}{d\lambda} = \frac{40\pi c_0 h}{\lambda_m^6 [\exp(hc_0/\lambda_m kT) - 1]}$$

$$+ \frac{8\pi c_0 h}{\lambda_m^5} \frac{c_0 h}{\lambda_m^2 kT} \frac{\exp(hc_0/\lambda_m kT)}{[\exp(hc_0/\lambda_m kT) - 1]^2} = 0,$$

where λ_m is the wavelength at which Ψ'_λ (or R_λ) has its greatest value. After simplification this gives

$$\left(1 - \frac{hc_0}{5\lambda_m kT}\right) \exp(hc_0/\lambda_m kT) = 1.$$

Replacing $hc_0/\lambda_m kT$ by x, we have

$$\left(1 - \frac{x}{5}\right) e^x = 1.$$

Using an iterative method or a table of natural logarithms we find that $x = 4.965$. Thus

$$\frac{hc_0}{k\lambda_m T} = 4.965.$$

Introducing numerical values for h and k as $h = 6.6252 \times 10^{-34}$ joule sec, $k = 1.3805 \times 10^{-23}$ joule/deg K we find that

$$\lambda_m T = 2893 \text{ micron degrees.} \tag{2.40}$$

This is the Wien displacement formula. Figure 2.7 shows the distribution of power radiated by a number of black body radiators, each of which is at a different temperature. The shape of each curve is in accordance with the Planck radiation formula 2.38, the area under each curve corresponds to the Stefan-Boltzmann relationship 2.12, and the displacement of the wavelength of peak emission is in agreement with the Wien displacement formula 2.40.

Figures 2.9 through 2.11 can be used for rough calculations of the spectral radiant emittance of most practical thermal sources. Figures 2.12 through 2.14 are useful for making the same type of calculations where it is desirable to know the number of photons emitted per unit time per unit area in each spectral interval of interest.

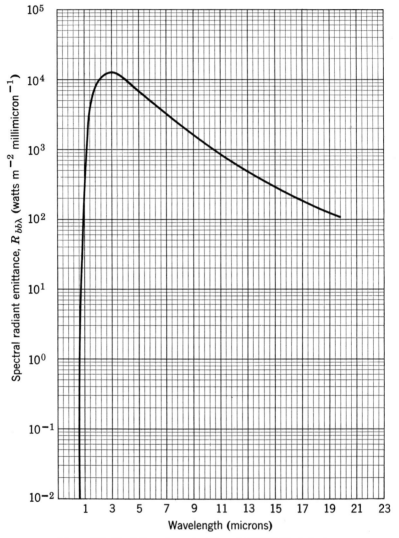

Figure 2.9. Radiation characteristics of 1000°K black body.

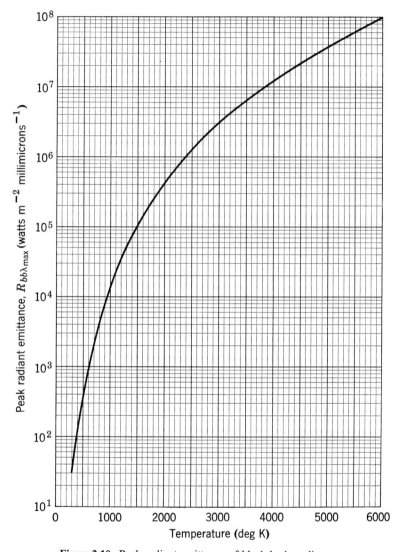

Figure 2.10. Peak radiant emittance of black body radiators.

34   ELEMENTS OF INFRARED TECHNOLOGY

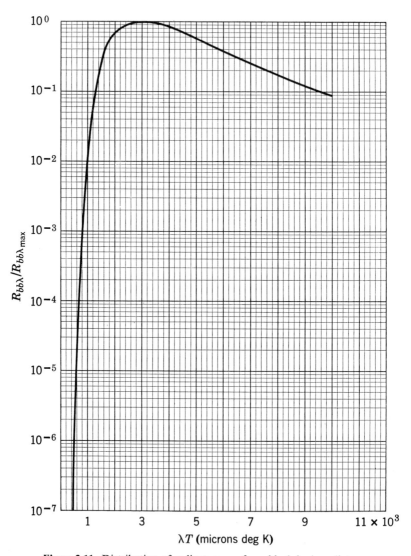

Figure 2.11. Distribution of radiant power from black body radiators.

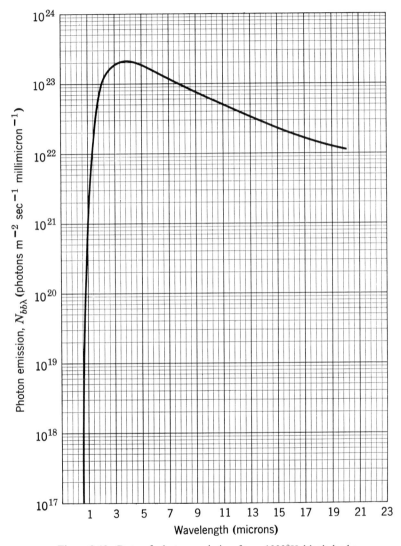

Figure 2.12. Rate of photon emission from 1000°K black body.

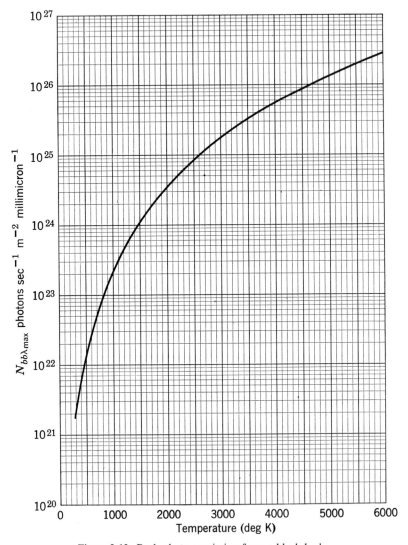

Figure 2.13. Peak photon emission from a black body.

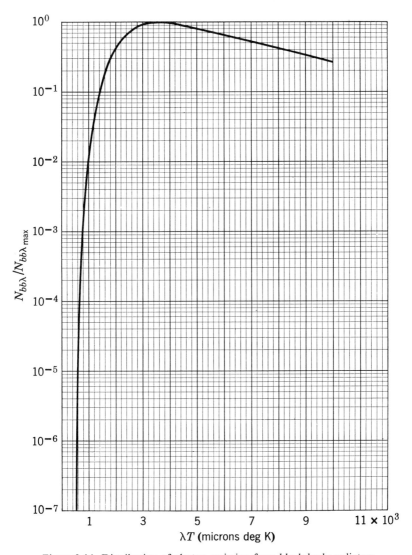

Figure 2.14. Distribution of photon emission from black body radiators.

2.4 THE EXCHANGE OF RADIANT POWER BETWEEN TWO BODIES

As stated previously, one of the important problems encountered in the field of infrared applications pertains to the power and spectral distribution of the radiation which is available at a detection system. There are two factors which influence the radiant power transferred from one body to another:

1. The position and attitude of the source and detector relative to each other, that is, the "view" each has of the other.
2. The goniometric emitting and absorbing characteristics of the surfaces of the two bodies.

The first of these is discussed in Section 2.4.1 and the second in Section 2.4.2. Several illustrative examples are presented in Section 2.4.3. We shall discuss those cases in which there is no absorption or scattering by the medium between the two bodies. The effect of attenuation by either of these mechanisms is discussed in Chapter 5. The discussion immediately following considers only bodies for which τ, the transmissivity, is zero. In addition we will discuss only that situation in which very little of the radiation reflected by one body will strike the other.

2.4.1. The Geometrical Effects

By definition the spectral radiant power emitted from an elementary area, dA_1, of one body at temperature T_1, having a spectral radiance $R_{\omega\lambda(\text{surface 1})}(T_1)$, into a solid angle, $d\omega$, is

$$d^4P_{\lambda(\text{surface 1})} = R_{\omega\lambda(\text{surface 1})}(T_1, \theta, \phi)\, dA_1\, d\omega. \qquad (2.41)$$

Here we use the notation $d^4P_{\lambda(\text{surface 1})}$ to signify that four integrations would be needed to obtain the radiant power $P_{\lambda(\text{surface 1})}$ leaving surface 1 (two over the surface area, one over azimuth angle, and one over elevation angle). In the general case $R_{\omega\lambda}$ is a function of θ, ϕ, T, λ, and position on the surface. We shall not consider the variation of $R_{\omega\lambda}$ with position on a surface. If each elemental area of this body emitted equally in all directions, the spectral power radiated from dA_1 per unit solid angle would be independent of the direction from which dA_1 is viewed. Actually, however, $R_{\omega\lambda}$ is found to vary with the direction from which the emitting surface is viewed. As an example, consider a candle—it cannot emit any light in the direction back along its body; hence its radiance is zero when viewed from that direction.

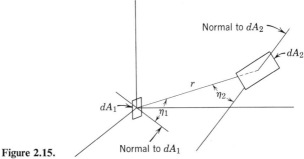

Figure 2.15.

The spectral radiant power emitted by dA_1 which is intercepted by the second body is

$$d^4P_{\lambda(\text{surface 1 to 2})} = R_{\omega\lambda(\text{surface 1})}(T_1, \theta, \phi)\, dA_1\, d\omega_2, \qquad (2.42)$$

where $d\omega_2$ is the solid angle subtended by the receiving body at dA_1. Equation 2.42 may be written

$$d^4P_{\lambda(\text{surface 1 to 2})} = \frac{R_{\omega\lambda(\text{surface 1})}(T_1, \theta, \phi)\, dA_1 \cos(\eta_2)\, dA_2}{r^2}, \qquad (2.43)$$

where $\cos(\eta_2)\, dA_2$ is the projected area of the receiver as "seen" from the emitting body,

η_2 is the angle between the normal to dA_2 and a line connecting dA_1 and dA_2, and

r is the distance between dA_1 and dA_2. (See Fig. 2.15.)

Similarly, the power emitted by the second body which is intercepted by the first is given by

$$d^4P_{\lambda(\text{surface 2 to 1})} = \frac{R_{\omega\lambda(\text{surface 2})}(T_2, \theta, \phi)\, dA_2 \cos(\eta_1)\, dA_1}{r^2}. \qquad (2.44)$$

However, 2.43 does not represent the spectral radiant power transferred from 1 to 2 and 2.44 does not represent the spectral radiant power transferred from 2 to 1. Some of the intercepted power may not be absorbed. The manner in which the absorptivity and emissivity affect the exchange of radiant power will be evident from the following discussion.

From definitions and Eq. 2.11 we may write for the radiant emittance of a black body,

$$R_{bb} = \sigma T^4 = \int_0^\infty R_{bb\lambda}(T)\, d\lambda = \int_0^\infty \left\{ \int_0^{2\pi} R_{bb\omega\lambda}\, d\omega \right\} d\lambda. \qquad (2.45)$$

By analogy we may write, when considering real surfaces,

$$R = \varepsilon_{\text{eff}}\sigma T^4 = \int_0^\infty \varepsilon_{\lambda\text{hem}} R_{bb\lambda}(T) \, d\lambda = \int_0^\infty \left\{ \int_0^{2\pi} R_{\omega\lambda} \, d\omega \right\} d\lambda$$

$$= \int_0^\infty \left\{ \int_0^{2\pi} R_{bb\omega\lambda}(T)\varepsilon_{\omega\lambda}(T, \theta, \phi) \, d\omega \right\} d\lambda, \quad (2.46)$$

where
$$\varepsilon_{\text{eff}}(T) \equiv \frac{R(T)}{R_{bb}(T)}, \qquad \begin{array}{l}\text{the effective hemispherical} \\ \text{emissivity,}\end{array} \quad (2.47)$$

$$\varepsilon_{\lambda\text{hem}}(T) \equiv \frac{R_\lambda(T)}{R_{bb\lambda}(T)}, \qquad \begin{array}{l}\text{the spectral hemispherical} \\ \text{emissivity, and}\end{array} \quad (2.48)$$

$$\varepsilon_{\omega\lambda}(T, \theta, \phi) \equiv \frac{R_{\omega\lambda}(T)}{R_{bb\omega\lambda}(T)}, \qquad \begin{array}{l}\text{the spectral goniometric,} \\ \text{or directional, emissivity.}\end{array} \quad (2.49)$$

The effective emissivity is the ratio of the total power emitted into a hemisphere by a surface to the total power emitted into a hemisphere by a black body at the same temperature. This is always less than or equal to unity. The spectral emissivity is the ratio of the spectral radiant power emitted between λ and $\lambda + d\lambda$ into a hemisphere by a real surface to that emitted by a black body within the same spectral interval. This ratio is always less than or equal to unity. The spectral goniometric emissivity is the ratio of the spectral radiant power emitted by a real surface between λ and $\lambda + d\lambda$ into the solid angle between ω and $\omega + d\omega$ for a specified value of θ and ϕ to that emitted into the same solid angle and in the same direction and within the same spectral interval by a black body. All these quantities can, in the general case, vary with temperature, but are unity for a black body.

The radiation emitted from a body originates within the volume of that body. However, most materials are so opaque to infrared radiation that a negligible portion of the radiant power leaving their surfaces originates more than a fraction of a millimeter below the surface. In treating the instances where this is not true, for example, sodium chloride, it is necessary to know the spectral transmissivity and its temperature dependence.

The following generalizations concerning the emissivity of surfaces are sometimes helpful:

1. The emissivities of most nonmetallic substances decrease with increasing temperature. They are typically greater than 0.8 at temperatures less than approximately 350°K, and are between 0.3 and 0.8 at refractory temperatures. Grain structure and color are more important than chemical composition in determining the emissivity of a surface.[6]

2. The emissivities of most metals are very low and approximately proportional to the absolute temperature, the proportionality constant

varying as the square root of the electrical resistance at a fiducial base temperature.[7,8] Unless particular care is taken to prevent oxidation, metals will exhibit a higher emissivity than predicted by this generalization.

We stated in Section 2.2.1 that in an isothermal enclosure, ε, the emissivity equals α, the absorptivity. This is also true under certain conditions encountered in practice. In order to set down explicitly the conditions under which the emissivity is equal to the absorptivity, we begin by noting that, using 2.47 and 2.48,

$$\varepsilon_{\text{eff}}(T_s) = \frac{\int_0^\infty \varepsilon_{\lambda\text{hem}}(T_s)R_{bb\lambda}(T_s)\,d\lambda}{\int_0^\infty R_{bb\lambda}(T_s)\,d\lambda},\qquad(2.50)$$

where T_s is the temperature of the emitting body.

The effective absorptivity of a surface is the ratio of the absorbed radiation, $\alpha_{\text{eff}}\mathcal{H}_i$, to the incident irradiance, $\mathcal{H}_i$. The spectral absorptivity, $\alpha_{\lambda\text{hem}}(T_i)$ of the body at temperature T_i, is defined as the ratio of the spectral absorbed radiation to the incident spectral irradiance, so that the effective absorptivity is

$$\alpha_{\text{eff}}[T_s,T_i] = \frac{\int_0^\infty \alpha_{\lambda\text{hem}}(T_i)\mathcal{H}_{i\lambda}(T_s)\,d\lambda}{\int_0^\infty \mathcal{H}_{i\lambda}(T_s)\,d\lambda}.\qquad(2.51)$$

Comparing 2.51 with 2.50 we see that Eq. 2.2 does not hold generally. Only when the incident radiation is black and when $T_i = T_s$ does 2.50 become equivalent to 2.51.

For those conditions under which the emissivity is equal to the absorptivity it is possible at least in principle to calculate the absorptivity of surface 2, the surface receiving the radiation. The spectral radiant power from dA_1 absorbed by dA_2 is, from 2.43, 2.44, and 2.49

$$d^4P_{\lambda(\text{surface 1 to 2})}[\text{absorbed}] = \frac{1}{r^2}\bigg\{\alpha_{\lambda(\text{surface 2})}(T_2,\theta,\phi)\varepsilon_{\lambda(\text{surface 1})}(T_1,\theta,\phi)\times$$
$$R_{bb\omega\lambda(\text{surface 1})}\,dA_1\cos(\eta_2)dA_2\bigg\}.\qquad(2.52)$$

Similarly,

$$d^4P_{\lambda(\text{surface 2 to 1})}[\text{absorbed}] = \frac{1}{r^2}\bigg\{\alpha_{\lambda(\text{surface 1})}(T_1,\theta,\phi)\varepsilon_{\lambda(\text{surface 2})}(T_2,\theta,\phi)\times$$
$$R_{bb\omega\lambda(\text{surface 2})}\,dA_2\cos(\eta_1)\,dA_1\bigg\}.\qquad(2.53)$$

Thus the net exchange of power between dA_1 and dA_2 is

$$d^4P_{\lambda\text{(surface 1 to 2)}}(\text{net}) = \frac{1}{r^2}\left\{ \alpha_{\lambda\text{(surface 2)}}(T_2, \theta, \phi)\varepsilon_{\lambda\text{(surface 1)}}(T_1, \theta, \phi) \right.$$

$$\times R_{bb\omega\lambda\text{(surface 1)}}(T_1)\cos(\eta_2) - \alpha_{\lambda\text{(surface 1)}}(T_1, \theta, \phi)\varepsilon_{\lambda\text{(surface 2)}}(T_2, \theta, \phi)$$

$$\times \left. R_{bb\omega\lambda\text{(surface 2)}}(T_2)\cos(\eta_1) \right\} dA_1\, dA_2. \quad (2.54)$$

The integration of this equation, for the most general case, is extremely difficult. In order to demonstrate its usefulness we shall first discuss the functional dependence of the emissivity and absorptivity with respect to the azimuth and elevation angles, that is, the goniometric distribution of the radiant power emitted and absorbed by a surface.

2.4.2 The Goniometric Emitting and Absorbing Characteristics of Surfaces

For black bodies the spectral radiance is independent of the azimuth angle and depends on the elevation angle θ in the following simple manner.

$$R_{bb\omega\lambda} = R_{bbn\omega\lambda}\cos\theta, \quad (2.55)$$

where $R_{bbn\omega\lambda}$ is the spectral radiance of a black body when viewed along the normal.

Equation 2.55 is called Lambert's cosine law. If the goniometric distribution of radiation emitted by a source obeys Eq. 2.55, the source is said to be Lambertian or perfectly diffuse. As a consequence of this law, glowing spheres may appear to be uniformly bright. Although the projected area near the rim represents more real area, that area is radiating less in the oblique direction, that is, where $\theta \to \pi/2$.

Since ideal black bodies are diffuse, that is, Lambertian, radiators, this fact helps in ascribing physical significance to the definitions given by Eqs. 2.47, 2.48, and 2.49. Consider a hemisphere, large by comparison to the dimensions of the black body source dA_s as shown in Fig. 2.16. Since in this case dA_c is always normal to a line from dA_s to dA_c,

$$d\omega_c = \frac{dA_c}{r^2} = \sin\theta\, d\theta\, d\phi.$$

Then the spectral radiant power striking the hemisphere, that is, emitted from the black body source is, from Eqs. 2.45 and 2.55,

$$R_{bb} = \sigma T^4 = \int_0^\infty R_{bbn\omega\lambda}\left\{ \int_0^{2\pi}\int_0^{\pi/2}\cos\theta\sin\theta\, d\theta\, d\phi \right\} d\lambda$$

$$= \pi\int_0^\infty R_{bbn\omega\lambda}\, d\lambda. \quad (2.56)$$

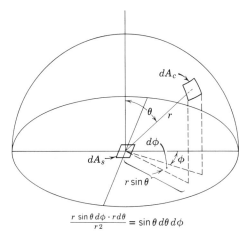

$$\frac{r\sin\theta\,d\phi\cdot r d\theta}{r^2}=\sin\theta\,d\theta\,d\phi$$

Figure 2.16.

Comparing this with 2.45 we see by means of 2.38 that

$$R_{bbn\omega\lambda}=\frac{1}{\pi}\int_0^{2\pi}R_{bb\omega\lambda}\,d\omega=\frac{R_{bb\lambda}}{\pi}=\frac{2c_0^2 h}{\lambda^5\left\{\exp\left(\dfrac{hc_0}{\lambda kt}\right)-1\right\}},$$

or when integrating over all wavelengths

$$\int_0^\infty R_{bbn\omega\lambda}\,d\lambda=\frac{\sigma T^4}{\pi}.\tag{2.57}$$

For a real radiating surface in the geometry shown in Fig. 2.16 the radiant power per unit source area is from 2.46

$$
\begin{aligned}
R=\varepsilon_{\mathrm{eff}}(T)\,\sigma T^4 &=\varepsilon_{\mathrm{eff}}(T)R_{bb}\\
&=\int_0^\infty R_\lambda\,d\lambda=\int_0^\infty \varepsilon_{\lambda\mathrm{hem}}(T)R_{bb\lambda}\,d\lambda\\
&=\int_0^\infty\left\{\int_0^{2\pi}R_{\omega\lambda}\,d\omega\right\}d\lambda\\
&=\int_0^\infty\left\{\int_0^{2\pi}R_{bb\omega\lambda}\varepsilon_{\omega\lambda}(T,\theta,\phi)\,d\omega\right\}d\lambda\\
&=\int_0^\infty\left\{R_{bb\omega\lambda}\int_0^{2\pi}\int_0^{\pi/2}\varepsilon_{\omega\lambda}(T,\theta,\phi)\sin\theta\,d\theta\,d\phi\right\}d\lambda\\
&=\int_0^\infty R_{bbn\omega\lambda}\left\{\int_0^{2\pi}\int_0^{\pi/2}\varepsilon_{\omega\lambda}(T,\theta,\phi)\sin\theta\cos\theta\,d\theta\,d\phi\right\}d\lambda.\tag{2.58}
\end{aligned}
$$

Thus

$$\varepsilon_{\lambda\mathrm{hem}}(T)=\frac{1}{\pi}\int_0^{2\pi}\int_0^{\pi/2}\varepsilon_{\omega\lambda}(T,\theta,\phi)\sin\theta\cos\theta\,d\theta\,d\phi.\tag{2.59}$$

It has been found experimentally[9,10] that a large number of real sources exhibit emissivities that are independent of direction, that is,

$$\varepsilon_{\omega\lambda}(T, \theta, \phi) = \varepsilon_{\omega\lambda}(T). \tag{2.60}$$

In this case 2.59 becomes

$$R = \varepsilon_{\text{eff}}(T)\,\sigma T^4 = \varepsilon_{\text{eff}}(T)\,R_{bb} = \int_0^\infty \left\{ \varepsilon_{\omega\lambda}(T) R_{bbn\omega\lambda} \int_0^{2\pi} d\phi \int_0^{\pi/2} \cos\theta\sin\theta\,d\theta \right\} d\lambda$$

$$= \pi \int_0^\infty \varepsilon_{\omega\lambda}(T) R_{bbn\omega\lambda}\,d\lambda, \tag{2.61}$$

and 2.60 becomes

$$\varepsilon_{\lambda\text{hem}}(T) = \varepsilon_{\omega\lambda}(T). \tag{2.62}$$

Although Lambert's cosine law is a good approximation for some surfaces, there are many for which Eq. 2.60 does not constitute a valid description of the goniometric distribution of emitted radiation. In the following discussion, which considers the electromagnetic nature of the emitted radiation, we shall outline the general theory explaining the angular distribution of emitted radiation. The emissivity depends on the chemical and physical condition of a surface, and, although it is often difficult to describe this condition precisely, it has been found that predictions based on this theory are in fairly good agreement with experiment. Consequently, they can be used to estimate emission characteristics of materials when experimental data are not available. For the purposes of the following discussion, materials will be grouped into two classes: dielectrics and electrical conductors.

Dielectric media. Figure 2.17 shows the angular dependence of the emissivity for typical dielectric surfaces.[11,12,13] The paramount feature of these curves is that the goniometric emissivity at any angle is less than that of a black body emitting in a perfectly diffuse, that is, Lambertian manner. The following discussion indicates that goniometric distributions of the type shown should be expected for dielectric surfaces.

It is shown in Section 3.4.3 that for radiation polarized parallel and for radiation polarized normal to the direction of the plane of incidence the reflectivities of dielectric materials are given by

$$\rho_{\parallel} = \left| \frac{\tan\,[\theta_i - \theta_r]}{\tan\,[\theta_i + \theta_r]} \right|^2$$

$$\rho_{\perp} = \left| \frac{\sin\,[\theta_i - \theta_r]}{\sin\,[\theta_i + \theta_r]} \right|^2,$$

where θ_i is the angle of incidence and
$\qquad \theta_r$ is the angle of refraction; both measured relative to a normal to the surface.

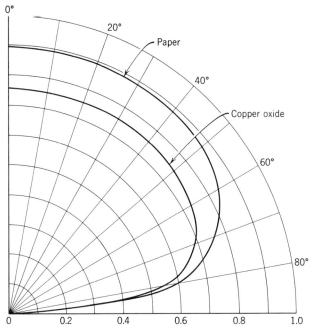

Figure 2.17. Goniometric emissivity of dielectrics. (Adapted from E. Schmidt and E. Eckert.[13])

These are called Fresnel's equations and they are derived on the basis of electromagnetic theory. The angles θ_i and θ_r are related by Snell's law:

$$n = \frac{\sin \theta_i}{\sin \theta_r},$$

where n is the index of refraction, a function of temperature and wave-length.

Assuming that:

1. The emitting surface is opaque, that is,

$$1 = \rho_\lambda + \alpha_\lambda;$$

2. The incident radiation is unpolarized so that

$$\rho = \tfrac{1}{2}\{\rho_\| + \rho_\perp\};$$

3. The spectral goniometric emissivity is equal to the spectral goniometric absorptivity, that is,

$$\varepsilon_{\omega\lambda} = \alpha_{\omega\lambda},$$

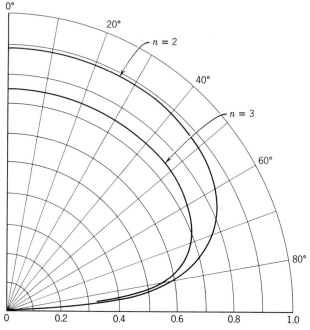

Figure 2.18. Polar plot of emissivity for various indices of refraction.

the angular dependence of the spectral goniometric emissivity can be given by[14]

$$\varepsilon_{\omega\lambda} = 1 - \frac{1}{2}\left\{\frac{\sin^2[\theta_i - \theta_r]}{\sin^2[\theta_i + \theta_r]} + \frac{\tan^2[\theta_i - \theta_r]}{\tan^2[\theta_i + \theta_r]}\right\}, \qquad n \neq 1 \quad (2.63)$$

where θ_i is now the angle from which the emitting surface is viewed.

The essential features of the curves shown in Fig. 2.17 can be deduced from Eq. 2.63. For radiation emitted normal to the surface of a dielectric

$$\varepsilon_{n\omega\lambda} = \frac{4n}{[1 + n]^2}, \qquad (2.64)$$

that is, the normal spectral emissivity of a dielectric source is equal to the normal spectral emissivity of a black body source diminished by a factor $4n/[1 + n]^2$. Figure 2.18 is a polar plot of calculated goniometric emissivity for several indices of refraction. It will be noted in this figure that the emissivity at any angle is always less than for black body emitters whose radiation is distributed according to Lambert's cosine law.

For many applications it is sufficiently accurate to assume that the goniometric emissivity of dielectrics is of the form

$$\varepsilon_{\omega\lambda} = \frac{4n}{[1 + n]^2} \cos^a \theta, \tag{2.65}$$

where $a < 1$ can be found experimentally.

If the wavelength and temperature dependence of n are neglected, Eq. 2.59 leads, by Eqs. 2.58 and 2.65, to

$$R_{\omega\text{diel}} = \frac{8\sigma T^4 n}{[a + 1][1 + n]^2} \tag{2.66}$$

Electrical conductors. According to electromagnetic theory, the reflectivity of a metallic surface for polarized radiation is given by[14]

$$\rho_{\parallel} = \frac{2n^2 \cos^2 \theta_i - 2n \cos \theta_i + 1}{2n^2 \cos^2 \theta_i + 2n \cos \theta_i + 1} \tag{2.67}$$

$$\rho_{\perp} = \frac{2n^2 - 2n \cos \theta_i + \cos^2 \theta_i}{2n^2 + 2n \cos \theta_i + \cos^2 \theta_i}. \tag{2.68}$$

It is shown in Section 3.2.2 that the index of refraction for a metal is approximately equal to

$$n = \sqrt{\frac{\mu\sigma\lambda c_0}{4\pi}}, \tag{2.69}$$

where c_0 is the speed of electromagnetic radiation in free space, σ is the electrical conductivity, μ is the absolute magnetic permeability, and λ the wavelength of the radiation. This relation is valid only for $\lambda > 10\ \mu$. In spite of this limitation, the goniometric emissivity calculated on the basis of Eqs. 2.67, 2.68, and 2.69 agrees quite well with experiment.

If we again assume that the transmissivity is zero, that unpolarized radiation is involved, and that the absorptivity equals the emissivity for all angles, then

$$\varepsilon_{\omega\lambda} = \left\{ 1 - \frac{1}{2} \left[\frac{2n^2 \cos^2 \theta_i - 2n \cos \theta_i + 1}{2n^2 \cos^2 \theta_i + 2n \cos \theta_i + 1} \right. \right.$$

$$\left. \left. + \frac{2n^2 - 2n \cos \theta_i + \cos^2 \theta_i}{2n^2 + 2n \cos \theta_i + \cos \theta_i} \right] \right\}. \tag{2.70}$$

If $n \gg 1$ and $\cos \theta_i \gg 1/2n[1 + \sqrt{3}]$ then Eqs. 2.67 and 2.68 become

$$\rho_{\parallel} = 1 - \frac{2}{n \cos \theta_i},$$

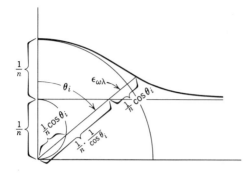

Figure 2.19. Goniometric emissivity of conductors. (After M. Jakob, *Heat Transfer*, Volume I, John Wiley & Sons, New York, Chapman & Hall, Limited, London, 1949, p. 48.)

and

$$\rho_{\perp} = 1 - \frac{2 \cos \theta_i}{n},$$

and a good approximation for Eq. 2.70 is

$$\varepsilon_{\omega\lambda} = \frac{1}{n} \left[\cos \theta_i + \frac{1}{\cos \theta_i} \right]. \tag{2.71}$$

Although Eq. 2.71 is not valid for $\theta \to \pi/2$, the values of the total radiant power emitted which are calculated by integration over 2π agree fairly well with experiment. The goniometric emissivities of conductors are shown in Fig. 2.19.

2.4.3 Examples

We shall now consider several illustrative examples.

Radiation from a point source to a finite area. In many practical instances the distance from the source to the collector may be so large that the dimensions of the source are negligible by comparison. If, in addition, the source radiates equally in all directions, that is, the emissivity does not depend on angle, then the source may be considered a point source.

Consider a point source, S, at the origin of the coordinate system, shown in Fig. 2.20. In order to determine the radiant power from the source incident on a finite area, we can proceed in the following manner. If dP_s represents the radiant power emitted into an elementary solid angle $d\omega$, then the radiant intensity E_s of the source is given by

$$E_s = \frac{dP_s}{d\omega}. \tag{2.72}$$

Conceivably E_s might be a function of the azimuth and elevation angles, that is, E_s could depend on the direction from which the source is viewed. But since a point source is considered to emit radiation in an isotropic manner, the radiant intensity can be given by

$$E_s = \frac{P_s}{4\pi} = \frac{\varepsilon_{\text{seff}}\,\sigma T_s^4}{4\pi}\,A_s, \tag{2.73}$$

where P_s is the total power emitted by the source,
    A_s is the source area,
    $\varepsilon_{\text{seff}}$ is the effective emissivity of the source,
    T_s is the source temperature.

It then follows that the radiation power $d^2P_{s\to c}$ incident on an element of the collector area dA_c, oriented at an angle η_c with respect to the normal to the elemental collector area dA_c at the point of incidence (see Fig. 2.20) at a distance r from the point source is

$$d^2P_{s\to c} = E_s\,d\omega = \frac{P_s}{4\pi}\frac{\cos\eta_c}{r^2}\,dA_c$$

$$= \frac{A_s\varepsilon_{\text{seff}}\,\sigma T_s^4}{4\pi}\frac{\cos\eta_c}{r^2}\,dA_c, \tag{2.74}$$

where $\cos(\eta_c)\,dA_c$ is the projected area of the collector "seen" from the source. In this and subsequent equations the subscripts s and c refer to source and collector, respectively. To determine the total radiant power incident on a collector of area A_c from the point source, Eq. 2.74 must be integrated, that is,

$$P_{s\to c} = \frac{\varepsilon_{\text{eff}}\sigma T_s^4 A_s}{4\pi}\int_{A_c}\frac{\cos(\eta_c)\,dA_c}{r^2}. \tag{2.75}$$

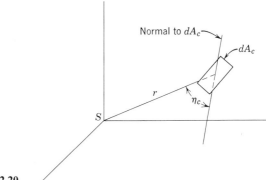

Figure 2.20

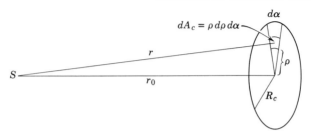

Figure 2.21.

If the collector is sufficiently small or sufficiently remote from the source, η_c and r do not vary over the collector area, then 2.75 reduces to

$$P_{s \to c} = \frac{\varepsilon_{\text{eff}} \sigma T_s^4 A_s}{4\pi} \frac{\cos (\eta_c)}{r^2} A_c. \tag{2.76}$$

As a second example, consider the power radiated from a point source normal to, and at a distance r_0 from, the center of a disc of radius R_c. (See Fig. 2.21.)

By 2.74 the power radiated from the point source to the elemental collector area is

$$d^2 P_{s \to c} = E_s \, d\omega_c = \frac{\varepsilon_{\text{eff}} \sigma T_s^4 A_s \rho \, \cos (\eta_c) \, d\rho \, d\alpha}{4\pi} \frac{}{r^2}. \tag{2.77}$$

But since

$$\cos (\eta_c) = \frac{r_0}{r} = \frac{r_0}{[\rho^2 + r_0^2]^{1/2}},$$

Eq. 2.77 becomes

$$d^2 P_{s \to c} = \frac{\varepsilon_{\text{eff}} \sigma T_s^4 A_s}{4\pi} \frac{r_0}{[r_0^2 + \rho^2]^{3/2}} \rho \, d\rho \, d\alpha, \tag{2.78}$$

which integrates to

$$\begin{aligned}
P_{s \to c} &= \frac{\varepsilon_{\text{eff}} \sigma T_s^4 A_s}{4\pi} r_0 \int_{\rho=0}^{R_c} \int_{\alpha=0}^{2\pi} \frac{\rho \, d\rho \, d\alpha}{[r_0^2 + \rho^2]^{3/2}} \\
&= \frac{\varepsilon_{\text{eff}} \sigma T_s^4 A_s}{2} \left\{ 1 - \left[1 + (R_c/r_0)^2 \right]^{-1/2} \right\}. \tag{2.79}
\end{aligned}$$

Radiation from a perfectly diffuse elemental source area to a finite area. Consider the configuration illustrated in Fig. 2.22. To determine the radiant power incident on the disc we see from Eq. 2.74 that the power from dA_s incident on $dA_c = \rho \, d\rho \, d\alpha$ is

$$d^2 P_{dA_s \to dA_c} = \frac{R_{\omega s} \, dA_s \, \cos (\theta_c) \rho \, d\rho \, d\alpha}{r^2}, \tag{2.80}$$

where $\qquad \cos \theta_c = \dfrac{h}{r} = \dfrac{h}{\left[h^2 + l^2 + \rho^2 + 2\rho l \cos \alpha \right]^{\frac{1}{2}}} \cdot$

Since by Eqs. 2.55, 2.57, and 2.48,

$$R_\omega = \int_0^\infty R_{\omega\lambda}\, d\lambda = \int_0^\infty R_{bb\omega\lambda}\varepsilon_{\omega\lambda}\, d\lambda \qquad (2.81)$$

$$= \int_0^\infty R_{bbn\omega\lambda}\varepsilon_{\omega\lambda} \cos \theta_s\, d\lambda$$

$$= \frac{\cos \theta_s \varepsilon_{\text{eff}} \sigma T_s^4}{\pi},$$

where $\qquad \cos \theta_s = \dfrac{1 + \rho \cos \alpha}{\left[h^2 + l^2 + \rho^2 + 2\rho l \cos \alpha \right]^{\frac{1}{2}}},$

then

$$P_{dA_s \to dA_c} = \frac{\varepsilon_{s\,\text{eff}} \sigma T_s^4\, dA_s h}{\pi} \int_{\rho=0}^{R_c} \int_{\alpha=0}^{2\pi} \frac{\rho[1 + \rho \cos \alpha]\, d\rho\, d\alpha}{[h^2 + l^2 + \rho^2 + 2\rho l \cos \alpha]^2} \qquad (2.82)$$

$$= \frac{\varepsilon_{s\,\text{eff}} \sigma T_s^4\, dA_s h}{2\pi} \left\{ \frac{h^2 + R_c^2 + l^2}{[h^2 + R_c^2 + l^2 - 4R_c^2]^{\frac{1}{2}}} - 1 \right\}.$$

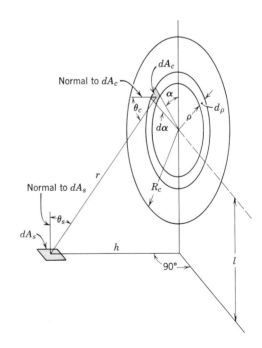

Figure 2.22. Adapted from E. R. G. Eckert and R. M. Drake, *Heat and Mass Transfer*, McGraw-Hill Book Company, New York, 1959, p. 399.

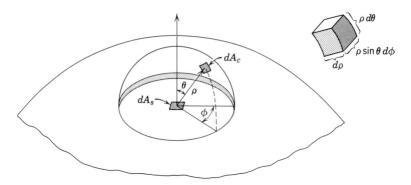

Figure 2.23

Let us consider another example. As indicated in Section 2.2.2 the radiant emittance R is equal to the energy density Ψ multiplied by one-fourth the speed of light. Proof of this statement follows.

In order to determine the relationship between the energy density and the radiant emittance, we shall consider the radiation emitted by a small hole in a heated cavity. Figure 2.23 shows a sphere of radius ρ above an aperture A_s, which has been provided in a heated cavity.

Only half of the radiation will be moving out through the aperture at any given time. Assuming that the radiation emitted through dA_s is perfectly diffuse, the radiation power from dA_s incident on an elemental area, $dA_c = \rho^2 \sin\theta \, d\theta \, d\phi$ is

$$d^5 P_c = \int_0^\infty R_{bb\omega\lambda} \sin\theta \, d\theta \, d\phi \, dA_s \, d\lambda \qquad (2.83)$$

$$= \sin\theta \cos\theta \, d\theta \, d\phi \, dA_s \int_0^\infty R_{bbn\omega\lambda} \, d\lambda,$$

or

$$d^4 P_c = R_{bbn\omega} \sin\theta \cos\theta \, d\theta \, d\phi \, dA_s.$$

The corresponding amount of energy, dU_c, within the elemental volume, $dV = \rho^2 \sin\theta \, d\rho \, d\theta \, d\phi$ is

$$dU_c = d^4 P_c t, \qquad (2.84)$$

where t is the length of time the radiation is in dV and is given by

$$t = d\rho/c_0, \qquad (2.85)$$

where c_0 is the velocity of light in free space. The elemental energy density due to the radiant power from dA_s is by definition

$$d\Psi_0 = \frac{dU_c}{dV} = \frac{R_{bbn\omega} \cos(\theta) \, dA_s}{\rho^2 c_0}. \qquad (2.86)$$

Upon integration this becomes

$$\Psi_0 = \int d\Psi_0 = \frac{R_{bbn\omega}}{c_0} \int \frac{\cos(\theta)\, dA_s}{\rho^2}$$

$$= \frac{R_{bbn\omega}}{c_0} \int_0^{2\pi} d\omega_s, \qquad (2.87)$$

where $d\omega_s$ is the solid angle subtended by the source at the collector area dA_c. Integrating 2.87 over the entire hemisphere and substituting $\pi R_{bbn\omega} = R_{bb}$ yields

$$\Psi_0 = \frac{2\pi R_{bbn\omega}}{c_0} = \frac{2R_{bb}}{c_0}. \qquad (2.88)$$

Since Ψ_0 represents the outward component of the total energy density, and therefore, on the average, represents one-half of the total energy density, the total energy density is given by

$$\Psi_{\text{tot}} = 2\Psi_0 = \frac{4R_{bb}}{c_0}. \qquad (2.89)$$

2.5 TIME DISTRIBUTION

The emission of radiation is accomplished by discontinuous, discrete processes and the observed radiation is the result of the execution of these processes by many individual radiating units. Thus, the observed characteristics and intensity of radiation are determined by some sort of average taken over the assemblage of radiating units. However, at any one instant in time, it is possible that the values of these quantities will differ from the values at some other instant. For example, if it were possible to examine individual radiating elements at some particular time, we would find that some would have just radiated, others would be on the verge of radiating, and still others would be in various conditions with regard to the completion of the emitting process. In other words, since the radiation characteristics of a source are the totality of many discrete, almost random processes, they also fluctuate in time. The fluctuations in the power and photon emission rate of a source form the subject of the two following sections. Two aspects of these spontaneous fluctuations in the emission from a source are considered. One, which pertains to the phase relationship between radiation emitted by the various radiating elements of which a source is composed, is called coherence. The second concerns the instantaneous variations in the power and/or rate at which photons are emitted by a source; these fluctuations are called photon noise.

2.5.1 Coherence

We have seen how the macroscopic properties of the radiation emitted from a source result from processes which are discontinuous and which fluctuate in time. The wave train emitted from one point on or in an incandescent source bears no predictable phase relationship to the train emitted from any other point. A body radiating in this manner is called an incoherent source. This situation is in contradistinction to that existing, for example, with radar. Although the radiation from a radar oscillator is the result of action by many electrons, they are constrained by the fields in the oscillator tube (for example, magnetron) to emit in phase with each other. The resulting radiation therefore forms a single continuous wave train. The amplitude and phase of this disturbance are determined uniquely at every point in space and time. This is the description of a coherent source.

Before discussing the possible advantages accruing through the use of a coherent source in an infrared system, it is appropriate to discuss first some of the elementary features characterizing an electromagnetic wave.

The wave contour of plane polarized, monochromatic radiation is represented by

$$A(t) = A_0 \cos \omega t = A_0 \cos 2\pi\nu t = A_0 \cos \left(\frac{2\pi c_0 t}{\lambda}\right). \qquad (2.90)$$

In Chapter 3 we shall see that Eq. 2.90 satisfies the wave equation which describes the propagation of an electromagnetic wave in any medium.

For a wave which is to advance in the $+x$ direction, 2.90 must be rewritten so that it becomes

$$A(t) = A_0 \cos \frac{2\pi}{\lambda} [c_0 t - x]. \qquad (2.91)$$

Figure 2.24 shows an advancing wave at two successive instants of time. Consider a source which consists of one radiating element. The wave

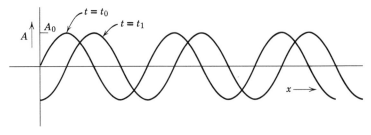

Figure 2.24

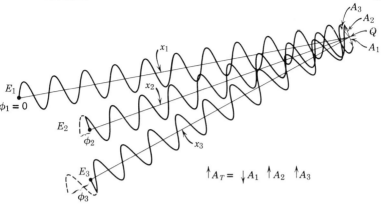

Figure 2.25. Summation of wave trains from three sources.

contour of its electromagnetic disturbance at a distance x from the source
can be given by

$$A(t) = A_0 \cos \left\{ \frac{2\pi}{\lambda} [c_0 t - x] - \phi \right\}, \tag{2.92}$$

where ϕ is the phase angle established with reference to some arbitrary
time at which t is set equal to zero. It is shown in Section 3.3 that the
average radiated power P transmitted by an electromagnetic wave, $A(t)$, is
proportional to the mean square value of $A(t)$, that is,

$$P \propto \overline{A^2} = \frac{1}{T} \int_0^T A^2(t) \, dt, \tag{2.93}$$

so that for a wave having the form described by 2.87 we find

$$P \propto \frac{A_0^2}{2}. \tag{2.94}$$

We can now ask the question, "What power would be radiated by a source
consisting of an assemblage of N such individual radiating elements?"
The answer, as we shall see, depends upon the extent to which these
elements act in phase. This is a measure of their coherence.

Consider N radiating elements emitting wave trains of the same frequency
and amplitude but of different phase relative to each other. By the superposition principle, the amplitude at the point Q (Fig. 2.25) due to the
contributions from all the radiating elements is given by

$$A_T = \sum_{j=1}^{N} A_j = \sum_{j=1}^{N} A_{0j} \cos \left\{ \omega t - \frac{2\pi x_j}{\lambda} - \phi_j \right\}, \tag{2.95}$$

where x_j is the distance between the point of observation Q and the jth radiating element, and ϕ_j is the phase of the jth element.

The average power, using 2.93, is given by

$$P_{av} \propto \frac{1}{T} \int_0^T \sum_{j=1}^N \sum_{k=1}^N A_{0j} A_{0k} \cos \left\{ \omega t - \frac{2\pi x_j}{\lambda} - \phi_j \right\}$$

$$\times \cos \left\{ \omega t - \frac{2\pi x_k}{\lambda} - \phi_k \right\} dt, \quad (2.96)$$

$$= \frac{1}{2} \sum_{j=1}^N \sum_{k=1}^N \frac{A_{0j} A_{0k}}{T} \int_0^T \left\{ \cos \left[2\omega t - \frac{2\pi}{\lambda} (x_j + x_k) - (\phi_j + \phi_k) \right] \right.$$

$$\left. + \cos \left[\frac{2\pi}{\lambda} (x_k - x_j) - (\phi_j - \phi_k) \right] \right\} dt. \quad (2.97)$$

If the ϕ_j's and ϕ_k's are not functions of time, then

$$P_{av} = \frac{1}{2} \sum_{j=1}^N \sum_{k=1}^N A_{0j} A_{0k} \left\{ \cos 2 \left[\frac{\pi (x_k - x_j)}{\lambda} + \frac{(\phi_k - \phi_j)}{2} \right] \right\}. \quad (2.98)$$

Now if the sources are coherent, that is, if $\phi_k = \phi_j$, then 2.98 becomes

$$P_{av} = \frac{1}{2} \sum_{j=1}^N \sum_{k=1}^N A_{0j} A_{0k} \cos \left[\frac{2\pi \Delta x_{kj}}{\lambda} \right], \quad (2.99)$$

where $\Delta x_{kj} \equiv x_k - x_j$.

Points can be found which satisfy the condition

$$\Delta x_{kj} = m\lambda, \quad (m = 0, 1, 2, \dots) \quad (2.100)$$

and at such points

$$P_{av} \propto N^2 A_0^2.$$

Similarly, there are points in space at $\Delta x_{jk} = m\lambda/4$, where the average power is zero. Whereas, if the source is incoherent, that is, the ϕ_j's and ϕ_k's randomly distributed, 2.98 becomes

$$P = \frac{1}{2} \sum_{j=1}^N \sum_{k=1}^N A_{0j} A_{0k} \cos \left\{ \frac{2\pi}{\lambda} [\Delta x_{kj}] + [\Delta \phi_{kj}] \right\}. \quad (2.101)$$

This can be expressed as two sums.

$$P = \frac{1}{2} \left\{ \sum_{\substack{j=1 \\ j=k}}^N A_{0j}^2 + \sum_{\substack{j=1 \\ j \neq k}}^N \sum_{k=1}^N A_{0j} A_{0k} \cos \left[\frac{2\pi}{\lambda} (\Delta x_{kj}) + \Delta \phi_{kj} \right] \right\}. \quad (2.102)$$

If N is large, the double sum in 2.102 is zero. For every choice of j and k another combination of j and k can be found which advances the argument by π. Since $\cos (\theta + \pi) = -\cos \theta$ these sets cancel in pairs. Thus only the first term remains which yields

$$P_{av} \propto N A_0^2.$$

For microwave sources where we have a single oscillator feeding into a number of radiating elements, it has been possible to construct arrays of N elements so that at a limited number of selected points in space, the radiated intensity is N^2 times as great as could be achieved with a single radiating element.

A system which has as its function the detection of electromagnetic radiation is able to gain an appreciable advantage if the radiation is coherent. This is because it is possible to establish at the receiver a condition which favors the reinforcement of the incident radiation by the creation of standing waves. Let us see how this may be accomplished.

An antenna which receives radiation described by Eq. 2.92 can be constructed so that its elements are a distance x apart, where x is given by

$$x = m\lambda \quad (m = 1, 2, 3, \ldots).$$

The interactions of a wave with the many elements of this array are made more efficient because of the following considerations:

1. The first cycle of a coherent wave which strikes the first element of the array sets up in this element and its feed lines an associated current-voltage wave. Each succeeding cycle of the wave serves to reinforce this current-voltage wave.

2. Additional elements of the array augment this process because their spacing is such that the voltage induced constructively interferes with the current-voltage wave already set up.

This is, of course, merely a way of setting up a condition of resonance. If the radiation incident on the elements is incoherent, the separate wave trains, of random phase, interfere destructively so that no reinforcement of the induced current-voltage wave can occur. We have chosen to illustrate this situation by suggesting, as a model, an antenna array. However, the discussion is equally valid if applied to other means of extracting energy from a wave, although in some cases this situation may be somewhat obscure.

2.5.2 Power Fluctuations Due to the Rate at Which Photons Are Emitted by a Source

Since the instantaneous power and the instantaneous rate at which photons are emitted by a source are the result of many individual processes, the rates of which fluctuate with time, these quantities also vary with time. The manner in which they fluctuate constitutes the subject of this section.

A very useful way to characterize a nonperiodic, fluctuating quantity, $p(t)$, is by using the concept of the noise spectrum $p(f)$ of that quantity. A convenient way of clarifying the physical significance of $p(f)$ is to consider it as a means of specifying the harmonic content of the fluctuating quantity. $p(f)$ states which frequencies are needed and what their magnitudes would have to be in order to synthesize the fluctuating quantity. It may be more precisely defined as the mean square of the deviation from the mean in the frequency interval between f and $f + df$, which thereby describes the frequency distribution of the fluctuating quantity.

By knowing $p(f)$ of the fluctuations in a quantity $p(t)$, the mean square deviation from the mean over all frequencies can be calculated by

$$\overline{[p(t) - \bar{p}]^2} = \int_0^\infty p(f)\, df, \tag{2.103}$$

where $\bar{p}$ is the time average of the fluctuating quantity.

It can be shown that if a randomly repeated event, such as the emission of a photon, occurs at an average rate $\bar{N}$, then the noise spectrum of the fluctuation $p(t)$ is given by

$$p(f) = 2\bar{N} \left| \int_0^\infty p(t) \exp{(j\omega t)}\, dt \right|^2. \tag{2.104}$$

We will now use Eq. 2.104 to compute the mean square deviation of both the power and the rate at which photons are emitted from a source.

The average rate of emission per unit area of the source per unit spectral interval is the average rate at which power is radiated per unit area per unit spectral interval divided by the energy for each individual emission event, that is,

$$\bar{N} = \frac{M(\nu, T)}{h\nu}\, d\nu, \tag{2.105}$$

where $M(\nu, T)$ is the spectral distribution function. Where M is the Planck distribution function, 2.105 becomes

$$\bar{N} = \frac{2\pi\nu^2/c_0^2}{\exp{[h\nu/kT]} - 1}\, d\nu. \tag{2.106}$$

We will assume here that the emission of a photon is an instantaneous random process so that the radiation power emitted can be described by

$$p(t) = h\nu\, \delta(t - t_0), \tag{2.107}$$

where $\delta(t - t_0)$ is the Dirac delta function defined by

$$\delta(t - t_0) = 0 \qquad t \neq t_0$$
$$\delta(t - t_0) = \infty \qquad t = t_0$$
$$\int_{-\infty}^\infty \delta(t - t_0)\, dt = 1.$$

Considering the radiation power as the fluctuating quantity, the noise spectrum of the power is obtained by substituting 2.106 and 2.107 into 2.104.

$$dp_P = \frac{4\pi A_s h^2 \nu^4 / c_0^2}{\exp\left[h\nu/kT\right] - 1}\, d\nu,$$

where A_s is the source area and we use the notation dp_P to denote a fluctuation in power in the frequency interval between ν and $\nu + d\nu$.

The noise spectrum for all spectral frequencies is

$$\begin{aligned}
p_P(f) &= \int_0^\infty dp_P \\
&= \frac{4A_s\pi h^2}{c_0^2} \int_0^\infty \frac{\nu^4\, d\nu}{\exp\left[h\nu/kT\right] - 1} \\
&= 7.66 A_s k\sigma T^5.
\end{aligned} \qquad (2.108)$$

Thus the mean square deviation from the mean in the fluctuation frequency interval between f and $f + \Delta f$ of the power emitted from a source is by 2.103

$$\overline{p_P^2} \equiv \overline{\left[p_P(t) - \bar{p}_P\right]^2} = 7.66 k A_s \sigma T^5 \int_f^{f+\Delta f} df \qquad (2.109)$$

$$= 7.66 k A_s \sigma T^5 \Delta f.$$

Strictly speaking, the emission of a photon is not a completely instantaneous, random process, but is governed by Einstein-Bose statistics, Eq. 2.32. An analysis which takes this into account yields for $p_P(f)$ the expression

$$p_P(f) = \frac{4A_s\pi h^2}{c_0^2} \int_0^\infty \frac{\nu^4 \exp\left[h\nu/kT\right]\, d\nu}{\{\exp\left[h\nu/kT\right] - 1\}^2}, \qquad (2.110)$$

which, when integrated, gives

$$p_P(f) = 8 A_s k\sigma T^5. \qquad (2.111)$$

If the source is a gray body having an effective emissivity, ε_{eff}, then the right-hand side of 2.110 and 2.111 should be multiplied by this factor.

2.5.3   Fluctuations in Number of Photons[15,16,17]

The average rate at which photons are emitted in a spectral interval $d\nu$ is

$$\bar{N} = \frac{2\pi\nu^2/c_0^2\, d\nu}{\exp\left[h\nu/kT\right] - 1},$$

and if the emission of a photon is again considered as an instantaneous, random event, the noise spectrum $p_N(f)$ of the emitted photons is

$$p_N(f) = \frac{4\pi A_s}{c_0^2} \int_0^\infty \frac{\nu^2 \, d\nu}{\exp\left[h\nu/kT\right] - 1} \tag{2.112}$$

$$= 3.042 A_s T^3 \times 10^{11}, \qquad A_s \text{ in cm}^2$$
$$T \text{ in deg K}$$

where we have used the subscript N to indicate a fluctuation in number. Thus the mean square deviation from the mean rate at which photons are emitted by a source in the fluctuation frequency interval between f and $f + \Delta f$ is

$$\overline{p_N^2} \equiv \overline{[p_N(t) - \bar{p}_N]^2} = 3.042 \, A_s T^3 \, \Delta f \times 10^{11}. \tag{2.113}$$

Again the use of Einstein-Bose statistics leads to a slightly different integral, namely,

$$p_N(f) = \frac{4\pi A_s}{c_0^2} \int_0^\infty \frac{\nu^2 \exp\left[h\nu/kT\right] d\nu}{\{\exp\left[h\nu/kT\right] - 1\}^2}, \tag{2.114}$$

or $$p_N(f) = 4.17 \, A_s T^3 \times 10^{11}. \tag{2.115}$$

If the source is a gray body, the right-hand side of 2.114 and 2.115 should be multiplied by the effective emissivity.

The limitation to the performance of infrared detectors due to fluctuations in power or number of photons from a radiating background is developed in Section 9.4.

2.6 LINE AND BAND SOURCES—INTRODUCTION

The power radiated in the infrared region of the spectrum by discontinuous sources may be ascribed to a number of mechanisms. The first of these is the movement of electrons from one discrete level of energy to another within the structure of a single atom. This is the mechanism responsible for the production of optical spectra. The changes in energy are relatively large so that the radiation produced is of short wavelength, that is, in the ultraviolet, visible, and very near infrared. Transitions of small energy difference occur with much smaller probability. For this reason the fabrication of a long wavelength pulsed source of high brightness in the infrared has been a formidable problem. A second cause responsible for the emission of discrete lines in the infrared is that of vibration of the atoms which make up a molecule. These vibrations produce electromagnetic radiation if they alter the electric moment of the molecule. Molecular structure has been extensively studied by investigation of the infrared absorption caused by molecular vibration. Still another mechanism, that of molecular rotation, produces photons of even lower energy. The wavelengths associated with these photons are comparatively long so that pure rotational spectra are found predominantly in the far infrared. Mechanisms such as the Raman effect[18] also produce infrared spectra. In the following discussion, emphasis is placed on those aspects of sources which are most practical for use with infrared systems.

2.6.1 Stationary States and Line Spectra

For many years physicists sought a theory capable of explaining the orderly arrangement of the spectral lines which they observed in their spectroscopes. The first attempts to explain spectral series utilized the notion of a "fundamental" plus various "overtones." Then Bohr presented his postulate of stationary states. According to this idea, the electrons in their classical orbits do not radiate continuously even though they are undergoing continuous centripetal acceleration. Instead, they persist in stationary states until such time as they are perturbed sufficiently to undergo an appreciable "jump" in energy. The difference in the two energies is then emitted or absorbed as a single quantum of energy, $\Delta \mathscr{E}$, whose frequency ν is given by the Einstein relation

$$\nu = \frac{\mathscr{E}_{\text{initial}} - \mathscr{E}_{\text{final}}}{h}, \qquad (2.116)$$

where $\mathscr{E}_{\text{initial}}$ and $\mathscr{E}_{\text{final}}$ are the energies of the initial and final states and h is Planck's constant. This concept clarified the well-known Rydberg formula

$$\frac{1}{\lambda} = RZ^2 \left[\frac{1}{n_2{}^2} - \frac{1}{n_1{}^2} \right], \qquad (2.117)$$

where R, the Rydberg constant, $= 2\pi m q^4 / c_0 h^3 = 1.09737 \times 10^7$ m^{-1}, $Z =$ atomic number, and n_2 and n_1, the principal quantum numbers, depend upon which lines and which series are being considered.

The theory of atomic or optical spectra is very complex and only a few of the simplest cases can be discussed using an approximation like that of Eq. 2.117. Strictly speaking, it is valid only for the circular orbits in hydrogenlike atoms. Although many extensions of the theory could be developed, we make no attempt to be complete. Rather, only a few refinements are mentioned so as to indicate some of the factors which have been considered by those workers who have correlated electronic energy transitions with specific spectral lines.

As a rather obvious example, it was recognized almost immediately that the electron and nucleus rotated about their common center of mass. This had the effect of requiring the use of a reduced mass $m' = mM/[m + M]$ (where m is the electronic mass and M is the mass of the nucleus) so as to compute a more accurate angular momentum. Some of the lines were found to be explainable if noncircular orbits were introduced. In addition, corrections were found necessary because of the magnetic moment associated with electron spin, for relativistic effects, and for the departures from a central-field attracting force with orbits of large eccentricity. Many of these modifications to Bohr's original theory cannot be adequately discussed using classical concepts.

2.6.2 Matter Waves and Wave Mechanics

In 1924 Louis DeBroglie suggested that all particles might have a wavelike nature in addition to a corpuscular nature. His speculations led him to the view that the wavelength associated with a particle would be related to its momentum, $\mathbf{p}$, by

$$\lambda = \frac{h}{|\mathbf{p}|}, \qquad (2.118)$$

where h is Planck's constant. These matter waves would have a wavelength far too short to be observable with macroscopic objects such as a baseball. On the other hand, the length of the waves would be comparable to the dimensions of the particle for bits of matter as small as an electron.

Following DeBroglie, a series of lectures given by Schrödinger laid the foundations for a wave mechanics. This work was continued by Born, Dirac, and others, and was given considerable impetus by the diffraction of electron waves reported by Davisson and Germer soon after DeBroglie's first paper on matter waves. Just as the classical development of atomic spectra could be touched on only briefly, so also our treatment of the subject of wave mechanics is confined to comments on its utility in dealing with phenomena on an atomic scale. Wave mechanics has shown itself to be an extremely useful tool for calculating the permissible energy levels within an atom or molecule. Schrödinger's wave scalar is related to the probability of finding the particle in a given position. In other words, suppose we have found a solution of the wave equation representing a standing wave in a particular region of space. We say that there is a high probability of finding the particle in those regions where the standing wave amplitude is large, and conversely, that there is small probability that the particle is occupying those regions where the standing wave has a node.

Using wave mechanics and numerical integration techniques, energy levels can be calculated with precision even for the penetrating orbits in heavy or complex atoms. Actually, it is somewhat paradoxical to speak of an orbit in conjunction with wave mechanics. However, the term "penetrating orbit" is often used to describe those electron configurations in which the wave function extends over a large range of distances from the nucleus. Under these circumstances, the situation is analogous to that of a classical electron which moves part of the time under the action of a central force and part of the time within an electron cloud due to the other orbital electrons which act as a partial shield for the nuclear charge. Since such a potential function is difficult to compute or to express analytically these configurations require laborious iterative procedures for their calculation.

The spectral frequency (or wavelength) has been either measured or calculated for all the lines of strong or intermediate intensity in the near infrared. Those who are interested in the wavelengths of the radiation emitted by specific discontinuous sources should consult the tabulated values for sources which have atomic spectra in the infrared region.[19]

2.6.3 Vibration-Rotation Spectra

It will be helpful in the discussion of vibration-rotation spectra to use a simple model of a diatomic molecule. In this model each atom is depicted as a sphere acted upon by a force along the line joining their centers. The relationship between this force and the distance of separation of the two spheres is shown in Fig. 2.26. At small distances the repulsive force is large; at very large distances the particles do not interact at all; and, in an intermediate range of distances the force passes through zero and becomes an attracting force.

It is reasonable to expect this type of dependence of the force upon distance because of the nature of the charge distribution in a typical atom.

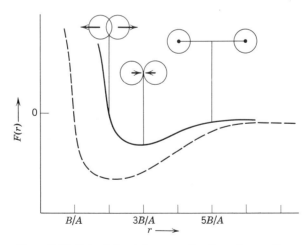

Figure 2.26. Potential and force as a function of separation.

At small values of r, the electron clouds are interpenetrating, and the attraction of each nucleus by the electrons of the other atom is reduced because the nuclei are partly inside the electron clouds. Since the nuclear charges are not as effectively screened during this interpenetration, they cause the net force to be one of repulsion. At slightly greater distances the mutual repulsion of the nuclei is overbalanced by the attraction of each nucleus by the electrons of the other atom. The resulting force is one of attraction, and is shown as an $F(r)$ less than zero in Fig. 2.26. At large distances this net force becomes insignificant. The dotted curve in Fig. 2.26 is a plot of the potential energy $V(r)$ of the system under discussion.

Experiment has shown that this curve can be represented by the sum of two simple terms, that is, $V(r) = -A/r + B/r^2$. Since potential functions in conservative systems are additive, the first of these terms may be considered as representing the long range attractive interaction and the second the short range repulsion. In this situation it turns out that

$$F(r) \equiv -\frac{d}{dr}[V(r)] = -\frac{A}{r^2} + \frac{2B}{r^3}. \tag{2.119}$$

Setting this equal to zero we find that a minimum of $V(r)$ occurs for $r = 2B/A$.

Expanding the function $F(r)$ about the point $r = 2B/A$ using Taylor's series, we have

$$F(r) = \frac{F(2B/A)}{0!} + \frac{F'(2B/A)(r - 2B/A)}{1!} + \frac{F''(2B/A)(r - 2B/A)^2}{2!} + \dots$$

For small displacements in the region of the minimum, the third term above and those of higher order may be neglected, leaving

$$F(r) = -\frac{A}{\left(\frac{2B}{A}\right)^3}[r - 2B/A]. \tag{2.120}$$

The fact that the restoring force is linearly dependent upon the displacement predicts that the atoms will execute simple harmonic motion.

For larger displacements the restoring force will be nonlinear and the wave form of the oscillatory motion will be complex. This is one of many situations we shall encounter where a wave train can be represented as composed of a number of sine waves of slightly differing frequencies. We therefore expect a broadening of vibrational bands where the radiation is of high intensity, that is, large amplitude.

The picture we have presented is typical of a situation which causes the appearance of resonance phenomena. A diatomic molecule is thus able to vibrate at a resonant frequency, ν, given by

$$\nu = \frac{1}{2\pi}\sqrt{\frac{k}{m^*}}, \tag{2.121}$$

where

$$k \equiv \frac{A}{\left(\dfrac{2B}{A}\right)^3}$$

and $m^* = m_1 m_2/[m_1 + m_2]$, the mean atomic weight of the two atoms involved. With HCl this frequency turns out to be approximately 9×10^{13} cps so that the wavelength is about 3.4 μ. A wave mechanical analysis of the linear oscillator reveals that the permitted frequencies are given by the condition

$$\mathscr{E}_{\text{permissible}} = (n + \tfrac{1}{2})h\nu_0, \qquad n = 0, 1, 2, 3, \ldots.$$

This is contrary to the implication of Section 2.3.2 that ν_0 is the frequency of an oscillator in the state of lowest energy. There we did not include the so-called half-quantum numbers. All the equations obtained in that section give results in agreement with experiment, however, since the additional $\tfrac{1}{2}h\nu_0$ adds a term to the Planck radiation formula which is independent of temperature.

The use of 2.121 aids in the calculation of the pure vibrational frequencies which can be generated by a diatomic molecule.

Superimposed upon this vibratory motion of the molecule are almost always effects due to its rotation. If the atoms are dissimilar, they will tend to form an electric dipole. With HCl, the chlorine tends to attract the electron from the hydrogen atom so that in the vicinity of the hydrogen there is an excess of positive charge and near the chlorine an excess of negative charge. Rotation of this pair then produces a variation in the electric field, the frequency of which corresponds to the speed of rotation. There is a quantum condition for this type of motion also, namely,

$$\mathscr{E}_{\text{permissible}} = \frac{J(J + 1)h^2}{8\pi^2 I_J}, \qquad J = 0, 1, 2, 3, \ldots \tag{2.122}$$

where J is the angular momentum quantum number and the I_J's are the rotational moments of inertia of the molecule about the two axes of rotation. The subscript J appears because the centripetal forces brought about by rotation causes a change in the spacing between the atoms and therefore a slight dependence of I upon J. Rotation about the line joining the atoms is not a permissible motion for a diatomic molecule. Diatomic

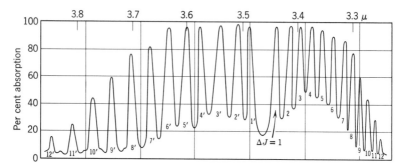

Figure 2.27. The principal absorption band of HCl in the near infrared. (After G. Herzberg, *Spectra of Diatomic Molecules*, Volume I, D. Van Nostrand Company, Princeton, New Jersey, 1950, p. 55.)

molecules obey selection rules requiring that $\Delta J = \pm 1$. In some polyatomic molecules $\Delta J = 0$ is also allowed by quantum mechanical selection rules.

When $\Delta J = \pm 1$ the difference in the energies associated with two adjacent spectral lines should be

$$\mathscr{E}_{\text{difference}} = \frac{2h^2}{8\pi^2 \bar{I}} \qquad (2.123)$$

where $\bar{I}$ is an average value of the various I_J's. Figure 2.27 shows that in this series of lines in the vibrational-rotational band of HCl no line appears for $\Delta J = 0$. The spacing of the innermost lines is thus $4h^2/8\pi^2\bar{I}$.

Pure rotational spectra have line spacings equal to those found within a vibrational-rotational band of the same type of molecule.

2.6.4 Polyatomic Molecules

In general, the vibrational spectra of polyatomic molecules are exceedingly complicated. This complexity occurs because the interactions among the atoms cause the motion of any one to excite motion on the part of the other atoms of which the molecule is composed. Very straightforward mathematics can be used to show that the number of possible and distinct vibrations can be specified by stating the number of degrees of freedom.

Molecules which contain more than two atoms are free to execute a great variety of motions. The number of vibrational degrees of freedom is given by $3N - 6$ where N is the number of atoms within the molecule. To see why this is so, we note that to specify completely the location of N atoms we need $3N$ coordinates. Six coordinates are needed to specify location and rotational orientation. If we consider only vibratory motion of the molecule, the center of mass does not move; hence 3 coordinates are sufficient to specify the position of the center of mass. Two more coordinates fix the orientation of some axis in the molecule. (For a linear molecule, this axis is most conveniently taken as the line joining the atoms. Since the atoms themselves are considered to be small and symmetrical, the energy of the system is not affected by rotation about this axis, hence $3N - 5$ coordinates are required to specify the vibratory motion of a linear molecule.) In nonlinear molecules one additional coordinate is needed to specify the orientation of the molecule about a specified axis, leaving $3N - 6$ coordinates dependent upon the vibrational state.

Suppose we consider a linear molecule composed of three atoms. It will have four vibrational degrees of freedom $[(3 \times 3) - 5]$. The interactions of the atoms are such that the various vibratory motions are likely to be coexistent. In order to analyze the behavior of a system of this kind, we must have recourse to the notion of normal coordinates and normal modes.

2.6.5 Normal Modes and Normal Coordinates

In Fig. 2.28 four types of motion are shown for the molecule AB_2. ν_1 is a bending mode that can occur in either of two planes. ν_2 is a mode in which the end atoms move in unison toward and away from the central atom. ν_3 is a mode in which the end atoms move in the same direction, whereas the central atom always moves in the direction opposite to their motion. If any one of these motions is initiated, it will persist; the atoms will execute simple harmonic motion (provided the amplitude is small), and the atoms will move in phase. That is, if they are started from positions at the tips of the arrows in any of the motions shown in Fig. 2.28, they will all return to these same positions after one period. It can be shown that any other motion of the system can be represented as the superposition of a number of these normal vibrations. A few of these have zero frequency and actually represent translational motion of the atom.

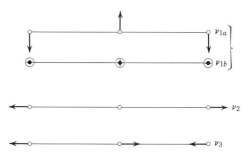

Figure 2.28. Vibrational modes for a triatomic linear molecule. (After G. Herzberg, *Infrared and Raman Spectra*, D. Van Nostrand Company, Princeton, New Jersey, 1945, p. 66.)

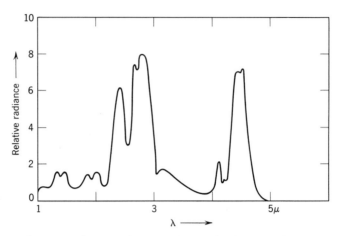

Figure 2.29. Spectrum of Bunsen flame. (After A. G. Gaydon, *The Spectroscopy of Flames*, John Wiley & Sons, New York, Chapman & Hall, Ltd., London, 1957, p. 166.)

The foregoing rather elementary example illustrates types of motion which can exist in even a very simple molecule. Since the acceleration of electric charge gives rise to the emission of electromagnetic radiation whose frequency, or wavelength, is related to the periodicity of the motion it can be seen that each of the motions shown in Fig. 2.28 gives rise to radiation of a characteristic frequency or wavelength. With complex molecules there are very many possible motions even when these have been resolved into their simplest form, that is, by the use of the concept of normal modes. Therefore, these molecules have many different lines or bands in their spectra, and each can be attributed, at least in theory, to one of the possible modes of motion.

As an illustration of the complexity of spectra that can arise from a simple low pressure Bunsen flame, see Fig. 2.29. Vibrational-rotational spectra are known for many types of molecules. These spectra need to be considered for several reasons. First, any combustion process generates band spectra. In burning hydrocarbons, large clouds of heated CO_2 and H_2O vapor are produced and these molecules radiate a series of lines or bands in the region between 0.69 and 15.5 μ[20]. Second, the absorption in these bands affects the performance of many infrared systems. The transmission of infrared through a long path in the atmosphere is often strongly dependent upon the concentration of CO_2 and H_2O in the vapor phase. These topics are discussed more completely in Chapter 5. Vibrational-rotational spectra are also important in some infrared system applications because of their contributions to the background radiation.

2.6.6   Line Breadth

All spectral lines have a finite width. The line width, δ, is defined as the wavelength difference between two points, one on each side of the wavelength of maximum intensity, at which the line intensity is half as great as at the wavelength of maximum intensity. (See Fig. 2.30.) Line broadening may be due to one or several of the causes discussed below.

Natural line width. According to classical theory, every line should have a finite width. An electron which is radiating energy would be expected to radiate a damped wave train of finite length. A wave train of this sort can be synthesized using a spectrum of undamped sine wave trains many cycles long but differing slightly in frequency. Similarly, from the wave mechanical point of view, there is an indeterminancy in the energy associated with a given wave function. The energy could be measured exactly if an infinite time were available for the measurement. However, one form of the uncertainty principle tells us that

$$(\Delta W)(\Delta t) \geq \frac{h}{2\pi}, \tag{2.124}$$

where ΔW and Δt are the uncertainties in measuring energy and time. The Δt applicable in atomic spectra may be considered to be the radiative lifetime, t_r, of the particular excited state under consideration. Then, if the decay is to the normal electron configuration of the atom, which state can be thought of as having an infinite lifetime,

$$\Delta W = \frac{h}{2\pi t_r},$$

and since

$$\Delta \nu = \frac{\Delta W}{h} = \frac{1}{2\pi t_r},$$

then

$$\delta = \Delta\lambda = \Delta\left(\frac{c_0}{\nu}\right) = -\frac{c_0}{\nu^2}\Delta\nu = \frac{\lambda^2}{c_0 2\pi t_r}.$$

If the transition is between two levels, each of which is an excited state for the atom, then

$$\delta = \Delta\lambda = \frac{\lambda^2}{2\pi c_0}\left(\frac{1}{t_{r_1}} + \frac{1}{t_{r_2}}\right). \tag{2.125}$$

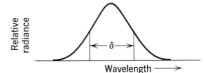

Figure 2.30

Doppler broadening. The width of a spectral line is also affected by the motion of the radiating atom along the line connecting the atom and the observer. According to Doppler's principle, atoms moving toward the observer will appear to radiate a wave of slightly higher frequency, that is, shorter wavelength, whereas atoms moving away will seem to radiate a lower frequency, or longer wavelength. The magnitude of the effect depends upon the velocity of the radiating atoms relative to the observer. In order to calculate the profile of a spectral line when only Doppler broadening is considered, we proceed as follows. If N_T represents the number of atoms emitting radiation at or near the discrete wavelength λ_0, then the number of atoms of this group which have a velocity between v_x and $v_x + dv_x$ is given by

$$dN \equiv N_{v_x} \, dv_x = N_T p(v_x), \qquad (2.126)$$

where N_{v_x} is the number of atoms per unit velocity interval at v_x,
$p(v_x)$ is the probability an atom has a velocity between v_x and $v_x + dv_x$. (Here v_x is chosen as the velocity component along the direction of observation.)

From Maxwell-Boltzmann statistics, (see 2.30), $p(v_x)$ is given by

$$p(v_x) = \sqrt{\frac{M}{2\pi R T}} \exp\left[\frac{-M}{2RT} v_x^{\,2}\right] dv_x, \qquad (2.127)$$

where M is the gram molecular weight,
R is the universal gas constant, and
T is the temperature.

Since v_x is the velocity of the radiating atom along the direction of observation, the relative change in the wavelength emitted by that atom is

$$\frac{\lambda - \lambda_0}{\lambda_0} = \frac{v_x}{c_0} . \qquad (2.128)$$

Combining 2.126, 2.127, and 2.128 we obtain

$$N_\lambda \equiv dN = N_T \sqrt{\frac{M}{2\pi R T}} \exp\left[\frac{-Mc_0^{\,2}}{2RT\lambda_0^{\,2}} (\lambda - \lambda_0)^2\right]. \qquad (2.129)$$

The spectral interval, δ, between those wavelengths where $N_\lambda = \frac{1}{2} N_{\lambda \max}$ (which occurs at $\lambda = \lambda_0$) is

$$\delta = \frac{2\lambda_0}{c_0} \sqrt{\frac{2RT}{M} \ln 2} . \qquad (2.130)$$

If T is expressed in degrees absolute ($^\circ$ K) and M in gram molecular weight, then δ is given by

$$\delta = 7.2 \times 10^{-7} \lambda_0 \sqrt{T/M}. \qquad (2.131)$$

Pressure broadening. Several factors combine to cause broadening of emitted spectral lines when many atoms are present in a confined space. The energy levels of an atom are shifted slightly when other atoms of a gas are in close proximity. This effect is small, however, by comparison with the effect due to interruption of the trains of radiation by collision. Each collision causes a discontinuity to appear in the radiated train of waves, either a change of phase or a large amount of damping. At high pressures where collisions are frequent, the pulses radiated between collisions consist of a smaller number of cycles.

It can be shown[21] that the spectrum of frequencies present in a wave train broadens as the number of cycles in the train decreases. Pressure broadening of a spectral line may be very large in a glow discharge or arc at high pressure. In these instances the lines may be so smeared that the wavelength distribution may begin to approximate that of a black body. This situation will exist when the mean free path of the radiating atoms becomes small by comparison with the dimensions of the radiating source. Radiative equilibrium is achieved only when the mean free path for internal absorption of the radiation in question is much smaller than the dimensions of the system. With systems whose dimensions are small compared to the mean free path for absorption of photons the actual radiation intensity from the source will be approximately the black body value diminished by the ratio of the dimensions of the system to the mean free path for the absorption of the photons.[22] This relationship is valid only for the continuum and not for the lines.

2.7 PRACTICAL SOURCES

It is frequently necessary to choose an infrared source for experiments to be performed in the laboratory. Precise spectral measurements demand a source of large spectral radiance in the wavelength interval of interest. On the other hand, if a system or component is being examined to determine its ability to resolve closely spaced objects, the source characteristic most important is large total radiance.

High temperature black body. To obtain large values of total radiance, the ideal source would be a black body maintained at a high temperature. Figure 2.31 shows a source which constitutes a good approximation to a black body operable at temperatures as high as 3000°K. A source of this kind can be made in the following way. One mil tungsten ribbon 7/8 in.

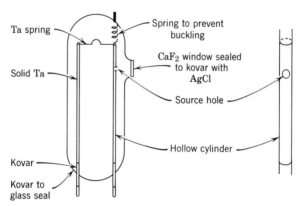

Figure 2.31. High temperature black body.

wide is rolled on a 1/8 in. diameter copper mandrel and seamed by a series of overlapping spot welds; a hole approximately thirty thousandths of an inch in diameter can then be cut in the foil and the copper dissolved out. The resulting hollow cylinder is mounted by inserting tapered tungsten rods at its ends. The figure shows other constructional details.

Nernst glower and Globar. The source described above is difficult to fabricate and not commercially available. Of the infrared sources in common use, two of the most useful are the Nernst glower and the Globar. The Nernst glower is a mixture of zirconium, yttrium, and thorium oxides in the form of a hollow rod approximately 25 mm long and 2 mm in diameter. This source may be obtained from the Stupakoff Ceramic Company under the trade name *Insulcon*. It operates at a temperature of 1800°K in air and must be supplied by a well stabilized power supply capable of providing 90 watts at 75 volts ac. The primary shortcomings of the Nernst glower are:

1. Negative temperature coefficient of resistance requiring a ballast tube in the supply circuit.
2. Low mechanical strength.
3. Susceptibility to temperature variations caused by air currents.
4. Low temperature of operation (1800°K).
5. Small size.

The Globar, a rod of bonded silicon carbide, can be obtained from the Perkin-Elmer Corporation. The most common form is that of a cylinder about 2 in. long and 3/16 in. in diameter. It also requires a ballasted supply delivering 180 watts at approximately 50 volts. It has an approximate lifetime of 250 hours when operated at 1500°K in air. Its disadvantages are:

1. Comparatively low maximum operating temperature.
2. Need for a ballasted supply.
3. The sublimation of material from its surface may cause difficulty when used near delicate optical surfaces.
4. Higher temperatures require water cooling of electrode attachment points.
5. Small size.

Conical cavity black body. Still another useful source, particularly for testing detectors, is described in an NBS report.[23] It consists of a conical cavity of half angle 15° machined in a copper cylinder which is allowed to oxidize during operation. (See Fig. 2.32.) Joulean heat is supplied to a noninductively wound element wrapped on the outside of the cylinder. Both heater and copper are encased in an insulating lava tube so that the end of the cylinder containing the conical cavity can be viewed through a small aperture. Such a source has a comparatively long life.

Welsbach mantle. A source can be prepared in the form of a cloth mantle impregnated with thorium oxide containing a small amount of cesium oxide. This arrangement is commonly seen on gasoline lamps and lanterns and is known as a Welsbach mantle. It can be operated at 2400°K. Its emissivity is rather low in the region between 1 and 6 μ but beyond 6 μ it approximates a black body.

Carbon arc. The carbon arc is also used as a source of infrared radiation. Its emission can be considered as approximately characteristic of a 3500°K black body when operated with a typical current of about 5 amp and a 40 volt drop across the gap. The addition of metallic salts to the carbon rods is a means of altering the wavelength distribution of the light emitted by the arc.

In general, all arcs are capable of serving as sources of very high radiance.

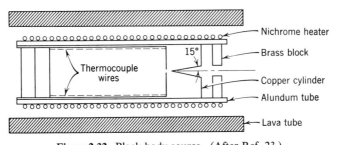

Figure 2.32. Black body source. (After Ref. 23.)

Unfortunately for those concerned with the utilization of infrared sources, the extremely high equivalent black body temperature has the effect of decreasing the usefulness of these sources due to the Wien shift of the radiation toward the ultraviolet and the consequent lowering of the fraction of the total energy emitted at long wavelengths.

Discharge arc. Electrical discharges in gases can also serve as sources. One of their unique advantages is the possibility of pulsed operation. The electronic character of the discharge permits modulation of the emitted radiation by modulation of the electrical power supplied. Several commercially available sources of this type are the low pressure mercury vapor lamp, the high pressure mercury vapor lamp, and the sodium lamp. The long wavelength limit of the aforementioned lamps is generally imposed by the transmission of their envelopes (usually quartz).

Although most discharge lamps emit little infrared radiation, their output being mainly in the ultraviolet and visible regions, the xenon lamp exhibits some output in the near infrared. Because of this, and its capability of being modulated at audio frequencies, it is used in infrared communication systems. Figure 2.33 depicts the spectrum of a xenon flashtube having characteristics similar to the modulated discharge lamps. Note the spectral lines superimposed upon the continuum.

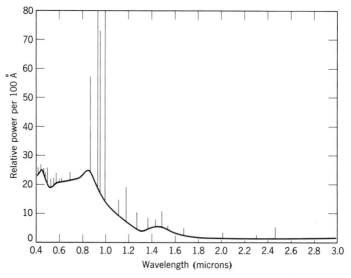

Figure 2.33. Spectral emission from Xe flashtube. [After R. J. Uhl, *M.I.T. Technical Memorandum* 7668-TM-3 (April, 1958) from data supplied by General Electric, Nela Park, Cleveland, Ohio.]

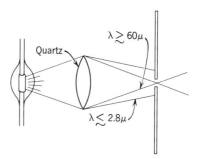

Figure 2.34

The maximum attainable radiance of electrical discharge sources is imposed ultimately by the fact that as they are modified so as to dissipate greater and greater amounts of electrical power, their line spectra broaden and approach that of a black body. As a result, they are subject to the same limitations as an incandescent solid. For example, a cesium discharge at 100 atmospheres is very nearly a black body so that the upper limit on its radiance is given by the Planck radiation law.

Long wavelength source. Rubens produced a good source of wavelengths greater than 60 μ by utilizing the marked wavelength dependence of the index of refraction in quartz. Figure 2.34 shows his arrangement. Radiation from an intense source was passed through a quartz lens shaped and oriented in such a way that those wavelengths greater than approximately 60 μ were focused on an aperture ($n_{60\mu} = 2.1$), whereas radiation in the wavelength interval between 2.8 and 60 μ was absorbed because of the natural opacity of quartz in this interval. The transmitted radiation of wavelength less than 2.8 μ was intercepted by the aperture plate because it was not properly focused on the aperture ($n_{2\mu} = 1.5$).

2.8 SOURCES ENCOUNTERED OUTSIDE THE LABORATORY

The sun. For certain types of experimental work, the sun, because of its very high intrinsic brightness, provides an excellent source of infrared radiation. The wavelength distribution in the solar spectrum, as observed through the atmosphere, is approximated by that of a black body at 5600°K.

Terrain. For approximate engineering calculations regarding the general level of background radiation, it will sometimes be convenient to assume that terrain is a gray body having an effective emissivity of 0.35. Its temperature is the temperature which would be indicated by a thermometer located in the same terrain. For example, suppose we wish to compute the normal radiance from a square meter of terrain whose

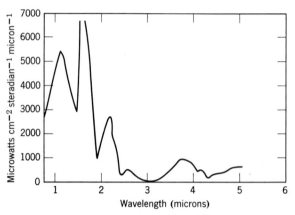

Figure 2.35. Distribution of sky background. [After E. E. Bell, et al., *Final Engr. Rpt.*, Contract No. 33(616)-3312, Ohio State University, Columbus (October 1957), AD 151221, p. 69.]

temperature is $50°F = 283°K$. Assuming terrain is a Lambertian radiator, we have

$$R_{n\omega} = \frac{\varepsilon\sigma T^4}{\pi} = \frac{(0.35)(5.67 \times 10^{-8})(283)^4}{\pi}$$

$$= 40.4 \text{ watts m}^{-2} \text{ steradian}^{-1}.$$

This calculation is concerned only with the radiation leaving the surface. It should be borne in mind that when an exact figure is needed, it will be necessary to refer to experimental data for the emissivity of the particular surface under consideration. Both the emissivity and the temperature depends on the nature of the vegetation and will, of course, be different for such surfaces as asphalt and concrete. These various surfaces come to different temperatures because each of them has an emissivity that varies differently with wavelength. Each, therefore, increases in temperature until it is radiating as much power at long wavelengths as it is absorbing from the sun at comparatively short wavelengths. For this reason infrared systems are able to distinguish such features as roads and trees when mapping the terrain.

Clouds. Clouds function quite efficiently as reflectors of radiation emanating from the sun and the earth. Figure 2.35 shows the spectral distribution of radiation reflected from clouds. Such a distribution seems plausible when one considers that the thermal radiation emitted by the sun should be expected to peak in the vicinity of 0.5 μ, whereas the terrain radiation will, in many cases, peak at about 9.6 μ. The lower surface of a cloud has radiating characteristics typical of a body at 0°C, whereas the upper surface of a cloud radiates as though it were a $-40°C$ black body. Although the radiance of cloud surfaces is low, the total amount of power they radiate can be large because of their very great size.

Hot gases.[24, 25, 26] Combustion processes involving hydrocarbons give rise to CO, CO_2, and water vapor. Because of the proximity of these products to the exothermic reaction, the molecules are generally in an excited state so that they are emitting vibrational and rotational spectra in the infrared. As an example, the main features of a Bunsen flame have been described by Plyler.[27] (See Fig. 2.29.) There are two very intense bands, one at 4.4 μ and another at 2.8 μ. The 4.4 μ band is associated with CO_2 vibration. The superposition of CO_2 and H_2O emission is responsible for the band at 2.8 μ. At the same time the carbon particles in the flame give off a black body continuum.

The complex processes involved in the emission from flames and hot gases constitute the material for many books. One of the complicating aspects of this topic is that rarefied vapors emit as line or band sources, but these undergo a gradual transition toward behavior as thermal

TABLE 2.2. **Normal spectral emissivities at 295 K**

Wavelength in Microns

Material	1.0	2.0	3.0	4.0	5.0	7.0	9.0	10.0	12.0	14.0
Aluminum	0.26	0.18	0.12	0.08	0.07	0.04	0.03	0.02	0.02	—
Cadmium	0.30	0.13	0.07	0.04	0.04	0.02	0.02	0.02	0.01	0.01
Chromium	0.43	—	0.30	—	0.19	—	0.08	—	—	—
Copper	0.10	—	0.03	—	0.02	—	0.02	—	—	—
Gold	0.62	—	0.03	—	0.02	—	0.02	—	—	0.02
Graphite	0.73	0.65	0.57	0.52	0.49	0.46	0.42	0.41	—	0.37
Iron	0.35	0.22	0.16	0.12	0.09	0.07	0.06	—	—	0.05
Lead	—	—	—	—	0.08	—	0.06	—	—	0.04
Magnesium	0.26	0.22	0.20	0.16	0.14	0.09	0.07	—	—	—
Molybdenum	0.42	0.18	0.12	0.10	0.08	0.07	0.06	0.06	0.05	—
Nickel	0.27	—	0.12	—	0.06	—	0.04	—	—	—
Silver	0.04	—	0.03	—	0.03	—	0.01	—	0.01	0.01
Tungsten	0.38	0.10	0.06	—	0.05	—	—	0.04	0.04	—
Steel										
untempered	0.37	0.23	0.17	—	0.11	0.07	0.07	—	—	0.04
Stellite	0.31	0.25	0.21	0.18	0.15	—	0.12	—	0.11	—
Brass,										
Trobridge	—	0.09	—	—	0.04	0.03	0.02	—	—	—

[Adapted from Table 6g-4, page 6–72, of *American Institute of Physics Handbook*, McGraw-Hill Book Co., Inc., New York (1957).]

radiators when the assemblage of radiating molecules becomes large and dense. For example, in hydrocarbon flames about 20 in. thick the radiance of the 4.4 μ CO_2 band approaches that of a black body.

Metal surfaces. A good figure for the hemispherical emissivity of a typical piece of sheet stock from a rolling mill is approximately 0.35 in the near infrared. It has been observed that the emissivity decreases for radiation of longer wavelength. There is also a slight increase of emissivity with increasing temperature. For a brief summary of emissivities, see Table 2.2.

Field infrared source. LaRocca and Zissis[28] have described black body sources useful in the field. The sources were metal cones coated internally with black enamel. One was heated by the ambient air; the other by a water jacket with provision for heating and circulating the water to maintain a variable but uniform temperature. It was estimated that the emissivity of such sources is 0.99 or greater.

2.9 OTHER SOURCES

Irasers and lasers. An intense source of infrared radiation can be devised using the "maser" principle. This type of source is very directional, is concentrated within a narrow spectral interval, and possesses a high degree of coherence. The word *maser* is an acronym for *m*icrowave *a*mplification by *s*timulated *e*mission of *r*adiation. The term *infrared maser*, which implies that both infrared and microwaves are involved, is therefore contradictory. The terms *iraser* (*i*nfrared *a*mplification by *s*timulated *e*mission of *r*adiation) and *laser* (*l*ight *a*mplification by *s*timulated *e*mission of *r*adiation) are properly applied to devices operating as amplifiers but are often also applied to devices operating as oscillators. Bloembergen[29] has proposed a maserlike device called an "infrared quantum counter" which has some advantages over conventional infrared detectors for certain special applications.

In this section we shall discuss irasers and lasers as sources of radiation. Such sources may function either as oscillators or as amplifiers to increase the intensity of other sources. In Chapter 8 we shall discuss irasers and narrow band quantum counters used as detectors, and in Chapter 9 their ultimate performance.

The principle of operation of the iraser and laser has been described by Schawlow and Townes[30] and by Dicke.[31] It is an extension of the maser principle first described by Gordon et al.[32] for a gas and later by Bloembergen[33] for a solid, based upon the concept of population inversion of atoms distributed among two or more energy levels.

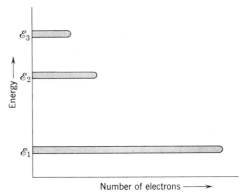

Figure 2.36. Distribution of electrons among energy states.

The operation of an iraser or laser source may be clarified by the following qualitative explanation. In Section 2.6 we mentioned that the electrons surrounding the nucleus of an atom are arranged in a number of discrete energy levels. The electrons may undergo transitions in energy between these levels. Transitions between certain pairs of levels are permitted by quantum mechanical selection rules; others are forbidden.

As was commented upon in Section 2.3.2, the question as to how particles are distributed in energy arises frequently in physics. Figure 2.36 shows a typical distribution of electrons among energy levels. The population distribution shown is typical of the so-called high energy tail of the distribution specified by a Fermi-Dirac distribution function. (See Eq. 2.31.) That is, on the average the number of electrons per level decreases as the energy increases. There will be many more electrons in the lower lying levels (close to the nucleus) than in the higher levels. Specifically, the average number per level is, to a good approximation, exponentially dependent upon the negative of the energy. Under equilibrium conditions the average population distribution does not change with time. However, the individual electrons undergo excitation to higher energy levels from which they may decay by one or more steps to the lower energy levels.

The rate at which electrons make transitions between energy levels depends upon the number of electrons in the initial state and upon the transition probability, which is determined by quantum mechanical considerations. If in the lower energy level there is an excess of electrons over the equilibrium value, more electrons will undergo a transition upward in energy per unit time than downward from the higher of the two energy levels to the lower. Conversely, if the higher energy level has a population greater than the equilibrium value, the rate of transitions downward in energy will exceed the rate of transitions upward in energy.

Thus far we have been discussing spontaneous transitions. In addition to those shifts from one energy level to another which occur spontaneously, there is another type of transition called an induced transition. The presence of radiation of wavelength $\lambda_{12} = hc_0/(|\mathscr{E}_2 - \mathscr{E}_1|)$ increases the probability that an electron at energy level 1 will jump to energy level 2, where levels 1 and 2 have energies which differ by an amount which satisfies the above relationship. These induced transitions can take place either upward or downward in energy. In the first case the process is accomplished by the annihilation of a photon. Induced downward transitions are not accompanied by the annihilation of a photon.

If a sufficiently intense source of radiation of the proper wavelength, that is, λ_{13}, is present, the so-called pumping radiation, it can induce enough transitions from a lower energy level to an upper energy level so that the upper level is more heavily populated than some level between the two. Such a situation, while it persists, is called a population inversion. It can be shown that the rate of induced transitions is proportional to the number of electrons in the level corresponding to the initial state of the transition. Therefore if photons of energy $|\mathscr{E}_3 - \mathscr{E}_2|$ are present, they will cause more transitions from the upper level to the intermediate level than vice versa. Since each transition from the upper to the intermediate level is accompanied by the emission of a photon and since the photon inducing the downward transition is not absorbed, there will be more photons emitted than absorbed. Thus an increase of the initial number of photons will occur at the expense of the pumping radiation.

To insure the continuing presence of photons whose energy is proper to stimulate the emission of the desired radiation, the processes just described must take place in a resonant cavity. The dimensions and shape of the

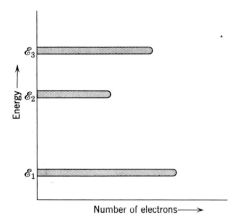

Number of electrons⟶   **Figure 2.37.** Population inversion.

cavity are somewhat critical. The cavity should be constructed so that very precise adjustments can be made with regard to its length so that a single frequency and mode can be selected. It must be possible to make the length of the cavity some arbitrary integral multiple of $\lambda/2$, where λ is the wavelength to be stimulated. The ends of the cavity must be highly reflective so that when an induced downward transition occurs, the resulting wave will traverse the cavity many times. By this action it will be able to induce other downward transitions before being absorbed by the walls, or passing out the ends of the cavity in the form of useful radiation. Each new downward transition is caused by the standing wave to occur in such a way that the radiation produced by it acts in phase with the wave already present. The cavity thus acts to bring about a high degree of coherence in the emitted radiation.

To sustain a single oscillatory mode in the cavity, the radiation must impinge on the end wall within a very small solid angle. The acceptance angle for radiation that can be amplified is of the order of $(\lambda/d)^2$, where d is the diameter of the source.[34] For $\lambda = 5\,\mu = 5 \times 10^{-4}$ cm and $d = 1$ cm the acceptance angle is about 2.5×10^{-7} steradian.

According to A. Gamba,[35] an assembly of excited molecules may emit a spectral line of width less than the "natural" width. This is a quantum-statistical effect arising from correlations between excited molecules and emitted quanta. The assembly of excited molecules within an iraser and the consequent emission of highly monochromatic radiation makes the iraser a desirable source for some applications. The iraser and laser can function as a continuous source only if the "pumping" radiation supplies electrons to the upper state at a rate which more than offsets their depletion by loss of radiation at the walls and through the ends of the cavity. If this condition is not satisfied, the device must be operated as an amplifier or as a pulsed source.

The greatest development effort has been devoted to ruby lasers operating in a pulsed manner. Synthetic ruby, which is aluminum oxide, Al_2O_3, doped with about 0.1 % chromium oxide, Cr_2O_3, is used to produce stimulated emission at 6943 Å, due to electron transitions within the Cr^{3+} ion.[36, 37] Although variations are found, a typical ruby laser will be described. Ruby rods, having dimensions of 5 cm by 0.5 cm, are used. The ends, which are flat to $\frac{1}{2}$ the wavelength of sodium light and parallel to 30 sec of arc, are coated with silver such that one end is totally reflecting and the other has a transmission of 1 to 5 %. The rod is mounted along the axis of a helical xenon flashtube. A 400 μf capacitor charged to 4 kv is discharged through the tube, producing a flash equivalent to a 5000°K source. Radiation at 4150 Å, 4750 Å, and 5600 Å is absorbed of sufficient intensity to invert the populations of electrons in the appropriate

Cr^{3+} levels. Induced emission occurs at 6943 Å with a peak power of 10 kw in a pulse duration of about 0.5 msec. The output power is less than 5% of the input power. The radiation emitted through the partially transmitting end is confined to a beam of one-third degree angular subtense.

It has been observed that breakdown within the rod occurs in a number of separate regions. Furthermore, the stimulated emission occurs as a series of apparently random microsecond pulses distributed over the total pulse period.[37] It has been postulated that each pulse is linked to emission from a single filamentary region.[38]

That the light emitted at 6943 Å is coherent has been determined by observing the diffraction pattern arising when a grating is interposed in the emitted beam. The line width is about 0.15 Å, although ruby rods can be found which are sufficiently strain free for the line width to be about 0.05 Å.

The emitted radiation has a brightness temperature of the order of 10^{10} °K. By focusing the radiation to an extremely small spot, it has been found possible to cause a plasma jet to be emitted from a carbon rod by a single pulse. In order to minimize overheating of the ruby rod by internal absorption from the lamp, cooled nitrogen gas is circulated within the housing which encloses it and the flash lamp. The reduced temperature also serves to reduce the rate of spontaneous downward transitions and narrow the natural width of the 6943 Å line.

Other solid materials[39, 40] used in pulsed operation include Sm^{++} doped CaF_2 emitting at 7082 Å and U^{+++} doped CaF_2 operating at 2.5 μ. These rare earth doped crystals operate at a pumping power of roughly 10% of that for ruby, and require cooling to cryogenic temperatures.

The necessity for low duty cycle pulsed operation limits the utility of the lasers discussed above. Javan[41] has reported a continuous wave iraser operating at 1.15 μ with a mixture of helium and neon gases contained in an 80 cm long tube excited by a 50 watt rf discharge. Exchange of energy by resonant collisions between excited He atoms in metastable states and Ne atoms in the ground state causes a population inversion within the Ne atoms. The observed line width is of the order of 10^{-6} Å. A Fabry-Perot interferometer provides a cavity capable of precise wavelength selection, the same function performed by the silvered ends of crystalline lasers. The output power is 15 mwatts.

In order to convey information in either pulsed or cw operation it is necessary to modulate the beam of stimulated radiation. It is a formidable technical problem to take full advantage of the information capacity of laser and iraser sources.

Recombination radiation.[42, 43] Some intrinsic semiconductors can be utilized for the production of nearly monochromatic infrared radiation by a process called radiative recombination. A free electron-hole pair can be produced by the absorption of a photon. (See Section 7.1.) Recombination radiation results from the inverse process, namely, the production of photons by direct recombination of an electron-hole pair. (See Section 6.8.) The wavelength of maximum spectral radiance occurs for $\lambda = hc_0/\mathscr{E}_i$ where $\mathscr{E}_i$ is the forbidden band gap. This phenomenon does not provide an intense source of radiation. However, judicious shaping of the specimen improves the utility by concentrating the radiant flux in a particular direction.[41]

Microwave oscillators. Tuned cavity resonators such as the klystron and magnetron can be used to generate very long wavelength infrared. Workers at Columbia University have been able to generate more than a kilowatt of peak power at a wavelength of 3000 μ (3 mm) using a pulsed magnetron.[45] The reported efficiency of a few per cent was far greater than that obtained by most of the others who have experimented in this field. The dimensions of the resonant cavity must be comparable to the wavelengths being generated and must be maintained constant within very small limits in spite of temperature fluctuations and mechanical stresses. The first requirement is alleviated somewhat by taking some high harmonic as the output frequency. Even so, a continuous output of a few milliwatts is difficult to attain at wavelengths of a few millimeters (a few thousand microns). One outstanding advantage of conventional radar oscillators for producing submillimeter waves is that they produce coherent radiation.

Incoherent oscillators. On the basis of classical electromagnetic theory, it can be shown that currents created in extremely small metallic particles will radiate in a portion of the spectrum where the wavelength is comparable to the dimensions of the particles. A number of different types of particles have been used. Several methods have been used to excite the particles: 1 mm diameter metal particles have been suspended in oil and passed through a spark gap,[46] highly charged mercury droplets have been dropped into a grounded pool of mercury so that the resulting charge deceleration produced the radiation,[47] and fine copper particles have been injected into the plasma stream of a spark discharge.[48] The intensity of these sources is extremely small and, in addition, the incoherence of the radiation limits their utility and renders them no more valuable as sources than incandescent solids. The intensity of the radiation can be increased if large numbers of particles are used, but since the radiation is incoherent, the interactions among particles cause them to behave like a dense gas and approach a black body distribution.

Cerenkov radiation. If a particle moving in an optically dense medium attains a velocity $v > c_0/n$ where n is the index of refraction of the medium and c_0 is the velocity of light in free space, it will emit radiation into a cone of half angle ϕ where $\phi = \text{arc cos } c_0/nv$. This radiation is called Cerenkov radiation. Unfortunately, several competing processes, nuclear collisions, and ionization losses along the path of the particle reduce the intensity of this type of radiation to very low levels.

The Raman effect. Under certain conditions the wavelength of monochromatic radiation may be changed when incident upon a scattering medium. This effect, called the Raman effect after its discoverer, sometimes produces radiation lying in the infrared portion of the spectrum. For a more thorough discussion of this, the reader is referred to Herzberg.[49]

Bremsstrahlung. Classical electrodynamics predicts that the deceleration of a charged particle will give rise to radiation. Such radiation has been observed and is called Bremsstrahlung (braking radiation). If the deceleration is small, the radiation may be in the infrared region, although normally it lies at much shorter wavelengths.

2.10 USEFUL APPROXIMATIONS FOR ENGINEERING CALCULATIONS

There will be many times when it would be helpful to find an approximate solution of an infrared problem. If, for example, a new type of system is being considered, a quick calculation will often help to decide between alternate modes of operation or choices of components. In some cases it may provide a basis for a decision that the proposed system is not feasible. Here and in Chapter 10 we shall list several convenient "rules-of-thumb" that are helpful in making such approximate calculations.

A convenient rule to use in estimating the spectral distribution of the energy emitted by a gray or black body is Wien's displacement law

$$\lambda_{\max} T = 2893 \ \mu \ \text{deg K}, \qquad (2.132)$$

where $\lambda_{\max}$ is the wavelength of maximum spectral radiance. Twenty-five per cent of the total power emitted by a black or gray body is of wavelength less than the $\lambda_{\max}$ computed from 2.132.

Another convenient rule is

$$\lambda_{1/2} T = 1800 \ \mu \ \text{deg K and } 5100 \ \mu \ \text{deg K}, \qquad (2.133)$$

where the $\lambda_{1/2}$'s are the wavelengths at which the spectral radiance is one-half of that at $\lambda_{\max}$. Approximately 3.8% of the total radiation lies below the lower $\lambda_{1/2}$ calculated from 2.133, whereas 35% of the power is radiated at wavelengths greater than the larger of the two $\lambda_{1/2}$'s.

Two types of radiation slide rules are available for computing the spectral distribution and the total radiant emittance or radiance of black or gray sources. These rules contain conversion scales of various kinds which

make them useful in other types of calculations such as comparisons of
the response of radiation detectors and the transmission of lens materials.*

REFERENCES

1. P. M. Morse, *Vibration and Sound*, McGraw-Hill Book Co., New York (1948), p. 389.
2. F. W. Sears, *Mechanics, Heat and Sound*, Addison-Wesley Press, Cambridge, Mass. (1950), p. 456ff.
3. D. ter Haar, *Elements of Statistical Mechanics*, Rinehart and Co., New York (1954).
4. R. W. Gurney, *Introduction to Statistical Mechanics*, McGraw-Hill Book Co., New York (1949).
5. J. E. Mayer and M. G. Mayer, *Statistical Mechanics*, John Wiley and Sons, New York (1940).
6. M. Michaud, Sc.D. Thesis, University of Paris (1951).
7. P. D. Foote, *J. Wash. Acad. Sci.* **5**, 1 (1915).
8. C. M. White, *Proc. Roy. Soc. (London)* **A123**, 645 (1929).
9. W. R. Wade, *NACA Technical Note 4206*, Langley Aero. Lab., Langley Field, Virginia (1958).
10. W. R. Wade, *NASA Memorandum 1-20-59L*, Langley Research Center, Langley Field, Virginia (1959).
11. E. Schmidt, *Gesundh-Ing. Berhefte Ser. 1*, No. 20, **70**, 885 and 947 (1927).
12. E. Schmidt, *Forsch. Ing.-Wes.* **5**, 1 (1934).
13. E. Schmidt and E. Eckert, *Forsch. Ing-Wes.* **6**, 175 (1935).
14. J. A. Stratton, *Electromagnetic Theory*, McGraw-Hill Book Co., New York (1941), p. 494.
15. J. R. Carson, *Bell System Tech. Jour.* **10**, 374 (1931).
16. R. C. Jones, *J. Opt. Soc. Amer.* **37**, 879 (1947).
17. R. B. Felgett, *J. Opt. Soc. Amer.* **39**, 970 (1949).
18. G. Herzberg, *Infrared and Raman Spectra*, D. Van Nostrand and Co., Princeton, New Jersey (1944).
19. *Massachusetts Institute of Technology Wavelength Tables*, John Wiley and Sons, New York (1956).
20. G. Herzberg, *Infrared and Raman Spectra*, D. Van Nostrand and Co., Princeton, New Jersey (1944).
21. S. Goldman, *Frequency Analysis, Modulation and Noise*, McGraw-Hill Book Co., New York (1948), p. 56.
22. R. F. Post, *Rev. Mod. Phys.* **28**, 338 (1956).
23. *The Procedures Used in the Study of Properties of Photoconductive Detectors*, Unpublished report by A. J. Cussen, National Bureau of Standards (1951), ATI 194147.
24. *Energy Transfer in Hot Gases*, U.S. Dept. of Commerce, NBS Circular 523.
25. A. G. Gaydon, *Spectroscopy and Combustion Theory*, Chapman and Hall, Ltd., London (1948).
26. A. G. Gaydon and H. G. Wolfhard, *Flames, Their Structure, Radiation, and Temperature*, Chapman and Hall, Ltd., London (1953).

* One of the rules designated GEN-15 is obtainable from the General Electric Company, 1 River Road, Schenectady, New York. A somewhat more elaborate rule can be purchased from the A. G. Thornton Company, Ltd., P.O. Box 3, Wythenshawe, Manchester, England. The Jarrell-Ash Company of Boston, Massachusetts, also distributes the British rule.

27. E. K. Plyler, *J. Res. Nat. Bur. Stds.* **40**, 113 (1948).
28. A. LaRocca and G. Zissis, *Rev. Sci. Instr.* **30**, 200 (1959).
29. N. Bloembergen, *Phys. Rev.* **104**, 324 (1956).
30. A. L. Schawlow and C. K. Townes, *Phys. Rev.* **112**, 1940 (1958).
31. R. H. Dicke, U.S. Patent 2,851,652 (Sept. 9, 1958).
32. Gordon, Zeiger and Townes, *Phys. Rev.* **99**, 1264 (1955).
33. N. Bloembergen, *Phys. Rev. Lett.* **2**, 84 (1959).
34. R. W. Gelinas, *Masers and Irasers*, RAND Report P-1585 (Dec. 30, 1958).
35. A. Gamba, *Phys. Rev.* **110**, 601 (1958).
36. T. H. Maiman, *Nature* **187**, 493 (1960).
37. R. J. Collins et al., *Phys. Rev. Lett.* **5**, 303 (1960).
38. J. F. Ready, personal communication.
39. P. P. Sorokin and M. J. Stevenson, *IBM J. Res. and Dev.* **5**, 56 (1961).
40. P. P. Sorokin and M. J. Stevenson, *Phys. Rev. Lett.* **5**, 557 (1960).
41. Javan, Bennett, Jr., and Herriot, *Phys. Rev. Lett.* **6**, 106 (1961).
42. J. Oberly, *J. Opt. Soc. Amer.* **47**, 439 (1957).
43. W. Van Roosbroeck and W. Shockley, *Phys. Rev.* **94**, 1558 (1954).
44. P. Aigrain, *Physica* **20**, 1010 (1954).
45. Columbia Radiation Laboratory Quarterly Reports (1949–1951).
46. A. Glagolewa-Arkadiewa, *Zeits. f. Phys.* **24**, 153 (1924).
47. F. R. Dickey, *Cruft Laboratory Tech. Report No. 123* (July 10, 1951).
48. H. A. Prime, *Research* **3**, 51 (1950).
49. G. Herzberg, *op. cit.*

3

The theory of the infrared optical characteristics of media

3.1 INTRODUCTION AND GENERAL DISCUSSION

'The theoretical and practical aspects of the generation of infrared radiation have been treated in the preceding chapter. The next phase of this topic concerns the disposition of the radiation between the time it leaves the source and strikes the detector. After its generation, the radiation will in many instances pass through an atmosphere in which it may be refracted, absorbed, or scattered; subsequently it may be refracted, reflected, diffracted, or polarized either by a target or a collector in an optical system. An understanding of how these optical phenomena affect the radiation is of extreme importance to students of infrared technology. The following sections treat the infrared optical characteristics of media.

These next sections are not meant as a definitive treatise on electromagnetic theory; instead, their function is to clarify the concepts and terminology that are concerned with the practical aspects of the optical properties of media. An attempt will be made first to establish, on the basis of electromagnetic theory and the wave equation, a theoretical foundation and a familiarity with the terms necessary for an understanding of the subject. With this foundation, various optical effects are treated as they apply to particular states and classes of matter.

3.2   ELECTROMAGNETIC THEORY

As was stated in Chapter 1, Maxwell, in 1862, was able to show analytically that light could be considered as a transverse wave phenomenon.
We intend to demonstrate how the wave equation, derived from electric
and magnetic field theory, describes the manner in which this wave motion
varies with space and time.

3.2.1   Maxwell's Equations

The basis for electromagnetic theory is contained in four differential
equations generally referred to as Maxwell's equations.[1] These four
equations state in mathematical terms that which had been determined
experimentally by Coulomb, Oersted, Ampere, Biot, Savart, Henry, and
Faraday. Indeed, all the information regarding electric and magnetic
fields was available by the middle of the nineteenth century, but it took the
genius of Faraday to suggest the concept of electric and magnetic fields.
Faraday was not a mathematician, and his concepts of these fields were not
accepted by most of his contemporaries. The first mathematical formulation of field theory was provided by Gauss who was inspired by the work
of Laplace and Poisson on gravitational fields. It remained for Maxwell
to state the correct mathematical formulation of Faraday's concepts, that
is, (in MKS units)* (the reader unfamiliar with vector notation is referred
to Wills[2])

$$q_v = \frac{\partial D_x}{\partial x} + \frac{\partial D_y}{\partial y} + \frac{\partial D_z}{\partial z} \equiv \text{div } \mathbf{D} \equiv \nabla \cdot \mathbf{D}. \tag{3.1}$$

$$0 = \frac{\partial B_x}{\partial x} + \frac{\partial B_y}{\partial y} + \frac{\partial B_z}{\partial z} \equiv \text{div } \mathbf{B} \equiv \nabla \cdot \mathbf{B}, \tag{3.2}$$

$$-\frac{\partial \mathbf{B}}{\partial t} = \mathbf{u}_x \left[\frac{\partial E_z}{\partial y} - \frac{\partial E_y}{\partial z} \right] + \mathbf{u}_y \left[\frac{\partial E_x}{\partial z} - \frac{\partial E_z}{\partial x} \right] + \mathbf{u}_z \left[\frac{\partial E_y}{\partial x} - \frac{\partial E_x}{\partial y} \right]$$

$$\equiv \text{curl } \mathbf{E} \equiv \nabla \times \mathbf{E}, \tag{3.3}$$

$$\mathbf{J} = \mathbf{u}_x \left[\frac{\partial H_z}{\partial y} - \frac{\partial H_y}{\partial z} \right] + \mathbf{u}_y \left[\frac{\partial H_x}{\partial z} - \frac{\partial H_z}{\partial x} \right] + \mathbf{u}_z \left[\frac{\partial H_y}{\partial x} - \frac{\partial H_x}{\partial y} \right]$$

$$\equiv \text{curl } \mathbf{H} \equiv \nabla \times \mathbf{H}, \tag{3.4}$$

* We will follow the terminology recommended in the *American Institute of Physics
Handbook*, McGraw-Hill Book Co., New York (1957).

where   **D** is the electric displacement vector in coulomb/m², which is
the sum of the electric field in the absence of the medium
plus the field resulting from the polarization of the medium
by the field,

q_v is the net electric charge per unit volume in coulombs/m³,
B is the magnetic induction in weber/m², which is the sum of the
magnetic field in the absence of the medium plus the field
resulting from the magnetization of the medium by the field,
E is the electric field in volt/m,
J is the current density in ampere/m²,
H is the magnetic intensity in ampere turns/m, and
$\mathbf{u}_x$, $\mathbf{u}_y$, and $\mathbf{u}_z$ are unit vectors in the x, y, and z-directions.

The first equation states that the net electric field emanating from an
elemental volume arises because of the electric charge in or on that
volume, that is, the charge distributed throughout the volume as well as
the charge distribution induced by polarization of the media within the
volume. The second equation is the fundamental equation for magnetic
induction. It states that the net number of lines of magnetic induction
emanating from any elemental volume is zero, that is, lines of magnetic
induction have no beginning or ending; they are closed loops. The third
equation states that a changing magnetic field gives rise to an electric field,
and the fourth equation indicates that the flow of current will produce a
magnetic field. Although these are usually stated as four independent
equations, it can be shown that if one assumes the conservation of charge
and the validity of the last two equations, the first two equations are not
independent.

The essence of Maxwell's contribution lay in his interpretation of
J in the fourth equation. Before Maxwell, this term had been interpreted
as the conduction current density. This interpretation led to a seemingly un-
reconcilable paradox that is evident from the following discussion. Taking
the divergence of both sides of the fourth equation, we see that div **J** = 0
since the divergence of the curl of any vector is identically zero. This result
is inconsistent with the equation of continuity, that is, div $\mathbf{J} = -\partial q_v/\partial t$.
From this Maxwell concluded that the correct interpretation of **J** must
also include a term which represents the displacement current, $\mathbf{J}_d$, that is,

$$\mathbf{J} = \mathbf{J}_c + \mathbf{J}_d, \tag{3.5}$$

where $\mathbf{J}_c$ is the conduction current.

From the continuity equation and the time derivative of Maxwell's
first equation, the displacement current is given by

$$\mathbf{J}_d = \frac{\partial \mathbf{D}}{\partial t}. \tag{3.6}$$

For a large number of substances the electric and magnetic properties are constant from point to point. Such substances are said to be homogeneous. If, in addition, the physical properties are the same in all directions, the substances are termed isotropic. For isotropic, homogeneous bodies, the electric displacement is proportional to the electric field, that is, $\mathbf{D} = \varepsilon\mathbf{E}$ (where ε is the absolute capacitivity also known as the absolute permittivity); the magnetic induction is proportional to the magnetic field, that is, $\mathbf{B} = \mu\mathbf{H}$ (where μ is the absolute permeability); the current density is proportional to the electric field, that is, $\mathbf{J} = \sigma\mathbf{E}$ (where σ is the electrical conductivity); and Maxwell's equations can be written

$$q_v = \varepsilon \operatorname{div} \mathbf{E}. \tag{3.7}$$

$$0 = \operatorname{div} \mathbf{H}. \tag{3.8}$$

$$-\mu\frac{\partial \mathbf{H}}{\partial t} = \operatorname{curl} \mathbf{E}. \tag{3.9}$$

$$\sigma\mathbf{E} + \varepsilon\frac{\partial \mathbf{E}}{\partial t} = \operatorname{curl} \mathbf{H}. \tag{3.10}$$

As stated previously, a changing magnetic field produces a changing electric field which in turn produces a changing magnetic field, and so on. Consequently, whenever an electric or magnetic disturbance takes place, energy will be continuously transferred back and forth between the electric and magnetic fields. It turns out that the magnetic energy is not located at the same place in the medium as the electrical energy from which it was derived, but is generated a little beyond it. The electrical energy resulting from that magnetic energy is displaced a little further in space. As a result, as the energy changes form from electrical to magnetic, and so forth, it is propagating through the medium in the form of an electromagnetic wave.

3.2.2 The Wave Equations and Optical Constants

In this section we intend to derive the differential equations which describe the transmission of an electromagnetic disturbance through any medium.[3]

By taking the curl of both sides of Eq. 3.9, we obtain

$$\operatorname{curl} \operatorname{curl} \mathbf{E} = -\mu\frac{\partial}{\partial t}(\operatorname{curl} \mathbf{H}). \tag{3.11}$$

However, from Eq. 3.10 and the fact that the curl of the curl of a vector equals the gradient of the divergence minus the Laplacian of that vector, Eq. 3.11 becomes

$$\nabla^2\mathbf{E} - \operatorname{grad} \operatorname{div} \mathbf{E} = \mu\sigma\frac{\partial \mathbf{E}}{\partial t} + \varepsilon\mu\frac{\partial^2 \mathbf{E}}{\partial t^2}. \tag{3.12}$$

By similar manipulations, we obtain

$$\nabla^2 \mathbf{H} = \mu\sigma \frac{\partial \mathbf{H}}{\partial t} + \varepsilon\mu \frac{\partial^2 \mathbf{H}}{\partial t^2}. \tag{3.13}$$

Equations 3.12 and 3.13 constitute the electromagnetic wave equations. These equations admit to a number of solutions depending upon the geometry. When radiation is emitted by an infrared source, the wave front approximates a spherical wave. If the detection system is sufficiently removed from the source, the wave front, being a very small section of a spherical wave, appears to be a plane. This type of wave is discussed below. The concepts developed are, of course, applicable to spherical or cylindrical wave fronts, but the more complex mathematics has a tendency to obscure the physical reasoning.

We will discuss the propagation of electromagnetic waves through (1) free space, (2) dielectric media, and finally, (3) electrical conductors.[4]

1. *Free space.* The term "free space" implies the following assumptions:
(a) $q_v = 0$; therefore div $\mathbf{E} = 0$.
(b) $\sigma = 0$.

On the basis of these assumptions, 3.12 becomes

$$\nabla^2 \mathbf{E} = \varepsilon_0 \mu_0 \frac{\partial^2 \mathbf{E}}{\partial t^2}, \tag{3.14}$$

where the subscript "0" refers to free space. A similar equation, that is,

$$\nabla^2 \mathbf{H} = \varepsilon_0 \mu_0 \frac{\partial^2 \mathbf{H}}{\partial t^2}, \tag{3.15}$$

can be written for $\mathbf{H}$, from 3.13.

Figure 3.1 shows an electric wave vector, called a plane polarized wave, which is always in the xy-plane. In optical terminology the plane normal to the electric vector and containing a vector drawn in the direction of

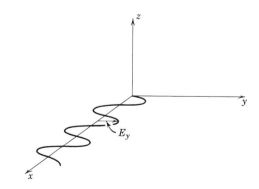

Figure 3.1

propagation is called the plane of polarization. In radio terminology the electric vector and the direction of propagation determine the plane of polarization. We, of course, use optical terminology.

By referring to Fig. 3.1, in which **E**, the electric vector, is in the xy-plane, and the direction of propagation is in the $+x$-direction, 3.14 becomes, since $E_x = E_z = 0$,

$$\frac{\partial^2 E_y}{\partial x^2} = \varepsilon_0 \mu_0 \frac{\partial^2 E_y}{\partial t^2} . \tag{3.16}$$

Although any function of the form $E_y = f[t \pm x/c_0]$ will satisfy Eq. 3.16, in particular we choose

$$E_y = E_{y0} \cos \left\{ \omega \left[t - \frac{x}{c_0} \right] \right\}, \tag{3.17}$$

where $c_0 = [\varepsilon_0 \mu_0]^{-1/2}$ is the speed of propagation in free space, that is, the speed of light (we will denote the speed of light in other media by c), and E_{y0} is the maximum amplitude of the electric field vector in the y-direction. Equation 3.17 describes a wave propagating through space with a speed c_0 and a frequency $\nu = \omega/2\pi = c_0/\lambda$, where λ is the wavelength of the disturbance in free space. For convenience in the mathematics, it has become accepted practice to use an exponential function instead of the trigonometric function as was used in Eq. 3.17; then Eq. 3.17 becomes

$$E_y = E_{y0} \exp \{2\pi j \nu [t - x/c_0]\}, \tag{3.18}$$

where $j \equiv \sqrt{-1}$, with the understanding that only the real part of 3.18 is to be considered.

2. *Dielectric media.* The statement that a substance is an uncharged dielectric involves the following assumptions.

   (a) $\sigma = 0$,

   (b) $q_v = 0$,

so that Eq. 3.12 becomes

$$\nabla^2 \mathbf{E} = \mu \varepsilon \frac{\partial^2 \mathbf{E}}{\partial t^2} , \tag{3.19}$$

in which μ and ε are the absolute permeability and capacitivity of the dielectric medium. An analogous equation can be given for **H**. Equation 3.19 has as a solution for the case depicted in Fig. 3.1,

$$E_y = E_{y0} \exp \{2\pi j \nu [t - x/c]\}, \tag{3.20}$$

in which $c = [\mu \varepsilon]^{-1/2}$.

The index of refraction n is defined as the ratio c_0/c or

$$n \equiv \frac{c_0}{c} = \sqrt{\frac{\mu \varepsilon}{\mu_0 \varepsilon_0}} = \sqrt{K_m K_e} , \tag{3.21}$$

where $K_e \equiv \varepsilon/\varepsilon_0$ is the relative capacitivity or dielectric constant, and $K_m \equiv \mu/\mu_0$ is the relative permeability.

Except for ferromagnetic materials K_m is approximately unity so that

$$n = \sqrt{K_e}. \tag{3.22}$$

The relation $n = \sqrt{K_e}$ is valid for radiation of long wavelength. In some materials this is valid for the infrared region of the spectrum and in other materials it is not valid excepting at wavelengths longer than those of the infrared region. For shorter wavelengths, the dielectric constant varies with frequency because the charges of which an atom or molecule are composed are not capable of moving in unison with the fields acting upon them. This aspect of propagation is discussed more fully in Section 3.5.1 which concerns dispersion.

3. *Electrical conductors.* In this case $q_v = 0 = \operatorname{div} \mathbf{E}$.

For such a medium Eq. 3.12 becomes

$$\nabla^2 \mathbf{E} = \mu\sigma \frac{\partial \mathbf{E}}{\partial t} + \varepsilon\mu \frac{\partial^2 \mathbf{E}}{\partial t^2}. \tag{3.23}$$

A similar equation can be obtained for $\mathbf{H}$.

Again, for plane polarized radiation and the geometry shown in Fig. 3.1, solutions of Eq. 3.23 take the form

$$E_y = E_{y0} \exp\{2\pi j\nu t - \Gamma x\}, \tag{3.24}$$

where Γ is called the propagation coefficient and is given by

$$\Gamma = [2\pi\nu j\mu\sigma - 4\varepsilon\mu\pi^2\nu^2]^{1/2},$$

$$= 2\pi\nu\sqrt{\varepsilon\mu}\left[1 + \left(\frac{\sigma}{2\pi\nu\varepsilon}\right)^2\right]^{1/4} \exp\left\{-\frac{j}{2}\left[\arctan\left(\frac{\sigma}{2\pi\nu\varepsilon}\right)\right]\right\}. \tag{3.25}$$

It is often convenient to express Γ in terms of a real and imaginary part in the following manner:

$$\Gamma = \frac{2\pi\nu j}{c_0}[n - jk]. \tag{3.26}$$

Since Γ^2 is in the second quadrant of the complex plane, n and k are both positive and given by

$$n \equiv \left[\frac{\varepsilon\mu}{\varepsilon_0\mu_0}\right]^{1/2}\left[1 + \left(\frac{\sigma}{2\pi\nu\varepsilon}\right)^2\right]^{1/4} \sin\left\{\frac{1}{2}\left[\arctan\left(\frac{\sigma}{2\pi\nu\varepsilon}\right)\right]\right\}, \tag{3.27}$$

and

$$k \equiv \left[\frac{\varepsilon\mu}{\varepsilon_0\mu_0}\right]^{1/2}\left[1 + \left(\frac{\sigma}{2\pi\nu\varepsilon}\right)^2\right]^{1/4} \cos\left\{\frac{1}{2}\left[\arctan\left(\frac{\sigma}{2\pi\nu\varepsilon}\right)\right]\right\}. \tag{3.28}$$

In this case Eq. 3.24 can be written

$$E_y = E_{y0} \exp\left[-2\pi\nu kx/c_0\right] \exp\left[j2\pi\nu(t - nx/c_0)\right]. \tag{3.29}$$

The factor $\exp\left(-2\pi\nu kx/c_0\right)$ describes the damping of a wave as it advances into a conductive medium. If k is large, the wave is damped out very quickly, and for that reason k is called the "absorption constant." By analogy with 3.18 and 3.20, we see that the Γ appearing in 3.24 can be expressed as

$$\Gamma = \frac{2\pi j\nu}{c^*} = \frac{2\pi j\nu}{c_0}\Gamma', \tag{3.30}$$

where

$$\Gamma' \equiv \frac{c_0}{c^*},$$

and $c^* \equiv c_0/(n - jk)$ is the complex velocity of light.

The similarity of this relationship to 3.21 has led some authors to refer to Γ' as the complex index of refraction. We see from 3.26 that it is given by

$$\Gamma' = n - jk. \tag{3.31}$$

Extending the analogy, 3.22 becomes

$$\Gamma' = \sqrt{K_e^*}, \tag{3.32}$$

where K_e^* is the complex dielectric constant.

We can now characterize the propagation of an electromagnetic wave by specifying two quantities n and k in much the same way that admittance and conductance characterize an electric circuit. It should be noted that for conductive media n and k are functions of the frequency of the radiation.

The manner in which the index of refraction n and the absorption constant k are affected by the properties σ, μ, and ε, as well as by the wavelength of the radiation, can best be seen by considering two limiting cases of Eqs. 3.27 and 3.28.

First let us consider $\sigma \ll 2\pi\varepsilon\nu$. This is the case for dielectrics or for radiation having an extremely short wavelength or a high spectral frequency. Here k is insignificant and the index of refraction is

$$n = \left[\frac{\varepsilon\mu}{\varepsilon_0\mu_0}\right]^{1/2} \approx [K_e]^{1/2}. \tag{3.33}$$

As before, this last approximation is justified on the basis that for all materials except ferromagnetic materials, $\mu \approx \mu_0 \approx 1$. The wave, in this case, propagates through the medium with no absorption and with the wavelength reduced by n.

For long wavelengths, that is, long radio waves, the velocity of propagation is given by $[K_e]^{1/2}$ as determined from the static dielectric constant.

As the wavelength decreases, this situation is no longer true; the index of refraction increases and the medium starts to absorb.

Let us now consider the case where $\sigma \gg 2\pi\nu\varepsilon$. This situation exists for metals or when the incident radiation has an extremely long wavelength. From Eqs. 3.27 and 3.28 we see that

$$k = n = c_0\left[\frac{\mu\sigma}{4\pi\nu}\right]^{1/2} = \left[\frac{\mu\sigma\lambda c_0}{4\pi}\right]^{1/2}, \tag{3.34}$$

so that for long wavelengths, σ controls the absorption characteristics of metals. Equation 3.34 is a restatement of the relationship first mentioned in Section 2.4.2. The reader is cautioned against substituting the dc value of the conductivity into the foregoing equations. The dependence of σ on the frequency is discussed in more detail in Section 3.5.1.

It is found that the optical characteristics of metals are described very well by the preceding simple theory in the spectral range having a wavelength greater than visible light. The theory is not applicable for short wavelengths because the capacitivity is not a constant but varies with the wavelength of the radiation involved.

3.3 ELECTROMAGNETIC POWER FLOW

The transmission of power through a medium by means of electromagnetic waves is discussed next.[5]

It can be shown that the energy stored per unit volume in a capacitor is $\frac{1}{2}\mathbf{D} \cdot \mathbf{E}$. A calculation of the energy per unit volume stored by a current-carrying inductor yields $\frac{1}{2}\mathbf{B} \cdot \mathbf{H}$. Consequently, the energy $d\mathscr{E}$ contained in an elemental volume dv of a medium through which an electromagnetic wave passes is

$$d\mathscr{E} = \tfrac{1}{2}\{\mathbf{D} \cdot \mathbf{E} + \mathbf{B} \cdot \mathbf{H}\}\, dv. \tag{3.35}$$

This energy passes through the elemental volume and its passage represents the flow of electromagnetic energy, that is (for a homogeneous, isotropic dielectric),

$$\mathscr{H} \cdot d\mathbf{a} = -\frac{d\mathscr{E}}{dt} = -\left\{\varepsilon\mathbf{E} \cdot \frac{\partial \mathbf{E}}{\partial t} + \mu\mathbf{H} \cdot \frac{\partial \mathbf{H}}{\partial t}\right\} dv, \tag{3.36}$$

where $\mathscr{H}$ is the power per unit area and
$d\mathbf{a}$ is the elemental area across which this power flows.

From Eqs. 3.9 and 3.10, Eq. 3.36 becomes

$$\mathscr{H} \cdot d\mathbf{a} = \{\mathbf{H} \cdot \nabla \times \mathbf{E} - \mathbf{E} \cdot \nabla \times \mathbf{H}\}\, dv. \tag{3.37}$$

For any vectors $\mathbf{M}$ and $\mathbf{N}$ the following identity is true:

$$\nabla \cdot \mathbf{M} \times \mathbf{N} = \mathbf{N} \cdot \nabla \times \mathbf{M} - \mathbf{M} \cdot \nabla \times \mathbf{N},$$

so that Eq. 3.37 becomes

$$\mathscr{H} \cdot d\mathbf{a} = \nabla \cdot \mathbf{E} \times \mathbf{H} \, dv,$$

or

$$\oint \mathscr{H} \cdot d\mathbf{a} = \int_v \nabla \cdot \mathbf{E} \times \mathbf{H} \, dv. \tag{3.38}$$

By Gauss' theorem the volume integral becomes an integral over the surface, in which case Eq. 3.38 becomes

$$\oint \mathscr{H} \cdot d\mathbf{a} = \oint (\mathbf{E} \times \mathbf{H}) \cdot d\mathbf{a}. \tag{3.39}$$

This can be true for any surface only if

$$\mathscr{H} = \mathbf{E} \times \mathbf{H}. \tag{3.40}$$

The vector $\mathscr{H}$, the cross product of $\mathbf{E}$ and $\mathbf{H}$, is often referred to as the Poynting vector. In order to compute the power, we must take only the real part of $\mathbf{E}$ and $\mathbf{H}$.

If we take $E_y = E_{y0} \cos \{\omega(t - x/c)\}$ (see Fig. 3.1) and compute $\mathbf{H}$ from 3.10, Eq. 3.40 becomes

$$\mathscr{H} = \frac{\mathbf{u}_x \varepsilon E_{y0}^2}{c} = \mathbf{u}_x \sqrt{\frac{\varepsilon}{\mu}} E_{y0}^2, \tag{3.41}$$

where $\mathbf{u}_x$ is the unit vector in the x-direction. Equation 3.41 indicates that the power is transmitted in the x-direction and has a magnitude proportional to the square of the amplitude of the electric vector. By a similar derivation it can be shown that the power is also proportional to the square of the amplitude of the magnetic vector.

3.4 MEDIA DISCONTINUITIES

In the foregoing sections we have seen how electromagnetic waves travel and how they convey power in free space or in any homogeneous media. There we treated only those cases in which μ and ε are not functions of the frequency of the radiation. In all cases we were concerned with media of infinite extent; in the following discussion we will be interested in the disposition of the electromagnetic energy when the wave travels from one medium into another.

The properties of a medium through which an electromagnetic wave is passing determine what fraction of the total energy is, on the average, associated with the electric field and, consequently, what fraction is associated with the magnetic field. If a wave is incident on the surface of a medium of a different index, the electric and magnetic properties of the new medium cause a reapportionment of energy between the two fields. This change in the relationship between the electric and magnetic fields occurs just as the wave crosses the boundary between the two media. Since no energy can be added to the incident radiation, the only way a new balance can be achieved between the two fields is for either some of the electric or some of the magnetic field energy to be rejected. However, if energy in either form is rejected, its fluctuations will give rise to an electromagnetic wave and hence radiation is rejected. This rejected electromagnetic energy is simply reflected radiation, and the fraction rejected in any spectral interval defines the reflectivity of that surface for that portion of the spectrum. The rejection of electromagnetic energy at the surface of a medium does not depend on the ultimate disposition of the energy entering the new medium. Often the energy transmitted across media discontinuities is absorbed within a few Ångstroms of the surface. The fraction of the electromagnetic energy rejected by reflection at a boundary between two media does not depend upon whether the wave is absorbed or transmitted in the second medium.

To describe the relationship between the incident, reflected, and transmitted energies, we must first know the appropriate boundary conditions that can be used with Maxwell's equations. The resulting relationships describing the reflectivity and transmissivity at an interface between two media are called Fresnel's equations.

3.4.1 Boundary Conditions

When an electromagnetic wave is incident on the surface of a homogeneous, uncharged material not carrying a current, the vectors **E**, **H**, **D**, and **B** conform to certain boundary conditions,[6] that is,

1. The tangential components of **E** and **H** are conserved.
2. The normal components of **D** and **B** are conserved.

Tangential components are those parallel to the interfacial plane; normal components are perpendicular to the plane. The first of these conditions means that the components of the electric field and magnetic intensity vectors which are in the plane of incidence are not changed as the wave passes from one medium to another. The second restriction on the electric displacement and magnetic induction vectors states that the components of these vectors which are perpendicular to the plane of incidence are not changed when the wave crosses a discontinuity.

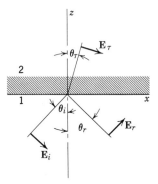

Figure 3.2. Refraction at a boundary.

3.4.2   Refraction and Snell's Law (see Fig. 3.2)

Equation 3.20 can be generalized to

$$E = E_0 \exp \left\{ 2\pi j \nu \left[t - \frac{\mathbf{u} \cdot \mathbf{p}}{c} \right] \right\}, \qquad (3.42)$$

where $\mathbf{u} \equiv \mathbf{u}_x + \mathbf{u}_y + \mathbf{u}_z$,

$\mathbf{p} \equiv$ the unit vector in the direction of propagation,

$\mathbf{p} \equiv \mathbf{u}_x \alpha + \mathbf{u}_y \beta + \mathbf{u}_z \gamma$ in which $\mathbf{u}_x$, $\mathbf{u}_y$, and $\mathbf{u}_z$ are unit vectors and α, β, and γ are direction cosines.

The electric field vector in Fig. 3.2 for the incident radiation is given by

$$E_i = E_{i0} \exp \left\{ 2\pi j \nu \left[t - \left(\frac{x \sin \theta_i - z \cos \theta_i}{c_1} \right) \right] \right\}, \qquad (3.43)$$

where $c_1 \equiv [\varepsilon_1 \mu_1]^{-\frac{1}{2}}$.

The transmitted electric vector can be described by

$$E_\tau = E_{\tau 0} \exp \left\{ 2\pi j \nu \left[t - \left(\frac{x \sin \theta_\tau - z \cos \theta_\tau}{c_2} \right) \right] \right\}, \qquad (3.44)$$

where $c_2 \equiv [\varepsilon_2 \mu_2]^{-\frac{1}{2}}$.

Since a homogeneous, isotropic uncharged dielectric not carrying a current is under consideration, wave fronts must match at the interface $-z = 0$. This can only be true if

$$\frac{\sin \theta_\tau}{c_2} = \frac{\sin \theta_i}{c_1} \quad \text{or} \quad \frac{\sin \theta_i}{\sin \theta_\tau} = \frac{n_2}{n_1}. \qquad (3.45)$$

Equation 3.45 is known as Snell's law of refraction.

3.4.3   Reflection

To determine the amount of incident radiation reflected by a surface, it is necessary to know the polarization of the radiation. By stating the direction of propagation, the direction of the electric vector is not uniquely defined, since there are an infinitude of vectors perpendicular to a single vector denoting the direction of propagation. For purposes of analysis, the array of electric vectors can be resolved into two orthogonal components which are generally taken to be parallel to and normal to the plane of incidence. If these orthogonal components are equal, the radiation is said to be circularly polarized, whereas the predominance of one component leads to elliptical polarization.

We will treat only the case where the electric vector is perpendicular to the plane of incidence. The results for the electric vector parallel to the plane of incidence will be stated and can be deduced from quite similar reasoning.

Referring to Fig. 3.2 we see that the reflected wave can be described by

$$\mathbf{E}_r = \mathbf{E}_{r0} \exp\left\{2\pi j\nu\left[t - \left(\frac{x \sin \theta_r + z \cos \theta_r}{c_1}\right)\right]\right\}. \tag{3.46}$$

Since the wave fronts are to match at $z = 0$, then from Eqs. 3.43 and 3.46 the angle of incidence must equal the angle of reflection.

The electric field vector, which is always parallel to the y-axis, can be described (for the incident, reflected, and transmitted or refracted radiation) as follows:

$$E_{yi} = E_{yi0} \exp\left\{2\pi j\nu\left[t - \left(\frac{x \sin \theta_i - z \cos \theta_i}{c_1}\right)\right]\right\}, \tag{3.47}$$

$$E_{yr} = E_{yr0} \exp\left\{2\pi j\nu\left[t - \left(\frac{x \sin \theta_i + z \cos \theta_i}{c_1}\right)\right]\right\}, \tag{3.48}$$

$$E_{y\tau} = E_{y\tau0} \exp\left\{2\pi j\nu\left[t - \left(\frac{x \sin \theta_\tau - z \cos \theta_\tau}{c_2}\right)\right]\right\}. \tag{3.49}$$

By substituting these expressions in Eq. 3.9, it can be shown that

$$E_{yi} \cos \theta_i = \mu_1 c_1 H_{xi}, \tag{3.50}$$

$$0 = H_{yi}, \tag{3.51}$$

$$E_{yi} \sin \theta_i = \mu_1 c_1 H_{zi}. \tag{3.52}$$

Similar equations can be written for the reflected and transmitted waves. Since the tangential components of **H** are conserved, that is, $H_{xi} - H_{xr} = H_{x\tau}$,

$$[E_{yi} + E_{yr}]\left[\frac{\varepsilon_1}{\mu_1}\right]^{\frac{1}{2}} \cos \theta_i = E_{y\tau}\left[\frac{\varepsilon_2}{\mu_2}\right]^{\frac{1}{2}} \cos \theta_\tau. \tag{3.53}$$

This can be seen by reference to Eq. 3.50 and similar equations derived for H_{xr} and $H_{x\tau}$.

Recalling that the tangential components of $\mathbf{E}$ are also conserved, that is, $E_{yi} = E_{yr} + E_{yr}$, and utilizing Snell's law, we find the amplitude reflection coefficient is given by

$$r_\perp \equiv \left[\frac{E_{yr}}{E_{yi}}\right]_\perp = \frac{\mu_1 \tan \theta_i - \mu_2 \tan \theta_r}{\mu_1 \tan \theta_i + \mu_2 \tan \theta_r}, \qquad (3.54)$$

where $r_\perp$ denotes the amplitude reflection coefficient when the electric vector is normal to the plane of incidence. If the media are nonmagnetic, Eq. 3.54 becomes

$$r_\perp = \frac{\sin\left[\theta_i - \theta_r\right]}{\sin\left[\theta_i + \theta_r\right]}. \qquad (3.55)$$

For normal incidence, that is, $\theta_i = 0$, Eq. 3.55 becomes

$$r = \frac{n_2 - n_1}{n_2 + n_1}. \qquad (3.56)$$

To compute the power reflection coefficient, ρ, the reflectivity of a surface, that is, $[E_{yr}/E_{yi}]^2$, we need only square either Eq. 3.55 or 3.56. Where r is complex

$$\rho_\perp = r_\perp r_\perp{}^*,$$

and

$$\rho_\parallel = r_\parallel r_\parallel{}^*,$$

where the asterisk denotes the complex conjugate. For radiation in which the electric vector is parallel to the plane of incidence

$$r_\parallel \equiv \left[\frac{E_{yr}}{E_{yi}}\right]_\parallel = \frac{\tan\left[\theta_i - \theta_r\right]}{\tan\left[\theta_i + \theta_r\right]}, \qquad (3.57)$$

which, for normal incidence, also reduces to $(n_2 - n_1)/(n_2 + n_1)$. Thus the amplitude reflection coefficient at normal incidence is independent of the polarization of the incident electric vector.

Just as $r \equiv [E_{yr}/E_{yi}]$, an amplitude transmission coefficient t is defined by

$$t \equiv [E_{y\tau}/E_{yi}].$$

Remembering that $rr^* + tt^* = \rho + \tau = 1$, ($\alpha = 0$ in an ideal dielectric), we can calculate the power transmission coefficient τ or transmissivity from either 3.55, 3.56, or 3.57.

$$\tau_\perp = \frac{\sin\left(2\theta_r\right) \sin\left(2\theta_i\right)}{\sin^2\left(\theta_i + \theta_r\right)}. \qquad (3.58)$$

$$\tau_\parallel = \frac{\sin\left(2\theta_i\right) \sin\left(2\theta_r\right)}{\sin^2\left(\theta_i + \theta_r\right) \cos^2\left(\theta_i - \theta_r\right)}. \qquad (3.59)$$

Again, for normal incidence the form of the expression is independent of the polarization so that 3.58 and 3.59 both reduce to

$$\tau = \frac{4n_1 n_2}{(n_1 + n_2)^2}. \tag{3.60}$$

In the general case of absorbing media, the mathematical treatment becomes rather complicated. The reason for this difficulty arises primarily because Γ is not only complex but the real and imaginary components are themselves vectors which are oriented in different directions in space. However, for long wavelengths and for normal incidence, the amplitude reflection coefficient of absorbing media such as metals can be calculated to be

$$r = \frac{(1 - n) - jk}{(1 + n) - jk}, \tag{3.61}$$

and the power reflection coefficient, that is, the reflectivity,

$$\rho \equiv rr^* = \frac{(1 - n)^2 + k^2}{(1 + n)^2 + k^2}. \tag{3.62}$$

3.5 ABSORPTION, DISPERSION, AND SCATTERING

The mechanisms responsible for the attenuation of an electromagnetic disturbance as it propagates through a medium are classified into two categories: absorption and scattering. Those attenuating mechanisms which convert the radiation to some other form of energy (or some other spectral distribution) are called absorption phenomena. On the other hand, scattering mechanisms are those which redirect the radiant energy from its original direction of propagation.

We shall treat each of these effects first from a phenomenological viewpoint and then through an approach that seeks to explain the observed facts on the basis of the motions of the atoms and molecules involved. We will start by deriving an expression for the attenuation in terms of the path length and the properties of the media. (See Fig. 3.3.) Consider a beam of collimated, monochromatic radiation of irradiance $\mathscr{H}_\lambda$ which is incident upon an elemental volume dv of an attenuating medium. It is reasonable to assume that the fraction of the radiation attenuated in passing through a thin layer dx of the medium is proportional to the thickness, that is,

$$\frac{\mathscr{H}_\lambda(x) - \mathscr{H}_\lambda(x + dx)}{\mathscr{H}_\lambda(x)} = \frac{d\mathscr{H}(\lambda)}{\mathscr{H}(\lambda)} \equiv -\sigma(\lambda)\, dx, \tag{3.63}$$

where $\sigma(\lambda)$, the attenuation coefficient, is the constant of proportionality, depending on the nature of the medium or media contained in the elemental

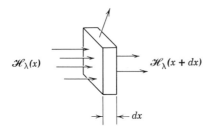

$\mathcal{H}_\lambda(x)$ $\mathcal{H}_\lambda(x + dx)$

dx

Figure 3.3. Attenuation of a beam.

volume dv as well as on the wavelength of the radiation. If we assume that the attenuation results from absorption and scattering and that these mechanisms act independently, Eq. 3.63 becomes

$$\frac{d\mathcal{H}(\lambda)}{\mathcal{H}(\lambda)} = -\{a + \beta\}\, dx, \qquad (3.64)$$

where a is the absorption coefficient, and
        β is the scattering coefficient.

Both these constants are functions of the wavelength of the radiation considered.

3.5.1 Absorption and Dispersion

It is convenient to discuss absorption in terms of the individual elements causing the absorption. Under suitable conditions, the absorption of each absorbing element is independent of the concentration of that element. Thus the absorption coefficient is proportional to the number of absorbing elements encountered per unit path length and hence is proportional to the concentration n_a of those elements. Mathematically this can be stated as

$$a = a' n_a, \qquad (3.65)$$

where a' (usually a function of wavelength) is the absorption coefficient per unit of concentration. Equation 3.65 is known as Beer's law.

The previous assumption concerning the independence of a' and n_a may be invalid in some cases. Changes in concentration, for example, may alter the nature of the absorbing molecular specie, or may cause interactions between the absorbing molecules.

In the same manner the scattering coefficient β may be written

$$\beta = \beta' n_s, \qquad (3.66)$$

where n_s is the concentration of scattering elements, and β' is the scattering coefficient per unit concentration.

Since a' and β' have the dimensions of area, they are referred to as the absorption and scattering cross sections. By using these definitions, Eq. 3.64 becomes

$$\frac{d\mathscr{H}(\lambda)}{\mathscr{H}(\lambda)} = -\{a'n_a + \beta'n_s\}\,dx, \qquad (3.67)$$

which integrates to

$$\mathscr{H}_\tau(\lambda) = \mathscr{H}_i(\lambda)\exp -\{a'n_a + \beta'n_s\}x, \qquad (3.68)$$

where $\mathscr{H}_i(\lambda)$ is the spectral irradiance at the front surface of the medium being considered. Equation 3.68 is known as the Lambert-Beers law.

In order to clarify the relationship between a number of such widely used terms as absorption coefficient a, absorption constant k, and extinction coefficient κ, we will, for the moment, neglect scattering so that we may write

$$\frac{\mathscr{H}_\tau(\lambda)}{\mathscr{H}_i(\lambda)} = \exp\{-ax\} \equiv \tau_a, \qquad (3.69)$$

where $\mathscr{H}_\tau(\lambda)$ is the transmitted spectral irradiance and
τ_a is the transmissivity as affected by absorption processes only.

From Eq. 3.29 we have

$$\frac{\mathscr{H}_\tau(\lambda)}{\mathscr{H}_i(\lambda)} = tt^* = \tau_a = \left|\frac{E_y}{E_{yi}}\right|^2 = \exp\left\{\frac{-4\pi\nu kx}{c_0}\right\}. \qquad (3.70)$$

Comparing Eqs. 3.69 and 3.70, we see that a, the absorption coefficient, is related to k, the absorption constant, by

$$a = \frac{4\pi\nu k}{c_0} = \frac{4\pi\nu}{c}\left(\frac{k}{n}\right) = \frac{4\pi}{\lambda}\left(\frac{k}{n}\right). \qquad (3.71)$$

The extinction coefficient, κ, defined as the ratio of the absorption constant to the index of refraction, is determined from Eq. 3.71 to be

$$\kappa = \frac{a\lambda}{4\pi}. \qquad (3.72)$$

Using the extinction coefficient, the complex index of refraction Γ' (see the discussion following Eq. 3.29) takes on a very simple form.

$$\Gamma' = n[1 - j\kappa].$$

Thus far, the discussion has concerned the relationships between various absorption parameters and the attenuation of radiation. The following is concerned with the basic mechanisms taking place in atoms or molecules that are responsible for absorption. We will see, by a simplified model, the relationship between the absorption coefficient and the basic properties of matter. We will also be led to the equation which accounts for dispersion and will show how the variation in n, the index of refraction, with wavelength, is related to the mass of the atoms in the material and to the binding forces between atoms. A relationship will then be derived between the absorption constant k and the same types of basic parameters. The absorption constant discussed here relates only to that associated with bound charges, for example, lattice absorption. The absorption computed below takes no account of such effects as free-carrier absorption and other absorbing mechanisms discussed in Chapter 4.

Each atom or molecule is composed of charges that can be displaced by an external electric field. The charges then act as if they were bound by a force proportional to the displacement from their equilibrium position. In addition, there is a damping force, proportional to the instantaneous velocity, which retards the motion of the charges. This damping force accounts for absorption.

The differential equation representing this physical situation is

$$m \frac{d^2 x}{dt^2} = qE(t) - k_1 x - k_2 \frac{dx}{dt}, \qquad (3.73)$$

where m is the mass of the charged particle,
  k_1 is the restoring force constant,
  k_2 is the viscous damping constant, and
  $E(t)$ is the magnitude of the electric field vector.

If the electric field which is interacting with the atoms varies sinusoidally with time, that is, if

$$E(t) = E_0 \exp (j\omega t), \qquad (3.74)$$

then 3.73 becomes

$$\frac{d^2 x}{dt^2} = \frac{q}{m} E_0 \exp (j\omega t) - \omega_0^2 x - \frac{k_2}{m} \frac{dx}{dt}, \qquad (3.75)$$

where $\omega_0 \equiv \left[\dfrac{k_1}{m} \right]^{1/2}$.

The solution of 3.75 can be stated as

$$x = \frac{q/m}{\omega_0^2 - \omega^2 + j\omega k_2/m} E_0 \exp (j\omega t). \qquad (3.76)$$

We therefore see that the charge oscillates with the same frequency as the applied radiation field. However, the amplitude of the vibration depends on the frequency of this field.

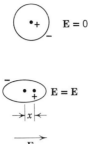

Figure 3.4. Atom polarized in an electric field.

The polarization of a material is a measure of how much the individual atoms of the material can be distorted by the application of an electric field. Application of a field and distortion of the atoms cause the field to have a different value inside the material. Since the degree of distortion and hence the change in the internal field is proportional to the applied field, we have

$$D = \varepsilon E = [1 + \chi_e]E\varepsilon_0 = \varepsilon_0 E + P,$$

where χ_e is the electric susceptibility, and
P is the polarization, a measure of the extent to which the field inside a material can be changed due to distortion of its atoms.

From this we see that

$$P \equiv \chi_e\varepsilon_0 E = [K_e - 1]\varepsilon_0 E, \tag{3.77}$$

where K_e is the relative capacitivity or dielectric constant, and
χ_e is proportional to the polarizability α of the atoms and the number of atoms per unit volume, N/V , that is,

$$\chi_e \equiv \left(\frac{N}{V}\right)\frac{\alpha}{\varepsilon_0} .$$

The polarizability α is also the induced electric moment per unit field. The electric moment M is defined as qx , where q is the magnitude of the charges being separated and x the distance of separation of the centers of mass of the charges. (See Fig. 3.4.) Then $\alpha \equiv M/E = qx/E$. From Eq. 3.77, the complex relative capacitivity $K_e{}^*$, referred to in connection with Eqs. 3.31 and 3.32, is given by

$$K_e{}^* = 1 + \chi_e = 1 + \left(\frac{N}{V}\right)\frac{qx}{\varepsilon_0 E} = 1 + \frac{\left(\frac{N}{V}\right)q^2/m\varepsilon_0}{\omega_0{}^2 - \omega^2 + j\omega k_2/m}$$

$$= [n - jk]^2, \tag{3.78}$$

where x is given by Eq. 3.76.

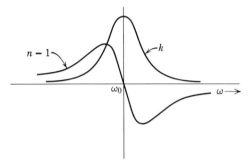

Figure 3.5. Frequency dependence
of $n - 1$ and k.

Where the second term is small compared to unity, we can equate the
imaginary parts of 3.78 and show how the absorption constant depends
upon the basic properties of the material and the wavelength of the
radiation, that is,

$$k \approx \frac{\left(\dfrac{N}{V}\right)(q^2/m^2)k_2\omega}{2\varepsilon_0[(\omega_0^2 - \omega^2)^2 + (\omega^2 k_2^2/m^2)]}. \tag{3.79}$$

We can also equate the real parts of 3.78 obtaining

$$n = 1 + \frac{\left(\dfrac{N}{V}\right)(q^2/m)[\omega_0^2 - \omega^2]}{2\varepsilon_0[(\omega_0^2 - \omega^2)^2 + (\omega k_2/m)^2]}. \tag{3.80}$$

We see that the index of refraction depends upon $\omega = 2\pi c/\lambda$ where λ is
the wavelength of the incident radiation. Thus the refraction of radiation
depends upon the wavelength, and this affords a means of separating
radiation into its component wavelengths. It is the magnitude of the
variation of n with λ, $dn/d\lambda$, which is termed dispersion and which measures
the ability of a prism to sort radiation according to wavelength.

Figure 3.5 shows $n - 1$ and k plotted as a function of ω and shows how
these quantities vary with frequency in the region of ω_0, the resonance
point. The behavior shown in this figure will be familiar to those readers
accustomed to working with resonance phenomena in electrical circuits.
The absorption constant k of a medium passes through a maximum in the
neighborhood of $\omega = \omega_0$ in analogy with the impedance of a parallel
resonant circuit which is a maximum at the resonant point. In solids
this frequency, ω_0, is known as the reststrahlen frequency. Section 4.3.3
discusses reststrahlen in dielectric materials. The index of refraction
follows a pattern analogous to that of the susceptance of an electric
circuit.

3.5.2   Scattering

In our discussion of scattering we shall adopt the procedure used with absorption. First we shall treat the relationship between the concentration and cross section of scattering particles and then consider the mechanisms, on an atomic scale, which give rise to cross sections of a specific magnitude.

For small particles suspended in a medium having different optical constants from those of the particles themselves, the radiation is sometimes caused to undergo an abrupt change in direction.[7] This change in direction, brought about by the absorption and reradiation of energy by the particles, is called scattering. The scattered energy is subtracted from the original beam, which is therefore attenuated.

Consider an elementary volume dv which is irradiated by monochromatic, parallel radiation having a spectral irradiance, $\mathscr{H}_i(\lambda)$. Consider that this volume contains a number, $N_s \equiv n_s\,dv$, of scattering particles which are capable of scattering radiative power out of the beam so that the power transmitted through the elemental volume has been reduced. The spectral radiance $R_{\omega\lambda}$ of radiation scattered by the particles in dv is proportional to the spectral irradiance and the number of scattering units within dv, that is,

$$\frac{\partial P_{\mathrm{scat}}(\lambda)}{\partial \omega} = E_{\mathrm{scat}}(\lambda) = \gamma_\lambda(\phi)N_s\mathscr{H}_i(\lambda) = \gamma_\lambda(\phi)n_s\,dv\mathscr{H}_i(\lambda)$$

$$= \gamma_\lambda(\phi)n_s\,dxP_i(\lambda), \qquad (3.81)$$

where $\gamma_\lambda(\phi)$ is the constant of proportionality,

ϕ is the scattering angle,

and where $\mathscr{H}_i(\lambda)$ da is the spectral power, $P_i(\lambda) = P_{\mathrm{trans}} + P_{\mathrm{scat}}$, incident on the elemental volume. Thus the power scattered is (see Fig. 3.6)

$$d^2P_{\mathrm{scat}}(\lambda) = P_i(\lambda)\gamma_\lambda(\phi)n_s\,d\omega\,dx = -d^2P_{\mathrm{trans}}(\lambda).$$

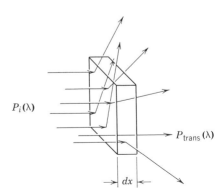

Figure 3.6

In the general case γ_λ is a function of ϕ and λ. Integration over ω yields

$$dP_{\text{trans}}(\lambda) = -n_s P_i(\lambda)\left\{2\pi \int_0^\pi \gamma_\lambda(\phi) \sin \phi \, d\phi\right\} dx, \qquad (3.82)$$

or

$$\frac{dP_{\text{trans}}(\lambda)}{P_i(\lambda)} = -n_s\left\{2\pi \int_0^\pi \gamma_\lambda(\phi) \sin \phi \, d\phi\right\} dx. \qquad (3.83)$$

Upon integration over x this becomes

$$\frac{P_{\text{trans}}(\lambda)}{P_i(\lambda)} = \frac{\mathscr{H}_r(\lambda)}{\mathscr{H}_i(\lambda)} = \exp\left\{-\left[n_s 2\pi \int_0^\pi \gamma_\lambda(\phi) \sin \phi \, d\phi\right]x\right\}. \qquad (3.84)$$

Comparing this with Eq. 3.68 and neglecting absorption, we see that

$$\beta'(\lambda) = 2\pi \int_0^\pi \gamma_\lambda(\phi) \sin \phi \, d\phi. \qquad (3.85)$$

For this reason $\gamma_\lambda(\phi)$ is generally called the differential scattering cross section.[8]

Having discussed the phenomenological aspects of scattering, let us examine the fundamental characteristics of media which affect the magnitude of scattering cross sections by considering a simple model.

The charges in the molecules which make up the scattering particle are caused to vibrate by the action of the incident electromagnetic wave. As with dispersion, the oscillating charge may be considered as a dipole having an equation of motion

$$m\frac{d^2x}{dt^2} = qE_0 \exp(j\omega t) - m\omega_0^2 x - m\frac{k_2}{m}\frac{dx}{dt}. \qquad (3.86)$$

The electric moment $M = qx$ will then be

$$M = qx = \frac{q^2}{m}\frac{E_0 \exp(j\omega t)}{(\omega_0^2 - \omega^2) + j\omega(k_2/m)}.$$

Since accelerated charges emit electromagnetic radiation, the dipoles that have been set in motion by the incident wave will emit radiation. The power scattered is related to the irradiance by

$$P_{\text{scat}}(\lambda) = \beta' N_s \mathscr{H}_i(\lambda),$$

or

$$\beta' = \frac{P_{\text{scat}}(\lambda)}{\mathscr{H}_i(\lambda)N_s}, \qquad (3.87)$$

where N_s is the number of scattering particles encountered. This defines β' as that area which a single scattering particle presents to the incident beam of radiation. It can then be shown[9] that

$$\beta' = \frac{32\pi R_e^{\,2}}{3} \frac{\omega^4}{[\omega_0^{\,2} - \omega^2]^2 + [\omega k_2/m]^2},\qquad (3.88)$$

where $R_e = q^2/8\pi\varepsilon_0 mc_0^{\,2}$, the classical radius of the electron. R_e has been found to be 2.818×10^{-15} meter. Where the damping is sufficiently small and $\omega/\omega_0 \ll 1$, 3.88 reduces to

$$\beta' = \frac{32\pi R_e^{\,2}}{3}\left(\frac{\omega}{\omega_0}\right)^4,\qquad (3.89)$$

or

$$\beta' = \frac{512\pi^5 c^4}{3\omega_0^{\,4}} R_e^{\,2} \frac{1}{\lambda^4}.\qquad (3.90)$$

Equation 3.90 is the Rayleigh scattering formula. Since the scattering cross section is inversely proportional to λ^4, the shorter wavelengths, that is, ultraviolet and visible, are scattered much more than the infrared ones. As a result the daylight sky appears blue because the shorter wavelengths from the sun are scattered toward the ground by the air molecules more than the longer wavelengths. On the other hand, the setting sun appears orange-red because its light, traversing a long, tangential path through the atmosphere, has its shorter wavelengths scattered, leaving the longer ones to reach our eyes.

The scattered radiation is not only changed in direction, it is also polarized. That is, the electric vector of the scattered radiation is not randomly oriented in the plane normal to the new direction of propagation. Instead, the electric vector of the scattered radiation will be normal to the plane containing both the incident and the scattered rays. This fact may be utilized as a means of discriminating against undesired scattered radiation in some types of infrared systems.

3.6   INTERFERENCE AND DIFFRACTION

The effects produced by interactions of wave fronts are so numerous that they are classified in two groups. Those effects due to a restriction of the wave front are called diffraction effects, whereas those caused by the combining of a number of separated beams, which originally came from the same source or coherent sources, are called interference effects.

Some very interesting and unusual phenomena manifest themselves when several electromagnetic waves occupy the same region in space and time. In 1678 Huygens, a Dutch scientist, formulated a law describing the interaction of two beams. This law, called the principle of superposition,[10] states essentially that at any point in space and time the total electric vector is the vector sum of the electric field associated with each beam. The amplitude, frequency, and all other characteristics of each beam are entirely uninfluenced by the other beam after having passed out of the region of crossing.

3.6.1 Interference

When waves from two sources move into the same region of space, they interfere. In this region of crossing, there are points at which the disturbance is zero and other points where the disturbance is greater than if there had been only a single wave. After each wave has passed through this region of interaction, it proceeds as if it had never encountered the other waves. Where the resultant disturbance is zero, the waves are said to have interfered destructively. Conversely, the points of greatly increased disturbance are examples of the effects of constructive interference.

When two infrared beams interact, interference effects occur and may sometimes affect the performance attainable with an infrared system.

The interference pattern arising in the case of a source-slit configuration shown in Fig. 3.7 was first investigated by Young.[11] Through his experiments, he was able not only to demonstrate the phenomena of interference and diffraction but was also able to measure numerical values of the wavelengths of the radiation involved.

Consider a source S behind the slits S_1 and S_2 that are located at a distance D from a screen oriented normal to a line from the source which

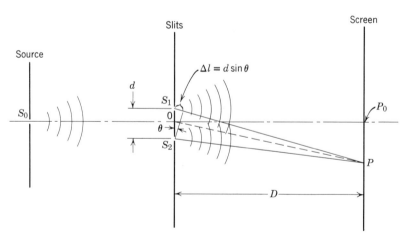

Figure 3.7. Interference due to two slits.

passes through the midpoint of a line joining the slits. We will consider the slits narrow by comparison with the wavelengths of the radiation being used. The wave fronts which spread out from S_1 and S_2 exhibit maxima and minima at regular intervals. As mentioned previously, we would therefore expect to find points of zero disturbance where the two waves had canceled each other and other regions where the resultant wave amplitude would be increased by the interaction.

At the point P, the irradiance will be a maximum if the path difference Δl is an even number of half waves, that is, $\Delta l = n\lambda/2$, $n = 0, 2, 4, 6, \ldots$. This will seem more plausible if we recall that the phase difference $\Delta\phi$ between two beams is given by

$$\Delta\phi = \frac{2\pi}{\lambda}(\Delta l). \qquad (3.91)$$

Consider the point P_0, which is on the screen and intercepted by a normal to the screen drawn from a point midway between the two slits. Here the path difference will be zero, $\Delta\phi$ will be zero, the waves will be in phase, and the resultant disturbance will have its largest possible value. Now consider the radiation reaching any arbitrary point P in the plane of the screen. Let the electric vector $\mathbf{E}_1$ associated with the radiation emanating from the first slit be given by

$$\mathbf{E}_1 = \mathbf{E}_0 \exp\left\{2\pi j\nu\left[t - \frac{\mathbf{u}\cdot\mathbf{p}_1}{c_0}\right]\right\},$$

where $\mathbf{u}$ is the vector connecting the point O with P,

   $\mathbf{p}_1$ is a unit vector in the direction of propagation (see Eq. 3.42), and

$$\mathbf{E}_2 = \mathbf{E}_0 \exp\left\{2\pi j\nu\left[t - \frac{\mathbf{u}\cdot\mathbf{p}_2}{c_0}\right]\right\},$$

represents the disturbance reaching P from S_2. It follows from 3.91 that

$$\Delta\phi = 2\pi\nu\left\{\frac{\mathbf{u}\cdot\mathbf{p}_2 - \mathbf{u}\cdot\mathbf{p}_1}{c_0}\right\} = \frac{2\pi}{\lambda}\{\mathbf{u}\cdot\mathbf{p}_2 - \mathbf{u}\cdot\mathbf{p}_1\},$$

and the resultant disturbance at the arbitrary point P is given by

$$\mathbf{E}_T = \mathbf{E}_1 + \mathbf{E}_2 = \mathbf{E}_0 \exp\left[2\pi j\nu t\right]\left\{\exp\left[-2\pi j\nu\left(\frac{\mathbf{u}\cdot\mathbf{p}_1}{c_0}\right)\right]\right.$$
$$\left. + \exp\left[-2\pi j\nu\left(\frac{\mathbf{u}\cdot\mathbf{p}_2}{c_0}\right)\right]\right\},$$

or

$$\mathbf{E}_T = \mathbf{E}_0 \exp\left\{2\pi j\nu\left[t - \frac{\mathbf{u}\cdot\mathbf{p}_2}{c_0}\right]\right\}\{1 + \exp\left[j\Delta\phi\right]\}. \qquad (3.92)$$

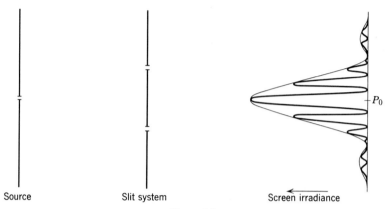

Source   Slit system   Screen irradiance

Figure 3.8

Since the average radiant power is proportional to $|E_T|^2$,

$$P_P \propto 2\,|E_0|^2\,(1 + \cos \Delta\phi). \qquad (3.93)$$

Figure 3.8 shows a plot of the irradiance as a function of $\Delta\phi$, that is, a function of the distance along the screen measured outward from P_0.

3.6.2 Diffraction

When radiation from a distant source falls upon an aperture in an opaque plate, the edges of the shadow image observed on a screen held at various distances behind the aperture are not infinitely sharp. Instead, if

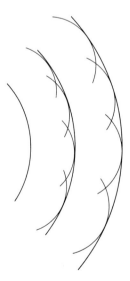

Figure 3.9

the screen is held very near the aperture, there are dark and light fringes around the edge of the shadow which are typical of Fresnel diffraction. As the screen is moved back further and further, these fringes become wider and move inward until they occupy the entire image area. This is an illustration of the transition from Fresnel to Fraunhofer diffraction. Finally, when the screen is far behind the aperture, the resulting shadow pattern bears no resemblance to the aperture, a situation typical of Fraunhofer diffraction.

Diffraction may be looked upon as a special case of interference. To facilitate the discussion of this effect, it is necessary to introduce Huygen's wavelets. According to Huygen's principle,[12] each point on a wave front may be considered a new source of spherical waves and the envelope of these waves constitutes the wave front at a later time. (See Fig. 3.9.) Diffraction arises because the radiation propagated by various small elements of the wave front interfere constructively or destructively, depending upon their position relative to each other and to the point of observation.

To illustrate this phenomenon, consider a plane monochromatic wave falling upon a small circular aperture. Figure 3.10a shows a plane wave parallel to an aperture plate. The wave is traveling from left to right and we will observe it through the aperture from the points P, Q, and R. These points, at the extreme right of Fig. 3.10a, may also be considered as defining a screen where the diffraction pattern would be displayed.

In the absence of the aperture plate the point P would receive its radiation from the point P_w on the plane wave. Similarly, the source of wavelets for Q is Q_w and the wavelets striking R emanate from R_w. R would receive no radiation from P or Q since we are considering a wave front which neither diverges nor converges.

We must now introduce the concept of Fresnel zones. Consider a circular portion, centered at P_w, of the wave front. It may be divided into a number of concentric, circular annular regions whose centers are also at P_w. The widths of the regions are chosen such that the radiation emanating from each region differs in phase, on the average, by π radians from the radiation arising from its nearest neighbors. It will seem more plausible that this can be done if we note that the edges of each zone are the loci of points equidistant from the chosen point of observation. From Eq. 3.91 we see that each individual zone therefore sends out radiation of approximately equal phase. This manner of division causes the widths of zones to become smaller as one proceeds outward from the center. However, the area of these zones becomes very gradually larger.

Consider now the radiation observed at the point P when the aperture plate is in position. It consists of contributions from each Fresnel zone.

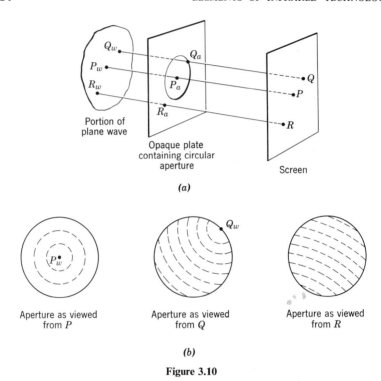

(a)

Aperture as viewed
from P

Aperture as viewed
from Q

Aperture as viewed
from R

(b)

Figure 3.10

In some cases the distance between P_a and P and the size of the aperture will be such that the aperture will contain an even number of zones. The radiation contributed by each of the zones will cancel in pairs so that the irradiance at point P will be zero. If, on the other hand, the number of zones within the aperture is odd, the contributions from the various zones cancel in pairs with one left over so that the irradiance at P is the same as it would be in the absence of the aperture. Thus we see that as we move the point P relative to the aperture or as we gradually change the aperture diameter, the point P is alternately light or dark.

We must now consider the irradiance at point Q. (See Fig. 3.10b.) If we now establish Fresnel zones about Q_a in the same manner as before, we see that only portions of the zones will be able to send their radiation through the aperture. There will thus be unequal contributions from each of the zones, so that the degree of cancellation depends strongly on just where Q is located relative to the aperture edge. As point Q moves so that Q_a travels across the aperture edge, there is a change in the irradiance due to variations in the degree of cancellation as portions of zones are covered and uncovered. This accounts for the diffraction fringes.

At R the wider central zones are completely obscured so that the irradiance arises as a result of the contributions from a large number of portions of narrow zones. If R is well removed from the center of the aperture, cancellation is so nearly complete that the irradiance is essentially zero. The sum of the contributions from the various zones fluctuates to a greater degree if the number of zones is small, that is, if we are considering apertures which are small by comparison with the wavelength. This illustrates the point mentioned previously that diffraction effects manifest themselves when we restrict the aperture of an optical system.

We saw in an earlier section that radiation passing through two slits produces an interference pattern on a screen. If the radiation from a source passes through two apertures small by comparison with the wavelength of the radiation, there will then be a diffraction pattern from each aperture in addition to the pattern resulting from the interference between apertures. This superposition of patterns may make it difficult to "resolve," that is, distinguish, the separate apertures. Figure 3.11 shows the combined interference-diffraction pattern arising from two apertures which are not very large compared to the wavelength of the radiation being used.

We have described qualitatively how diffraction effects occur and have indicated why restriction of the wave front can be expected to make this special type of interference easier to observe. The mathematics needed to specify and sum the contributions from the Fresnel zones in a practical case are quite complicated. A thorough discussion involves the derivation of an analog of Green's theorem, a powerful tool, but one of limited usefulness in the remainder of this book. Therefore, we merely note that the essential step in computing the diffraction pattern is the evaluation of the integral

$$\int \exp \left[\frac{-2\pi j}{\lambda} (r + r_1) \right] da, \qquad (3.94)$$

over the area of the aperture,

where r is the distance from the source to the aperture element da, located at Q_A, and
r_1 is the distance from da to the point Q at which we wish to determine the irradiance.

If we are considering a plane wave parallel to the aperture, r is the same for all elements and can be taken outside the integral. The integral represents the sum at Q of the amplitudes of the spherical wavelets of equal amplitude and phase starting from all elements in the aperture. It is the interference of these wavelets which produces the diffraction pattern.

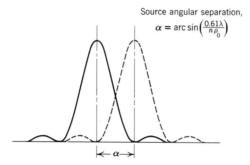

Source angular separation,

$$\alpha = \arcsin\left(\frac{0.61\lambda}{n\rho_0}\right)$$

Figure 3.11. Diffraction image of two slit sources.

The circular aperture. A case of considerable practical importance is that of the Fraunhofer diffraction produced by a circular aperture. (See Fig. 3.12.) This is the diffraction which might be observed when viewing a distant object using an optical system of small aperture. Our treatment of this case follows that of Slater and Frank.[13] For a distant source, r can be taken outside the integral in Eq. 3.94 so that

$$\int \exp\left[\frac{-2\pi j r_1}{\lambda}\right] da = \int \exp\left[\frac{-2\pi j}{\lambda}(lx + my)\right] da, \qquad (3.95)$$

where l and m are the direction cosines of the direction from the center of the aperture to any point Q on the observation plane; x and y are co-ordinates of the element da in the aperture plane measured from the center of the aperture.

If we introduce the polar coordinates $x = \rho \cos \theta$, $y = \rho \sin \theta$, and take advantage of the symmetry of the problem, we need only perform the integration for a set of elements which extend along some radius vector in

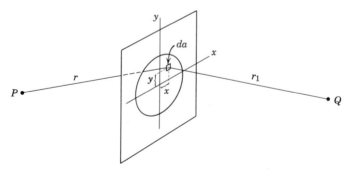

Figure 3.12. Geometry for diffraction calculation.

the aperture. It is most convenient to choose the one which extends in the $+x$-direction so that $m = 0$. Then 3.95 becomes

$$\int_0^{2\pi} d\theta \int_0^{\rho_0} \exp\left[\frac{2\pi j \rho l \cos\theta}{\lambda}\right] \rho \, d\rho, \qquad (3.96)$$

where ρ_0 is the radius of the aperture.

Integrating the second integral by parts, we get

$$\int_0^{2\pi} \left\{ \frac{\rho_0 \exp\left[\dfrac{2\pi j \rho_0 l \cos\theta}{\lambda}\right]}{\dfrac{2\pi j l \cos\theta}{\lambda}} - \frac{\exp\left[\dfrac{2\pi j \rho_0 l \cos\theta}{\lambda}\right] - 1}{\left(\dfrac{2\pi j l \cos\theta}{\lambda}\right)^2} \right\} d\theta.$$

This turns out to be equal to

$$\frac{\rho_0 \lambda}{l} J_1 \left\{ 2\pi \rho_0 \left(\frac{l}{\lambda}\right) \right\}, \qquad (3.97)$$

where J_1 is the first order Bessel function of the first kind.

We can deduce the variations in the irradiance from our knowledge of the properties of this function. The irradiance will be a maximum at $l = 0$, directly in line with the center of the aperture. It will be zero at $2\pi \rho_0 l / \lambda = 3.832$, 7.016, 10.173, $13.324, \ldots$, or $\rho_0 l / \lambda = 0.610$, 1.117, 1.621, $2.122, \ldots$ etc., with maxima at points between these values. This accounts for the fringes.

Whenever radiation passes through an optical system, it may be refracted or reflected, and, since it is restricted by some aperture in the system, it will also be diffracted. This may, in some instances, be the factor which limits the ability of the system to form separate images of closely spaced objects. This ability is known as the resolution capability of the system. The angular resolution of a system which uses a circular aperture can be obtained by utilizing 3.97. The angular separation, α, between two objects which can just barely be resolved is given by[14]

$$\alpha = \text{arc sin} \left[\frac{0.61\lambda}{n\rho_0}\right], \qquad (3.98)$$

where n is the index of refraction of the medium in object space.

From Eq. 3.98 it can be seen that infrared systems are capable of much greater resolution than radar systems of the same aperture. By the same token, infrared systems require a larger aperture than visible systems capable of resolving the same fineness of detail.

REFERENCES

1. J. A. Stratton, *Electromagnetic Theory*, McGraw-Hill Book Company, New York (1941).
2. A. P. Wills, *Vector Analysis*, Prentice-Hall, Englewood Cliffs, New Jersey (1931).
3. G. P. Harnwell, *Principles of Electricity and Electromagnetism*, McGraw-Hill Book Company, New York (1938), p. 533.
4. J. C. Slater and N. H. Frank, *Electromagnetism*, McGraw-Hill Book Company, New York (1947), p. 90.
5. H. H. Skilling, *Fundamentals of Electric Waves*, John Wiley and Sons, New York (1948), p. 131.
6. J. A. Stratton, *op. cit.*, p. 185.
7. H. C. Van de Hulst, *Light Scattering by Small Particles*, John Wiley and Sons, New York (1957).
8. W. E. K. Middleton, *Vision Through the Atmosphere*, University of Toronto Press, Toronto (1958), p. 16.
9. J. C. Slater and N. H. Frank, *op. cit.*, p. 160.
10. J. Strong, *Concepts of Classical Optics*, W. H. Freeman and Company, San Francisco (1958), p. 27.
11. F. A. Jenkins and H. E. White, *Fundamentals of Physical Optics*, McGraw-Hill Book Company, New York (1937), p. 52.
12. J. K. Robertson, *Introduction to Physical Optics*, D. Van Nostrand Company, New York (1947), p. 60.
13. J. C. Slater and N. H. Frank, *op. cit.*, Chapter XIV.
14. F. W. Sears, *Optics*, Addison-Wesley Press, Cambridge, Mass. (1949), p. 257.

4
Optical properties of media

4.1 INTRODUCTION

In the design of an infrared system, it is necessary to consider the optical properties of the media which lie in the path of the radiation as it makes its way through the system. For example, the designer must be aware of the many considerations involved in the choice of window materials, filters, and other optical elements, and must be able to take proper account of the optical parameters of the detector itself. In addition, he must be cognizant of the transmission characteristics of the atmosphere through which the radiation must pass.

Chapter 3 dealt with the physical theories underlying the optical properties of matter. This chapter will discuss the optical properties of the various states of matter, presenting some of the practical aspects to be considered in choosing optical components. The final section describes the measured characteristics of a number of available materials.

4.2 CLASSES OF MATTER—THEIR OPTICAL PROPERTIES

4.2.1 Solids

Several mechanisms account for the transmission characteristics of solid materials. The most important of these for our purposes include fundamental absorption, which is responsible for reststrahlen, characteristic absorption, intrinsic and extrinsic absorption in semiconductors, and absorption by free carriers. In polycrystalline substances, scattering may appreciably attenuate the radiation.

Fundamental absorption relates to the radiation attenuation arising because of the interaction between the electromagnetic waves and the

119

lattice, and is thus sometimes called lattice absorption. In amorphous materials, such as glass, there is only a short range order and we should, perhaps, more properly speak of absorption due to structure rather than the absorption due to the lattice. In certain materials of high ionicity resonance effects, called reststrahlen, occur. (See the discussion later in this chapter.)

Characteristic absorption becomes evident when we observe the absorption bands due to molecular vibrations and rotations. The infrared absorption bands of plastic films are typical examples of characteristic absorption.

In semiconductors it is convenient to distinguish between intrinsic and extrinsic semiconductors. (See Section 6.4.) At this point it may be sufficient to remark that intrinsic absorption causes excitation of electrons across the forbidden band. Extrinsic absorption causes transitions of electrons between impurity states and the conduction or valence band. On the other hand, the interactions between free, or almost free, electrons and radiation in a solid account for another type of absorption, termed free-carrier absorption.

Finally, scattering in polycrystalline materials of low optical absorption arises when radiation impinging upon intercrystalline boundaries suffers an abrupt change in its direction, causing it to be scattered out of the beam. Lattice imperfections which cause local modifications in the index of refraction also give rise to scattering.

Dielectrics. The optical properties of dielectrics are important because of their use as windows, filters, and lenses. Dielectrics typically possess rather good transmission in some portions of the infrared. By comparison with metals and semiconductors their reflectivity (at an air interface) is low. Their thermal and electrical conductivities are usually low and they are frequently brittle. In general, classical electromagnetic theory is quite adequate for explaining the observed optical behavior of nonconductors.

In Section 3.5.1 we pointed out the functional relationships between the wavelength of the radiation and the optical constants of dielectric materials. These relationships are very useful in that they permit fairly accurate estimates of the optical behavior of dielectric materials. For single crystal dielectrics, fundamental absorption accounts for most of the experimentally measured attenuation. Characteristic absorption in dielectrics is limited to the short wavelength visible, ultraviolet, and x-ray regions of the spectrum. At room temperature and lower, the number of free electrons in a dielectric is so small that free-carrier absorption is insignificant. The scattering of infrared by polycrystalline dielectrics depends upon both the crystallite size and the average density. Basically, the scattering depends upon a change in the index of refraction at the boundaries between individual crystals. The amount of scattering therefore depends upon the index of refraction of the material. The resolution capabilities of an optical system may be reduced by the use of polycrystalline dielectrics even when the intrinsic transmission (that is, the transmission of the individual crystallites) is good. Properties of some specific dielectrics are given in Section 4.3.4.

Glasses. The strong water absorption band at about 2.8 μ causes the common glasses to be unsuitable for use in those optical systems that are required to function at wavelengths longer than about 2.7 μ. Corning's high silica glass No. 7900, known as Vycor, is suitable for use as far as 3.3 μ, whereas fused silica can be used for wavelengths as great as 4.5 μ. Work has been under way for many years to prepare suitable glasses for use beyond 4.5 μ. To avoid the use of silicates, a wide variety of other compounds have been tested. Workers at Ohio State University have developed some germanium dioxide and lead tellurate glasses[1] which cut off at about 6 μ. The National Bureau of Standards has prepared glasses containing a mixture of barium and titanium compounds in addition to silicates.[2,3] The Battelle Memorial Institute[4] has investigated compounds which contain rather a large number of components in order to solve simultaneously the problems associated with the attainment of good infrared transmission, ease of working, and the increasing of resistance to devitrification.

One of the most promising materials which has been made available for use in the intermediate infrared is the series of calcium aluminate glasses first studied at NBS and further developed by Bausch and Lomb.[5] One of these, identified as RIR 12, is shown in Fig. 4.11. The spectral transmissions, indices of refraction, and a brief tabulation of other properties of a few types of glass are presented in Section 4.3.4.

Plastics. Plastics form another class of dielectric materials. Their use as optical elements in infrared systems is somewhat restricted for two reasons.

The first characteristic of plastics which limits their usefulness relates to the multiplicity of absorption bands in the infrared. Plastics are composed of strings or chains of molecular groups exhibiting rotational and vibrational bands. These characteristic bands are so numerous that in many regions they overlap. However, in some plastics there are spectral intervals in which the transmission characteristics are good, but care must be taken so that the "windows" of a particular plastic correspond to spectral intervals of interest. The second undesirable characteristic of plastics is that although there are some plastics which exhibit relatively good transmission characteristics over broad spectral regions, they do so only for thin samples. Unfortunately, most plastics cannot be used as unsupported optical elements in thicknesses sufficiently small to insure good transmission. For thicker samples absorption increases markedly because of the deepening and widening of absorption bands.

In Section 4.3.4 we shall show the spectral transmission characteristics of two representative plastics.

Metals. Metals are characterized by the fact that both their absorptivity and reflectivity are large. The concentration of free electrons is so great that they account for a large part of the optical absorption in metals. As a result, metals appear opaque in the infrared.

It has previously been shown in Eq. 3.62 that for normal incidence

$$\rho = \frac{(n - 1)^2 + k^2}{(n + 1)^2 + k^2}. \tag{4.1}$$

Where σ/ν is very large, that is, for long wavelengths it can be seen from Eq. 3.34 that

$$n = k = c_0 \sqrt{\frac{\mu\sigma}{4\pi\nu}}.$$

Both n and k are large compared to unity, so that 4.1 becomes

$$\rho \approx 1 - \frac{2}{n} = 1 - \frac{2}{c_0} \sqrt{\frac{4\pi\nu}{\mu\sigma}}. \tag{4.2}$$

Equation 4.2 is known as the Hagen-Rubens relation and, in general, is in agreement with experiment for $\lambda > 10\ \mu$. For the limiting case of a perfect conductor we see that $\rho = 1$. Since for most metals σ is very large,

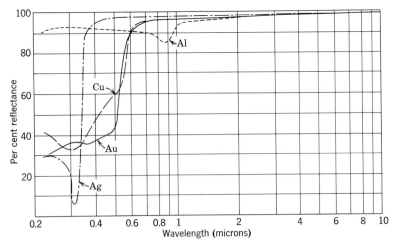

Figure 4.1. Spectral reflectance of metal films. (After G. Hass.[6])

the reason for their high reflectivity is evident. The reflectivity of metal films has been very carefully investigated by Hass.[6]

Figure 4.1 shows the reflectivity of some common metal films. It is interesting to note that Au, although it reflects well in the infrared, has an appreciably lower reflectivity in the visible, making it a good choice for reflecting optical surfaces in systems which are required to be "solar blind." Reflectivities of a few bulk metals are shown in Table 4.1.

TABLE 4.1. Reflectivity of metals (%), normal incidence*

Wavelength (microns)	Copper	Gold	Nickel	Steel
0.25	25.9	38.8	37.8	32.9
0.36	27.3	27.9	48.8	45.0
0.45	37.0	33.1	59.4	54.4
0.50	43.7	47.0	60.8	54.8
0.60	71.8	84.4	64.9	55.4
0.70	83.1	92.3	68.8	57.6
0.80	88.6	94.9	69.6	58.0
1.00	90.1	—	72.0	63.1
2.00	95.5	96.8	83.5	76.7
3.00	97.1	—	88.7	83.0
4.00	97.3	96.9	91.1	87.8
9.00	98.4	98.0	95.6	92.9

* Adapted from *Handbook of Chemistry and Physics*, 37th ed., Chemical Rubber Publishing Co., Cleveland (1955), p. 2691.

Semiconductors. A discussion of the physics of semiconductors is given in Chapter 6. In this section we will discuss only their optical properties.[7] Those who are not acquainted with semiconductor physics may find it helpful to first read Chapter 6.

From both an academic and a practical point of view, the optical properties of semiconductors in the infrared are extremely interesting and complex. Semiconductors are interesting because they continue to transmit radiation in a spectral interval where most glasses absorb strongly; in addition, their high indices of refraction adapt them for use as refracting elements in optical systems. The complexity arises primarily because semiconductors represent a transition between dielectrics and metals. Furthermore, the number of free charge carriers, and to a lesser extent, carrier mobility, is subject to control by varying either chemical composition, crystal structure, or temperature. The number of carriers can also be changed abruptly by electrical or optical injection.

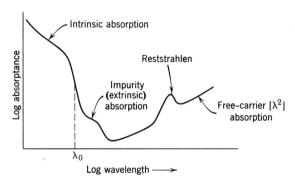

Figure 4.2. Spectral absorption in a semiconductor.

Whereas most dielectric materials transmit in the visible and absorb in the infrared, most semiconductors are opaque in the visible but transparent in a large part of the infrared.

Figure 4.2 shows the relationship between wavelength and absorption for a typical semiconductor. In the region where λ is smaller than λ_0, the incident photons have enough energy to excite electrons across the forbidden band of energies which is characteristic of the material. This process of intrinsic absorption is so efficient that even comparatively thin sections of the material are completely opaque. The wavelength λ_0 is called the absorption edge of the material, and it can easily be located by the relation

$$\lambda_0 = \frac{hc_0}{\Delta\mathscr{E}},$$

where $\Delta\mathscr{E}$ is the width of the forbidden energy gap of the semiconductor. At wavelengths slightly greater than λ_0 the absorption falls to very low values. The reason for this behavior is that the atoms in a valence crystal are nearly neutral so that in the first approximation the vibrations of these atoms do not give rise to dipole moments and hence abstract very little energy from the beam. In impure materials which contain energy levels within the forbidden band, absorption may occur as a result of the optical excitation of these levels. It may not be possible to observe this behavior unless the material has been cooled to prevent thermal excitation. In addition, in p-type materials, certain absorption bands are observed at wavelengths greater than λ_0. They are almost always temperature sensitive, being related to the mean square thermal displacement of the atoms. Furthermore since they are usually not altered by the introduction of impurities, they seem to be linked to lattice vibrations, termed "phonons." Although the primary bonding in semiconductors is considered to be covalent, some types of semiconductors exhibit a certain degree of ionicity. Reststrahlen can be observed in these materials. At still longer wavelengths, the absorption curve rises appreciably. This behavior can be accounted for by considering the interactions between the radiation and free electrons or holes. Although the laws pertaining to the conservation of energy and momentum forbid a complete transfer of energy from the radiation to perfectly free electrons, any real lattice is not perfectly periodic and is also vibrating because of thermal agitation. The resultant slight interaction between the lattice and the electrons or holes is such as to permit the absorption of radiation by them.

To see how the free carrier absorption varies with wavelength, consider Eq. 3.71.

$$a = \frac{4\pi v k}{c_0} = \frac{4\pi v k}{c_0}\left(\frac{n}{n}\right).$$ (4.3)

From Eqs. 3.27 and 3.28, kn is given by

$$kn = \frac{\varepsilon\mu}{\varepsilon_0\mu_0}\frac{\sigma}{4\pi v \varepsilon}.$$

Substituting this into Eq. 4.3 yields

$$a = \frac{\mu\sigma}{\mu_0\varepsilon_0 c_0 n} \approx \frac{\sigma}{\varepsilon_0 c_0 n},$$ (4.4)

by virtue of the assumption $\mu \approx \mu_0 \approx 1$.

The next step is to see how the conductivity varies with wavelength. Consider the equation of motion for a free carrier having an effective mass m_e, acted upon by a sinusoidally varying electric field, that is,

$$qE_0 \sin \omega t = \left\{\frac{d}{dt}\left[m_e \bar{v}\right]\right\}_{net},$$ (4.5)

where q is the electric charge of the carrier,

$E_0 \sin [2\pi c_0 t/\lambda]$ represents a sinusoidally varying electric field, and $\bar{v}$ is the average carrier velocity.

The factor $\left\{\frac{d}{dt}(m_e \bar{v})\right\}_{net}$ is composed of two terms. One pertains to the rate at which $m_e \bar{v}$ increases with the application of the electric field; the other relates to the rate at which $m_e \bar{v}$ decreases when the field is removed. Consider the latter. If the average momentum upon application of an electric field is $(m_e \bar{v})_0$, then removal of the field causes $(m_e \bar{v})$ to decrease with time in an exponential manner.

$$(m_e \bar{v}) = (m_e \bar{v})_0 \exp (-t/\tau_r),$$

where τ_r is the characteristic time required for a new velocity distribution to be attained; it is referred to as the relaxation time and is of the order of 10^{-12} sec for germanium.

Thus

$$\frac{d(m_e \bar{v})}{dt} = -\frac{1}{\tau_r}(m_e \bar{v}).$$

Including both terms in 4.5 results in

$$\frac{qE_0}{m_e} \sin \omega t = \frac{d\bar{v}}{dt} + \frac{\bar{v}}{\tau_r}. \tag{4.6}$$

The steady state solution of this equation is

$$\bar{v} = \frac{qE_0 \tau_r}{m_e[1 + \omega^2 \tau_r^2]} \{\sin \omega t - \omega \tau_r \cos \omega t\}. \tag{4.7}$$

Since the mobility μ_e is given by $\mu_e \equiv \bar{v}/E$ where $E = E_0 \sin \omega t$, the average mobility $\bar{\mu}_e$ can be calculated to be

$$\bar{\mu}_e = \frac{1}{T} \int_0^T \mu_e \, dt = \frac{q\tau_r}{m_e[1 + \omega^2 \tau_r^2]T} \int_0^T \{1 - \omega \tau_r \cos \omega t\} \, dt$$

$$= \frac{q\tau_r}{m_e[1 + \omega^2 \tau_r^2]}. \tag{4.8}$$

The conductivity $\sigma = n_e q \bar{\mu}_e$ is then

$$\sigma = \frac{n_e q^2 \tau_r}{m_e[1 + \omega^2 \tau_r^2]} = \frac{n_e q^2 \tau_r}{m_e\left[1 + \dfrac{4\pi^2 c_0^2 \tau_r^2}{\lambda^2}\right]} = \frac{\sigma_0}{\left[1 + \dfrac{4\pi^2 c_0^2 \tau_r^2}{\lambda^2}\right]}, \tag{4.9}$$

where n_e is the carrier concentration and σ_0 is $n_e q^2 \tau_r / m_e$, the zero frequency conductivity.

From 4.4 we see now that the absorption coefficient becomes

$$a = \frac{\sigma_0}{\varepsilon_0 c_0 n} \left[1 + \frac{4\pi^2 c_0^2 \tau_r^2}{\lambda^2}\right]^{-1}. \tag{4.10}$$

For very long wavelengths the expression in brackets is essentially equal to unity and the absorption is nearly independent of wavelength. (This condition is encountered at about 2 or 3 mm wavelength in a typical germanium sample.) At somewhat shorter wavelengths (but still well beyond λ_0), $[(2\pi c_0/\lambda)\tau_r]^2$ begins to determine the magnitude of the expression inside the brackets. In this range of wavelengths we observe the so-called λ^2 absorption as indicated in Fig. 4.2. At very short wavelengths, the free carrier absorption is negligible as is indicated by 4.10, and below λ_0 it is also masked by the very much larger intrinsic absorption. As seen later in the chapter, the free carrier absorption is sometimes markedly increased by heating the semiconductor. An example of the effect of heating on free carrier absorption is cited in Section 4.3.

In impure semiconductors absorption may be caused by the excitation of electrons from donor centers to the conduction band, or from the valence band to acceptor centers. Since these centers are near the band edges, the excitation energy required is small so that the absorption bands are observed at wavelengths past the absorption edge. A rather exact value of the energy can be determined by considering the electron as bound in a hydrogenlike configuration to the impurity center, which can be thought of as being immersed in a medium whose dielectric constant is equal to that of the crystal.[8]

There is still another type of absorption which may produce bands considerably more intense than the free carrier absorption at wavelengths past the absorption edge. This absorption results from transitions which take place within an energy band of the crystal. These effects are sometimes described as transitions between "heavy" and "light" holes and are found, for example, in p-type germanium.

4.2.2 Liquids

The infrared optical characteristics of liquids are of limited importance in infrared technology. Their very large absorption in the infrared generally precludes their use as refracting elements. Actually, "water cells," glass vessels containing water, are used in projection optical systems to remove the infrared component of high temperature radiation. The infrared reflectivity of most liquids is appreciable. This factor may adversely affect the performance of some systems because it produces "glint." Glint is the shimmering or fluctuation of the radiation reflected from a large number of small specularly reflecting areas as, for example, from the surface of the ocean.

The infrared spectral transmission characteristics of organic liquids are extremely complex. They are quite similar, in this regard, to plastics. Since few unique and useful properties have been found for organic liquids, they are not widely used in infrared technology.

Another extremely important aspect of the infrared optical characteristics of liquids becomes evident when the liquids are dispersed as small droplets in gaseous media. In this state they become very effective scattering agents. It will be more convenient to discuss this topic when we treat the transmission characteristics of the atmosphere in the next chapter.

4.2.3 Gases

The transmission of infrared through gases has concerned spectroscopists for many years. Much valuable information regarding the properties of gas atoms, their collision processes, and other interactions has been obtained. We, however, shall be concerned with the transmission of infrared through gases only as it relates to infrared applications. Our interest in the transmission properties of gases stems from the fact that in many applications the radiation has to pass through a long path in the atmosphere.

The reflectivity of pure gases and the effects upon the transmissivity of gas samples due to changes in their indices of refraction are both negligible. Their absorption is characteristic absorption due to vibrations and rotations of molecules as discussed in Chapter 2. In Chapter 5 transmission curves are presented for two pure gases which have the largest effect on infrared systems: water vapor and carbon dioxide.

4.3 CHOICE OF SPECIFIC MATERIALS

Until now we have been discussing the characteristics of classes of matter. In this section we present an abbreviated set of data on some specific materials which may be used as windows, domes, and filters in infrared applications. A brief description of the properties of the most commonly used materials may be helpful in several ways. It will perform the function referred to at the beginning of this chapter, that is, it will help the reader to make at least a tentative selection of a material for a specific application. Before making a final decision, the user may wish to obtain more complete data on those materials which seem best suited for his purposes. Ballard, McCarthy, and Wolfe[9] have prepared a very thorough report on infrared optical materials which should be of great assistance. The book by Moss[10] provides a more thorough discussion of specific semiconductors than can be given in the space available here. The reader who is interested in the reflectivity and emissivity of materials will find *Emittance and Reflectance in the Infrared*[11] an excellent bibliography.

4.3.1 Requirements for Windows

Infrared transmitting windows are needed in many different places in an infrared system, on the detector envelope, for example. Since the detector envelope application is, in many ways, the most demanding, we will discuss its usage to illustrate the type of requirements.

Mounting the sensitive element of a detector in an enclosure with an infrared transmitting window protects the element from mechanical damage and the effects of the atmosphere which surrounds it. Cooled detectors must have an enclosure with a window to permit vacuum encapsulation and prevent frosting. The window must have a high transmissivity (and therefore a very low absorptivity or emissivity) over the entire spectral interval to which the detector responds. If this condition is met, the window will efficiently admit the useful radiation from the source to be detected and will itself emit only very small amounts of radiation in the wavelength interval which can be seen by the detector. Both reflection and scattering losses should be low. In many instances it will be possible to obtain windows which transmit 95% of the incident radiation in the wavelength interval of interest. For cooled detectors, window material should have good sealing properties to glass or to some other material which can be shaped to make a sturdy, vacuum-tight detector housing. Glass-to-glass type seals require a close match between the thermal expansion coefficients of the materials being sealed. If the expansion coefficients differ greatly, use of a graded seal becomes necessary and construction difficulties are greatly increased. Some materials which have good optical properties can be joined to the detector envelope only by the use of epoxy resin. The use of cement leads to difficulties because of poor resistance to thermal shock and because of outgassing during storage.

Windows must meet many requirements in the way of chemical stability. If they are glasses, they should not undergo changes upon aging which cause them to devitrify or to change their spectral transmission. They should not be affected by moisture; they should be nontoxic; they should lend themselves to being worked or recast, and should be easy to cut or shape. Some of the materials we shall discuss have a tendency to "cold flow," an undesirable quality. Such materials undergo a very gradual but appreciable change of shape when subjected to stress by gravity or when used as part of an evacuated container. In addition to all the requirements mentioned, the cost should be low. We shall discuss materials which are used extensively as windows in Section 4.3.4.

4.3.2 Requirements for Domes

The component used at the front of the enclosure for the optical portion of an infrared system is often referred to as an irdome, or simply, dome. Materials which can be utilized for the fabrication of domes must meet a rather restrictive set of conditions.

To be suitable, the material must possess the proper spectral transmission characteristics. Just as with detector window materials, this requirement usually means that the dome must transmit the entire spectrum likely to be emitted by the source to be detected. If the detector used in the system responds also to radiation outside this interval, the material should also be transparent in these regions of the spectrum. The dome will then have an absorptivity, α, and an emissivity, ε, which are low for all detectable wavelengths. (See Chapter 2.) If its thermal radiation is small, the photon fluctuations which occur within the electrical bandwidth of the system will also be small and will constitute only a small false radiation signal. (See Sections 2.5.2, 2.5.3, and 9.4 for discussions of photon fluctuations.)

Some materials which exhibit good spectral transmission curves at normal room temperature deteriorate markedly in transmission at higher temperatures because of an increase in free carrier absorption. Domes are often heated aerodynamically if they are mounted on aircraft or missiles and used at high speeds, so this factor may be an important one for air-borne applications. The dome material should not possess discontinuities of a sort which cause the signal radiation to be scattered. From the earlier discussions in this chapter, we see that the material should be either in the form of a single crystal or its index of refraction should not undergo sharp discontinuities at the crystalline boundaries.

Irdomes need not ordinarily meet such stringent requirements as windows in regard to their coefficient of expansion. In most applications the seal between the dome and the remainder of the optical housing need not be vacuum tight. The dome may often be clamped in place against an "O" ring or gasket, rather than being fused or permanently bonded to other portions of the housing. For this same reason the vapor pressure of dome materials is relatively unimportant.

The material should be hard enough to resist abrasion caused by flying dust. Irdomes are exposed to the scouring action of particles of grit and dirt when mounted on equipment which is used near ground level or which must be in standby condition near ground level. For this reason an "eyelid" is sometimes used to protect an irdome during standby.

Although infrared detector windows are usually small, irdomes are often 3 to 6 in. in diameter and may occasionally be much larger, perhaps several feet in diameter. Small pieces of material are sometimes used to construct irdomes in the form of a mosaic, but this procedure has obvious shortcomings.

Finally, dome materials must be able to resist attack by salt water and corrosive gases in the atmosphere, and should be chemically stable even after prolonged exposure to intense solar radiation. Windows may be mounted in a sealed, dry, clean environment within the optical housing, but domes are, by their very nature, exposed to a harsh environment.

4.3.3 Other Optical Components

Infrared systems also use such components as mirrors, lenses, prisms, filters, and antireflection coatings. The choice of materials for these purposes is discussed in this section.

Deciding which metallic films to use for mirror coatings is usually simple. Gold has some advantage over silver and aluminum because it has a smaller reflectivity in the visible. (See Fig. 4.1.) In applications where the solar background is a problem, the use of gold reduces its magnitude. In many instances filters are utilized to eliminate the visible radiation and thus quality of the coating is of greater importance than the particular metal chosen.

The considerations involved in the choice of lens materials are somewhat similar to those which determine the choice of a window material. However, the coefficient of expansion is important only for immersed optics. Here the detector element is in intimate contact with the lens material; the detector is immersed in the lens. (See Section 10.2.7.) If the element must be cooled, it is likely to separate from the lens unless their expansion coefficients are well matched. Such separation greatly increases the reflection losses because of the two newly formed interfaces. If the material used for an immersion lens has the proper index of refraction, and if the detector element remains in intimate contact with the lens, the lens may serve the additional purpose of reducing the reflection losses which would otherwise have occurred at the front surface of the element. However, it may then become necessary to reduce reflection losses at the front surface of the immersion lens.

Antireflection coatings. For obvious reasons it is extremely important that windows, lenses, etc., transmit as large a fraction of the incident radiation as possible. By the use of dielectric films the transmission of a silicon window, for example, can be increased from 55% at 5 μ to nearly 100% by proper coating methods. (See Fig. 4.23.)

We have seen in Section 3.2.2 that ideal dielectrics exhibit a unique property by which they can change the phase of an electromagnetic wave passing through them without absorbing energy from the radiation. In practice this theoretical prediction is fulfilled with very useful and interesting consequences when dielectric thin films are deposited on reflecting surfaces. Through the use of such coatings the reflectivity of components of optical systems can be greatly reduced.

The treatment of the optical properties of dielectrics in Chapter 3 can be modified and extended to account for the improvements which can be achieved by the use of these reflection reducing films. Consider a thin dielectric film of thickness d and index of refraction n_2. Let it be deposited on a substrate having an index of refraction n_1 in an environment which

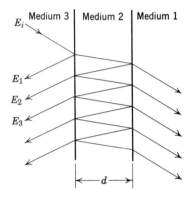

Figure 4.3

has an index of refraction n_3. (See Fig. 4.3.) The amplitude reflection coefficient for radiation traveling from medium 3 to 2 is r_{32} and from medium 2 to 1 is r_{21}.

Let radiation whose electric field vector E_i is described by

$$E_i = E_0 \exp \left\{ 2\pi j v \left[t - \frac{\mathbf{r} \cdot \mathbf{p}}{c_3} \right] \right\}, \qquad (4.11)$$

be incident on the boundary between dielectric media 3 and 2 in the direction shown. Then

$$E_1 = E_i r_{32}, \qquad (4.12)$$

specifies the amplitude of the radiation reflected from the first interface. In the same way,

$$E_2 = E_i[1 - r_{32}]r_{21}[1 - r_{23}] \exp (-j \, \Delta\phi), \qquad (4.13)$$

where for normal incidence

$$\Delta\phi = \frac{4\pi d n_2}{\lambda} \ .$$

Similarly,

$$E_3 = E_i[1 - r_{32}]r_{21}^2 r_{23}[1 - r_{23}] \exp (-2j \, \Delta\phi). \qquad (4.14)$$

So that the amplitude of the mth reflected beam is given by

$$E_m = E_i[1 - r_{32}]r_{21}^{m-1} r_{23}^{m-2}[1 - r_{23}] \exp [-j(m - 1) \, \Delta\phi]. \quad (4.15)$$

By recourse to the principle of superposition the amplitude of the reflected radiation may be obtained by summing the separate reflected amplitudes, that is,

$$E_r = \sum_{m=1}^{\infty} E_m = E_i\{r_{32} + r_{21}[1 - r_{32}][1 - r_{23}] \exp\left[-j(m-1)\,\Delta\phi\right]$$

$$\times\ [1 + r_{21}r_{23} \exp\{-j\,\Delta\phi\} + r_{21}{}^2 r_{23}{}^2 \exp\{-2j\,\Delta\phi\}$$

$$+ \ldots r_{21}^{m-2} r_{23}^{m-2} \exp\left[-j(m-2)\,\Delta\phi\right] + \ldots]\}, \tag{4.16}$$

$$= E_i\left\{r_{32} + \frac{r_{21}[1 - r_{32}][1 - r_{23}] \exp\left[-j\,\Delta\phi\right]}{1 - r_{21}r_{23} \exp\left[-j\,\Delta\phi\right]}\right\}. \tag{4.17}$$

Now from Eq. 3.53 $r_{kl} = -r_{lk}$ so that 4.17 yields the amplitude reflection coefficient

$$r \equiv \frac{E_r}{E_i} = \frac{r_{32} + r_{21} \exp\left[-j\,\Delta\phi\right]}{1 + r_{21}r_{32} \exp\left[-j\,\Delta\phi\right]}, \tag{4.18}$$

and the reflectivity is then

$$\rho = |r|^2 = \frac{r_{32}{}^2 + 2r_{32}r_{21} \cos \Delta\phi + r_{21}{}^2}{1 + 2r_{32}r_{21} \cos \Delta\phi + r_{32}{}^2 r_{21}{}^2}. \tag{4.19}$$

In order that the reflectivity be zero, the numerator of 4.19 must vanish, that is,

$$r_{32}{}^2 + 2r_{32}r_{21} \cos \Delta\phi + r_{21}{}^2 = 0. \tag{4.20}$$

This is possible only if

$$\Delta\phi = \pm(2l - 1)\pi, \qquad l = 0, 1, 2, 3, \ldots$$

and

$$r_{32} = r_{21}.$$

The consequence of the first of these two conditions is that

$$d = \frac{(2l - 1)\lambda}{4n_2}. \tag{4.21}$$

For $l = 1$ we get the well-known formula for a quarter wave plate,

$$d = \frac{\lambda}{4n_2}, \tag{4.22}$$

whereas the second condition, $r_{32} = r_{21}$, requires that

$$\frac{n_3 - n_2}{n_3 + n_2} = \frac{n_2 - n_1}{n_2 + n_1}, \tag{4.23}$$

or

$$n_2 = \sqrt{n_3 n_1}.$$

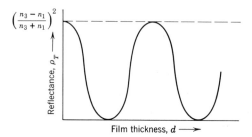

Figure 4.4. Reflectance as a function of film thickness for $n_2 = \sqrt{n_1 n_3}$.

We therefore see that an antireflective dielectric coating deposited on a material of a given index of refraction to be used in a specified environment should have an index of refraction which is the geometric mean of the indices of refraction of the substrate and environment, and should have a thickness one-fourth of the wavelength within the coating material.

Figure 4.4 shows the dependence of reflectance on film thickness. The properties of the film are strongly dependent upon the angle at which the radiation is incident. It is desirable in many instances that the coating exhibit its antireflection properties over as broad a spectral region as possible. The dependence of τ upon wavelength is least when $\Delta\phi$ has the smallest value possible, but even then the transmissivity is 100% at only a few discrete wavelengths. In order to extend the spectral interval over which the coating is effective, multiple coatings are often used. Cabellero[12] has studied this problem theoretically and the discussion concerning two coatings, as in the following outline, is quite similar to her treatment. Consider Fig. 4.5.

The problem of analyzing the case of two dielectric films resolves itself into consideration of the real interface 4–3 having an amplitude reflection

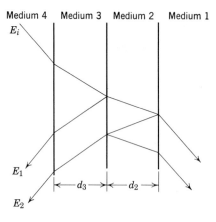

Figure 4.5

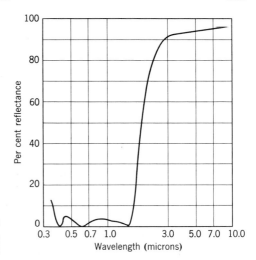

Figure 4.6. Antireflection coatings on aluminum. [After G. Hass, *J. Opt. Soc. Amer.* **46**, 31 (1956).]

coefficient r_{43} and a pseudointerface having a complex amplitude reflection coefficient Z where

$$Z = \frac{r_{32} + r_{21} \exp\left[-j\,\Delta\phi\right]}{1 + r_{32}r_{21} \exp\left[-j\,\Delta\phi\right]} = |Z| \exp\left[-j\,\Delta\theta\right],$$

and where $\Delta\phi \equiv 4\pi\, d_2 n_2/\lambda$. In this manner the problem can be treated as the one film case.

Using the notation $\Delta\psi = 4\pi\, d_3 n_3/\lambda$ the conditions for the composite reflectivity to be zero are

$$\Delta\psi + \Delta\theta = \pm(2s - 1)\pi, \qquad s = 0, 1, 2, 3 \dots$$

and

$$r_{43} = |Z| = \left[\frac{r_{32}{}^2 + 2r_{32}r_{21}\cos\Delta\phi + r_{21}{}^2}{1 + 2r_{21}r_{32}\cos\Delta\phi + r_{21}{}^2 r_{32}{}^2}\right]^{\frac{1}{2}}. \qquad (4.24)$$

Graphical or analytical methods may be used to determine from these conditions the optimum thicknesses and optimum indices for the two films involved. The extension to an even larger number of films can be carried out in an analogous manner. Figure 4.6 shows the reflection from an aluminum surface having three superimposed coatings of SiO-Al-SiO. It can be seen that the reflection is reduced over a comparatively broad spectral interval.

If antireflection coatings are used to minimize reflection at a particular wavelength, the conditions for minimum reflectivity will also be satisfied for other wavelengths which satisfy the conditions imposed by Eq. 4.21.

That is, those wavelengths such that

$$\lambda_{\text{sideband}} = \frac{4dn_2}{2l - 1}.$$

In multilayer antireflection films the sidebands are numerous and overlapping.

Georg Hass[13] has succeeded in producing "black mirrors" through the use of multiple coatings of dielectric materials. These surfaces have a very low reflectivity in the visible portion of the spectrum but have a very high reflectivity in the infrared. A mirror of this kind could be used as part of the optical system for an infrared source. Its high reflectivity in the infrared would render it capable of projecting a collimated beam of infrared yet it would not easily be revealed due to its low reflectivity in the visible.

Filters—interference. These same principles can be employed to prepare filters which transmit selectively in the infrared.[14] They are called interference filters because they depend for their operation upon interference of the electromagnetic waves, either constructive or destructive, depending upon the wavelength and its relationship to the thickness and indices of refraction of the films used. As with antireflection coatings, there are a number of wavelengths for which the transmission is a maximum. These side bands occur at wavelengths which satisfy the condition expressed by Eq. 4.21 for values of l larger than one. The width of the spectral interval transmitted can be made quite narrow, of the order of one-tenth micron, but the particular interval transmitted depends rather strongly upon the angle at which the radiation is incident on the film. Figure 4.7 shows the characteristics of a family of commercially available infrared interference filters.

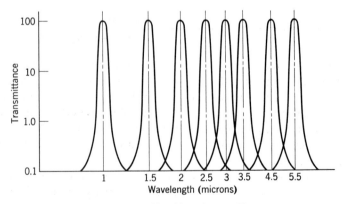

Figure 4.7. Family of interference filters.

Filters—reflection (reststrahlen). Another phenomenon observed with dielectric materials, called reststrahlen, can be used to isolate a narrow spectral interval. As radiation impinges upon the atoms which make up a dielectric, the atoms are set into oscillation by the electric field. This action is more pronounced with crystals in which the bonding is almost completely ionic. In that case there is a more complete charge transfer, and a specific sign of electrical charge can be identified in each region. Furthermore, ionic bonds are more rigid and resonances are more easily observable so that a characteristic oscillation is induced. By analogy with the ω_0 used in Eq. 3.75 and following, the frequency of this oscillation is given by

$$\nu = \left[\frac{k}{\dfrac{m_a m_b}{m_a + m_b}} \right]^{\frac{1}{2}}, \tag{4.25}$$

where $k \equiv$ the restoring force constant,
$m_a \equiv$ the mass of the ions of one type of electrical charge, and
$m_b \equiv$ the mass of the ions of opposite charge.

As an example, MgO, when irradiated, reemits a band of wavelengths sharply peaked in the vicinity of 21 μ. Other compounds are used to produce sources peaked at different wavelengths and slabs of these materials are sometimes arranged so as to produce narrower peaks by successive reflection. Strong[15] shows the construction of a reststrahlen apparatus using successive reflections from four crystals.

Filters—Christiansen. A filter which selectively transmits narrow spectral intervals in the infrared can also be prepared by using a finely

TABLE 4.2. **Christiansen filter peaks for various crystals***

Crystal	Christiansen Peak	Crystal	Christiansen Peak
LiF	11.2 μ	RbBr	65
NaCl	32	RbI	73
NaBr	37	CsCl	50
NaI	49	CsBr	60
KCl	37	TlCl	45
KBr	52	TlBr	64
KI	64	TlI	90
RbCl	45		

*Adapted from Sawyer.[16]

powdered dielectric. In these filters, called Christiansen filters, a transparent dielectric powder suspended in a fluid is placed between two parallel plates. In that particular small range of wavelengths for which the fluid and powder have the same index of refraction, the filter transmits as a plane parallel plate. At all other wavelengths for which the indices differ, the radiation is scattered and thereby attenuated. In the infrared, the alkali halides have regions of anomalous dispersion in which the index of refraction passes through unity. Air can therefore be used as the fluid and the filter will transmit approximately all the radiation in the region of anomalous dispersion. The transmission decreases rapidly on either side of this region. The wavelength at which the peak transmission occurs in alkali halide Christiansen filters is shown in Table 4.2.[16]

Filters—semiconductor. As pointed out in Section 4.2.1, semiconductors are characterized by an extremely high absorptivity at wavelengths less than λ_0 and a very low absorptivity at longer wavelengths. In most cases λ_0 marks a sharp transition between these two regions. (See Fig. 4.2.) This characteristic can be utilized to reject short wavelength radiation such as solar radiation in those systems where the radiation would otherwise cause extremely deleterious effects. In addition, semiconducting filters may be used in conjunction with interference filters and antireflection coatings to isolate a rather narrow spectral interval and simultaneously eliminate the side bands usually associated with interference filters. The short wavelength side bands of the interference filter are effectively eliminated by the fundamental absorption referred to above, whereas the long wavelength side bands are reduced to a considerable extent by free carrier absorption. The antireflection coating peaked at the wavelength interval of interest serves to increase the transmittance to a value near 100%.

Since the width of the forbidden band in semiconductors is somewhat temperature dependent, λ_0 may exhibit a temperature dependent shift.

Filters—gelatin. Gelatin filters are seldom used in infrared systems because of the difficulty of preparing gelatins which absorb in the visible and transmit infrared. A good source of information about such filters is in the booklet *Wratten Filters*, Eastman Kodak Publication No. B-3.

4.3.4   Available Materials

When considering the reflection, absorption, and transmission properties of optical materials, care must be taken to differentiate between reflectivity and reflectance, absorptivity and absorptance, and transmissivity and transmittance. To make the distinction between these terms clear, consider an optically flat piece of material as shown in Fig. 4.8.

Let P_i be the radiation power incident on the material. The power

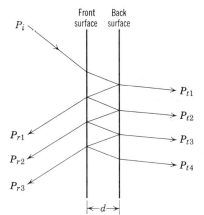

Figure 4.8

reflected from the front surface is

$$P_r = P_{r1} + P_{r2} + P_{r3} + \ldots + P_{rn} + \ldots \tag{4.26}$$

$$= \{\rho + \rho(1 - \rho)^2 \exp(-2ad) + \rho^3(1 - \rho)^2 \exp(-4ad) + \ldots\} P_i$$

$$= P_i \left\{ \rho + \frac{(1 - \rho)^2 \rho \exp(-2ad)}{1 - \rho^2 \exp(-2ad)} \right\},$$

where ρ is the reflectivity,
  a is the absorption coefficient, and
  d is the thickness.
Then the reflectance ρ_t is

$$\rho_t \equiv \frac{P_r}{P_i} = \rho + \frac{(1 - \rho)^2 \rho \exp(-2ad)}{1 - \rho^2 \exp(-2ad)}. \tag{4.27}$$

Similarly, the power transmitted is

$$P_t = P_{t1} + P_{t2} + P_{t3} + \cdots P_{tn} + \ldots$$

$$= \{(1 - \rho)^2 \exp(-ad) + (1 - \rho)^2 \rho^2 \exp(-3ad)$$

$$+ (1 - \rho)^2 \rho^4 \exp(-5ad) + \ldots\} P_i$$

$$= \frac{P_i(1 - \rho)^2 \exp(-ad)}{1 - \rho^2 \exp(-2ad)}, \tag{4.28}$$

and the transmittance τ_t is

$$\tau_t \equiv \frac{P_t}{P_i} = \frac{(1 - \rho)^2 \exp(-ad)}{1 - \rho^2 \exp(-2ad)}. \tag{4.29}$$

Since $1 - \rho_t - \tau_t = \alpha_t$, the absorptance is found to be

$$\alpha_t \equiv \frac{P_a}{P_i} = \frac{(1 - \rho)[1 - \exp(-ad)]}{1 - \rho \exp(-ad)}. \tag{4.30}$$

TABLE

Material	Trans. Limit (microns)	Index of Refraction 2.2 μ	4.3 μ	Young's Modulus (10^6 psi)	Knoop Hardness	Density (gms/cm^3)
Optical glasses	2.7	1.5	1.7	7–10	300–600	2.3–4.6
Fused silica	4.5	1.43	1.37	10.1	470	2.20
RIR 2	4.7	1.75	—	15.2	~600	~3
850324 (thorium glass) Bu. Std. Desig. F-158	4.7	1.80	—	—	—	4.62
915213 (flint glass) Bu. Std. Desig. A- 2059	4.8	1.85	—	7	—	6.01
Sapphire (Al$_2$O$_3$)	5.5	1.73	1.68	53	1370	3.98
RIR 20	5.5	1.82	1.79	12–14	542	5.18
RIR 12	5.7	1.62	—	15.2	594	3.07
Rutile (TiO$_2$)	6	—	2.45	—	880	4.26
Lithium fluoride (LiF)	6	1.38	1.34	11	110	2.6
Periclase (MgO)	6.8	1.71	1.66	36	690	3.59
Strontium titanate (SrTiO$_3$)	7.0	2.23	2.19	—	620	5.13
Irtran-1 (MgF$_2$)	8	1.37	1.35	—	576	3.18
Tellurium (Te)	—	—	$\begin{cases} 4.93 \perp \text{to } c \\ 6.37 \parallel \text{to } c \end{cases}$	—	—	6.24
Calcium fluoride (CaF$_2$)	9	1.42	1.41	15	158	3.18
Arsenic trisulfide (As$_2$S$_3$)	12	2.38	2.35	2.3	109	3.20
Barium fluoride (BaF$_2$)	13.5	1.46	1.45	8	82	4.89
Silicon (Si)	15*	3.44	3.42	19	1150	2.33
Irtran-2 (ZnS)	15	2.26	2.25	14	354	4.11

4.3.

Melting Point (°C)	Thermal Expansion Coefficient ($10^{-6}/°C$)	Solubility gr/100 ml of H_2O at 20°C	Soluble in	Comments
700	4–10	0.00	HF	Easily cut, polished; nontoxic, nonhygroscopic.
1667	0.55	0.00	HF	Excellent mechanical and thermal properties; nontoxic.
~900	8.3	0.00	—	Available in production quantities. Preliminary studies indicate that no surface protection is needed.
780	8.3	—	Slightly in HNO_3	Contains thorium. Conventional surfacing techniques apply.
430	9.8	—	Dissolves in 1% HNO_3	Conventional surfacing techniques apply. Available in random slabs of the order of 2″ × 2″ × 1½″.
2030	$\begin{cases}5.0 \perp \text{to } c \\ 6.7 \parallel \text{to } c\end{cases}$	0.00	NH_4 salts	Excellent mechanical and thermal properties; nontoxic. Can be fused directly to 7052 and 7520 glass.
760 ~900	9.6 8.3	— —	1% HNO_3	Good mechanical and optical properties; nontoxic.
1825	9	0.00	H_2SO_4	Nonhygroscopic; nontoxic.
870	36	0.27	HF	Scratches easily; nontoxic; noncorrosive.
2800	13	0.00	NH_4 salts	Nonhygroscopic; nontoxic; noncorrosive. A surface scum forms if stored in air.
2080	9.4	—	—	Of special interest because its refractive index ≈ $\sqrt{5}$. See text.
1396	16	Small	—	Emissivity very low even at 800°C.
450	16.8	0.00	—	Poisonous. $\perp$ denotes E vector normal to c-axis. $\parallel$ denotes E vector parallel to c-axis.
1403	25	0.002	NH_4 salts	Scratches easily; not resistant to thermal or mechanical shock; nontoxic.
196	26	0.00	—	Nonhygroscopic; noncorrosive.
1280	—	0.17	NH_4Cl	Slightly hygroscopic; nontoxic.
1420	4.2	0	$HF + HNO_3$	Resistant to corrosion. *Long wavelength limit depends upon impurity concentration as well as thickness and temperature. Very pure specimens are transparent even into the microwave region. Must be highly polished to reduce scattering losses at the surfaces.
800	7.9	0	Slightly in HNO_3, H_2SO_4	

TABLE 4.3

Material	Trans. Limit (microns)	Index of Refraction		Young's Modulus (10⁶ psi)	Knoop Hardness	Density (gms/cm³)
		2.2 μ	4.3 μ			
Cadmium telluride (CdTe)	15	—	2.56**	—	—	—
Indium antimonide (InSb)	16	—	3.99***	6.21	—	5.78
Germanium (Ge)	25*	4.09	4.02	14.9	—	5.33
Sodium chloride (NaCl)	25	1.53	1.52	5.8	17	3.16
Silver chloride (AgCl)	30	2.01	2.00	2.9	9.5	5.53
KRS-6 (Thallium bromide thallium chloride)	30	2.20	2.19	3.0	35	7.19
KRS-5 (Thallium bromide thallium iodide)	45	2.62	—	—	—	—
Cesium bromide (CsBr)	48	1.66	1.66	2.3	—	4.44
Cesium iodide (CsI)	60	1.75	1.73	0.8	—	4.53
Kel-F	4	—	—	—	—	—
Lucite	5.5	—	—	—	—	—

If the material is opaque, that is, $\exp(-ad) = 0$, then from 4.29 and 4.30 we see that

$$\rho = \rho_t,$$

and

$$\alpha_t = 1 - \rho_t.$$

When reflection, transmission, and absorption data are obtained experimentally, the measured values are the reflectance, ρ_t, absorptance, α_t, and transmittance, τ_t.

In Sections 4.3.1, 4.3.2, and 4.3.3 we have discussed the general requirements materials must satisfy if they are to be used in various parts of an infrared system. We will now describe the characteristics of available materials and attempt to indicate their value as materials for infrared components.

Table 4.3 is a compilation of the more important properties of commonly used materials.

An optical property of general interest is that of refractive index. Indices of a number of materials are shown in Fig. 4.9. As was stated earlier, the indices of refraction of semiconductors are typically higher than those of

(*Continued*)

Melting Point (°C)	Thermal Expansion Coefficient (10⁻⁶/°C)	Solubility gr/100 ml of H₂O at 20°C	Soluble in	Comments
1045	4.5	0.00	—	**At 10 μ. Transmission is 38% in the range 1 to 10 μ.
523	4.9	0.00	CP4	***At 8 μ.
940	6.1	0.00	Hot H₂SO₄ Aqua regia	*Same note as for silicon above.
803	44	35.7	Glycerin, H₂O	Corrosive; hygroscopic.
458	30	0.00	NH₄OH, KCN	Corrosive; nonhygroscopic.
424	51	0.01	HNO₃ Aqua regia	Toxic; cold flows; nonhygroscopic.
415	60	0.02	HNO₃ Aqua regia	Toxic; cold flows; nonhygroscopic; high reflection loss.
636	48	124.3	—	Soft, scratches easily; hygroscopic.
621	50	44	—	Soft, scratches easily; hygroscopic.
—	—	—	—	Soft, easily scratched; becomes milky if overheated.
Distorts at 72	110–140	—	—	Soft, easily scratched.

dielectrics. Although this permits the fabrication of thinner refracting elements for a given power, there is a compensating disadvantage in the greater reflection losses.

Glasses. Figure 4.10 shows the spectral transmission of three types of glass designated as types C-601, F-158, and A-2059 by the National Bureau of Standards during development work and now available as Types 827250, 850324, and 915213 from Bausch and Lomb. Figure 4.11 shows the spectral transmission of Corning type 0160, and Bausch and Lomb types RIR 2, RIR 12, and RIR 20. The spectral transmission of their types RIR 10 and RIR 11 are not shown because the differences between the curves for these glasses and those in Figs. 4.10 and 4.11 are comparatively minor.

In general, it can be said that some type of glass is the best choice for both detector window and irdome if the detector to be used in the system responds only to wavelengths shorter than about 2.8 μ, for example, lead sulfide. As can be seen from Figs. 4.10 and 4.11, glasses may also be suitable for wavelengths as long as 4.7 μ, but these newer glasses are somewhat less satisfactory because they are not stable. Some samples

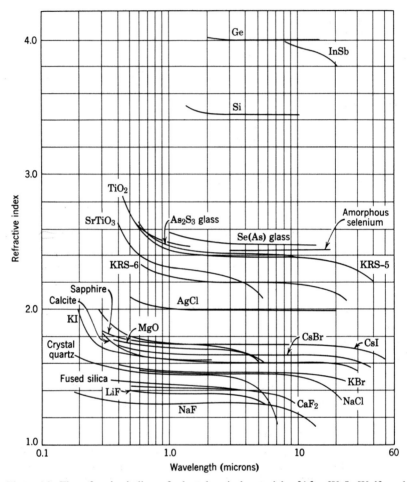

Figure 4.9. The refractive indices of selected optical materials. [After W. L. Wolfe and S. S. Ballard, *Proc. Inst. Radio Engrs.* **47**, 1540 (1959).]

tend to darken and discolor with age so that their transmission is reduced. Still other samples tend to devitrify, and the resulting structure sometimes gives rise to excessive losses by scattering.

Sealing problems are less severe with glasses than for any other materials. If proper facilities are available, highly specialized requirements can be met with regard to shape and size.

The transmission characteristics of this material are not strongly temperature dependent, so that it functions well as windows and domes when elevated in temperature as by aerodynamic heating.

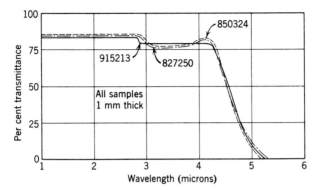

Figure 4.10. Spectral transmission of three special glasses. (Courtesy of Bausch & Lomb Optical Co., Rochester, New York.)

Fused quartz (SiO_2). This material is the amorphous form of silica and is sometimes referred to as transparent fused silica or vitreous silica. Its infrared properties are as good as those of crystalline SiO_2 and it is much cheaper.

The transmission characteristics of fused silica are shown in Fig. 4.12. The reflection loss for two surfaces in air is only 3.2% at 2 μ.

In the temperature range 20° to 900°C the thermal expansion of fused silica is extremely low, $5 \times 10^{-7}/°C$. This property makes it very difficult to seal quartz directly to other materials. A graded seal requiring three different glasses (G.E. designation GSC-1, 3, and 4) is required to seal fused silica to Pyrex. Vycor (Corning type 7900), which is 96% fused

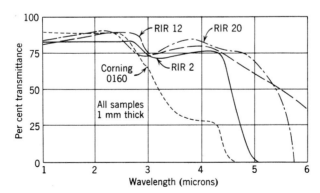

Figure 4.11. Spectral transmission of special glasses. (Courtesy of Corning Glass Works, Corning, New York, and Bausch & Lomb Optical Co., Rochester, New York).

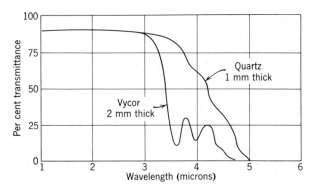

Figure 4.12. Spectral transmission of fused silica and Vycor. (Courtesy of Corning Glass Works, Corning, New York.)

quartz, has transmission properties quite similar to fused quartz below 2.5 μ; but above this wavelength Vycor does not transmit as well. (See Fig. 4.12.)

Both quartz and Vycor can be obtained in large pieces. They cut and grind well and when heated to 420°C the transmission decreases by only 3%. Since these are amorphous materials, they are sometimes classified as glasses.

Because of its hardness, its chemical inertness, and outstanding resistance to thermal and mechanical shock, quartz makes an excellent dome material for systems which utilize short wavelengths.

Sapphire (Al_2O_3). Sapphire single crystals have been grown by Linde Air Products. They are colorless and insoluble in water. As seen in Fig. 4.13, these crystals make excellent window material at wavelengths as long as 6 μ. At 4 μ, 12% of the incident radiation is reflected by two surfaces in air.

Sapphire is anisotropic and has a thermal coefficient of expansion of 5.0 × 10^{-6}/°C perpendicular to the c-axis and 6.7 × 10^{-6}/°C parallel to the c-axis. It can be sealed directly to Corning type 7520 and 7052 glasses. The melting point of sapphire is quite high, approximately 2030°C, and it is extremely hard. As a result it must be cut, ground, and polished with diamond or boron carbide tools and abrasives.

Sapphire single crystals have been made up to 6 in. in diameter. The thermal conductivity of sapphire at liquid nitrogen temperature is extremely high so that it provides an excellent means of conducting heat from a photosensitive element to a heat sink.

In addition to its usefulness as a window material, sapphire, because of its chemical inertness and availability in large pieces, also makes excellent irdome material.

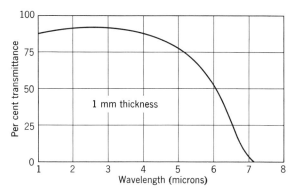

Figure 4.13. Spectral transmission of sapphire. (Courtesy of Linde Air Products, New York.)

Ruby, which is also aluminum oxide but with chromium oxide which colors it, has infrared transmission properties similar to sapphire.

Rutile (TiO$_2$). This material is also called titania. It is available from Linde Air Products as a hard, colorless, single crystal insoluble in water. Figure 4.14 shows the transmission properties of rutile. About 32% reflection loss is encountered at 2 μ for two surfaces.

Rutile possesses a somewhat unique property in that it has an index of refraction approximately equal to the square root of 5, which is nearly equal to the geometric mean of several common detector materials and air. Thus it can be used to minimize reflection losses. (See Section 4.3.3.)

Lithium fluoride (LiF). This material is available from Harshaw Chemical Company, Cleveland, Ohio, and Optovac, Inc., North Brookfield,

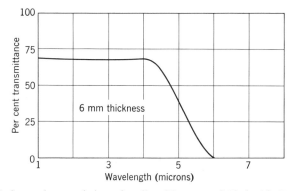

Figure 4.14. Spectral transmission of rutile. (Courtesy of Linde Air Products, New York.)

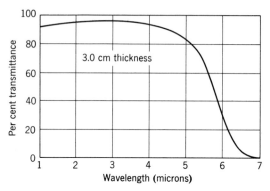

Figure 4.15. Spectral transmission of lithium fluoride. (Courtesy of Harshaw Chemical Company, Cleveland, Ohio.)

Massachusetts. It is a single crystal and is fairly soluble in water. The transmission characteristics, given in Fig. 4.15, show a cutoff wavelength of about 6 μ. At 4 μ, two surfaces will reflect 4.4% of the incident radiation. The low reflection losses result from the very low index of refraction this material exhibits.

It is difficult to seal lithium fluoride to glasses because of its large thermal coefficient of expansion. Castings are available in sizes up to 6 in. in diameter by 4 in. long.

Lithium fluoride cannot be used as a dome material because it scratches easily and because it is soluble in water. It is useful as a window or lens material only in dry, well-protected enclosures.

Periclase (MgO). Rough pieces of periclase are available from Norton Abrasive Company. It is a colorless crystal, insoluble in water. However, the surface upon prolonged exposure to the atmosphere exhibits a white scum which decreases its transmission.

The transmission properties of periclase are given in Fig. 4.16. At 4 μ, an 11.6% reflection loss for two surfaces is evident. Its forbidden energy gap is so great that free carrier absorption is negligible even at temperatures as high as 400°C.

The thermal coefficient of expansion is 1.38×10^{-5}/°C averaged over the temperature range from 20°C to 1000°C.

Coatings of SiO can be used to protect periclase irdomes from attack by the atmosphere. Irdomes made from this material have usually been made in the form of mosaics because of the difficulty of obtaining large crystals.

Strontium titanate (SrTiO₃). This material is available as a clear, single crystal. The transmission characteristics of strontium titanate are shown in Fig. 4.17. The reflection loss for two surfaces is 20% at 4 μ.

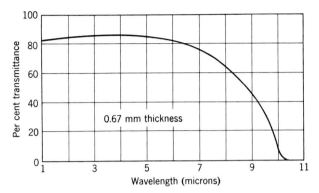

Figure 4.16. Spectral transmission of periclase. (Courtesy of Norton Company, Worcester, Mass.)

Strontium titanate is used primarily as an immersion lens for detector optics. As with rutile, its index of refraction is ideal for minimizing reflection losses at the front surfaces of some common detector materials.

Irtran-1. Irtran-1, magnesium fluoride, is an excellent infrared transmitting material developed by Eastman Kodak Company, Rochester, New York. This material is insoluble in water and appears to be resistant to thermal and mechanical shock. The spectral transmission properties of Irtran-1 are shown in Fig. 4.18. The reflection loss at 5 μ is approximately 8 % for two surfaces in air.

The transmission properties of Irtran-1 are not unduly affected by temperatures as high as 800°C. This fact, coupled with its other characteristics, makes this material an excellent choice for irdomes. It has the

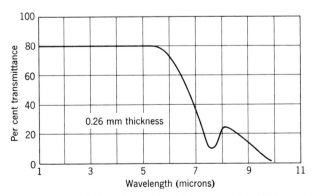

Figure 4.17. Spectral transmission of strontium titanate. (Adapted from S. B. Levin et al., *J. Opt. Soc. Amer.* **45,** 737 (1955).

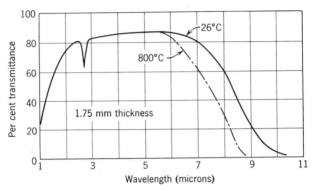

Figure 4.18. Spectral transmission of Irtran-1. (Courtesy of Eastman Kodak Company, Rochester, New York.)

additional advantage that it transmits microwave radiation very well. Its insertion loss is less than that of a similar window made from natural mica. It is therefore possible to consider using it when a combination irdome-radome is needed. This material may also be useful for lenses or as a substrate for interference filters.

Tellurium (Te). Tellurium is an element which has a hexagonal crystal structure. It is so soft that it can be scratched by a cotton swab. Its poor mechanical properties limit its use to situations in which it can be carefully protected. As a result of its lattice arrangement, the electrical and optical properties are anisotropic. It is transparent from $3.5\ \mu$ to greater than $8.0\ \mu$. (See Fig. 4.19.) The large value of the refractive index causes

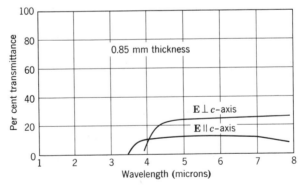

Figure 4.19. Spectral transmission of tellurium. (Courtesy of Minneapolis-Honeywell Regulator Company, Minneapolis, Minnesota.)

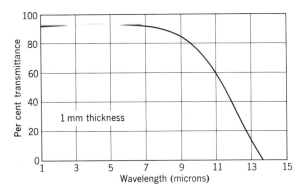

Figure 4.20. Spectral transmission of calcium fluoride. [After Calingaert, Heron, and Stair, *SAE Journal* **39**, 448 (1936).]

reflection losses to be appreciable, about 75% for two surfaces at 6 μ. These losses can be reduced by the use of antireflection coatings.

Large single crystals of this material have been grown by the Czochralski method. Optical polishing results in a conductive layer which can be removed by etching. Tellurium is insoluble in water.

Calcium fluoride (CaF₂). Calcium fluoride, or fluorite, is commercially available (Linde Air Products Company and Optovac, Inc.) as colorless single crystals which are fairly insoluble in water. The spectral transmission properties of this material are shown in Fig. 4.20.

Crystal surfaces of calcium fluoride scratch easily. The high value of the thermal coefficient of expansion, together with the low thermal conductivity, make this material unable to withstand thermal shock. It is also quite susceptible to damage by mechanical shock. A considerable difference exists between synthetic fluorite and natural fluorite which is mined. The synthetic crystals are less stable chemically and become cloudy after they have been exposed to the atmosphere for a few weeks. The naturally occurring crystals can be exposed to the atmosphere for many months, even years, before showing this effect.

This material cannot be used in domes for the reasons cited, but it makes an excellent window material in those applications where it can be shielded from mechanical and thermal damage.

Arsenic trisulfide (As₂S₃). This material can be procured from Servo Corporation of America, Fraser Glass Company, and The American Optical Company. It is an amorphous, red material which is practically insoluble in water. Figure 4.21 shows its transmission as a function of wavelength. The reflection loss is 28.5% at 10 μ for two surfaces in air.

Arsenic trisulfide is a rather soft material which can be cut and polished easily. The thermal coefficient of expansion, $2.46 \times 10^{-5}/°C$, is close to

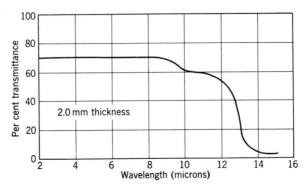

Figure 4.21. Spectral transmission of arsenic trisulfide. (Courtesy of Servo Corporation of America, New Hyde Park, Long Island, New York.)

that of aluminum and as a result this material is often used in aluminum mounts.

Arsenic trisulfide is available in pieces up to 5 in. in diameter. One of its shortcomings is that it "cold flows." As a result, it tends to change its shape very slowly when mounted so that some force such as gravity acts upon it in the same direction for an extended period. Although convenient for use in the laboratory, its softness and tendency to cold flow limit its use in field applications.

Barium fluoride (BaF_2). Barium fluoride is obtainable as a synthetic single crystal from Harshaw Chemical Company, Cleveland, Ohio. It is soluble in water. The excellent transmission properties of this material are shown in Fig. 4.22. The reflection loss is 7.7% at 8 μ for

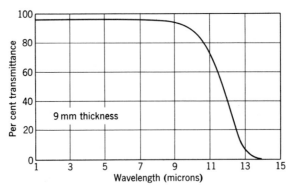

Figure 4.22. Spectral transmission of barium fluoride. (Courtesy of Harshaw Chemical Company, Cleveland, Ohio.)

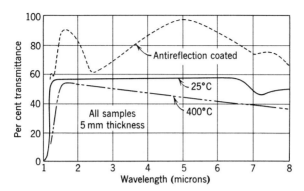

Figure 4.23. Spectral transmission of silicon. (Courtesy of Texas Instruments, Dallas, Texas.)

two surfaces. Its softness and solubility confine its use to the laboratory.

Silicon (Si). Single crystals of pure Si are available from Baird-Atomic, Inc., Cambridge, Massachusetts, Raytheon Manufacturing Company, Waltham, Massachusetts, and Texas Instruments, Inc., Dallas, Texas, among others. This material is insoluble in water and most acids, but will dissolve in a mixture of hydrofluoric, nitric, and acetic acids (CP-4).

Figure 4.23 shows the spectral transmission characteristics of silicon. The reflection loss for this material is quite high, 46% at 10 μ for two surfaces. SiO is used to reduce the reflection loss at the short wavelengths up to about 4 μ, whereas ZnS is used for reflection reducing films in the long wavelength region.

Silicon has a high resistance to mechanical and thermal shock. The transmission properties of silicon are less affected than are those of germanium by increasing temperature. This material is excellent for irdomes either as a single crystal or in polycrystalline form. As with many semiconductor materials, it is a useful filter for rejecting visible radiation.

Irtran-2. Irtran-2, zinc sulfide, is a convenient infrared optical material for applications requiring good transmission characteristics at wavelengths as long as 14 μ. It can be used over the temperature range -200 to $800°C$. Its thermal expansion coefficient is such that it can be readily sealed to glass, a great convenience if it is to be used as a window material. The material is not highly resistant to damage by scratching.

Irtran-2 can be thoroughly cleaned by a wide variety of acids and organic solvents without danger that it will be dissolved by them. For example, it resists attack by water, steam, nitric or sulfuric acids, dilute potassium hydroxide, dilute ammonium hydroxide, ethyl ether, trichloroethylene, chloroform, trichlorbenzene, acetone, and benzene. It is often desirable to

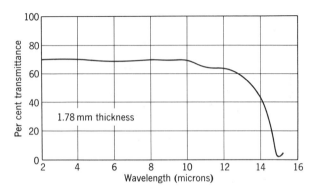

Figure 4.24. Spectral transmission of Irtran-2. (Courtesy of Eastman Kodak Company, Rochester, New York.)

use these solvents to clean the inner surfaces of detector envelopes just before the final sealing process to reduce outgassing. The spectral transmission of Irtran-2 is shown in Fig. 4.24.

Cadmium telluride (CdTe). Cadmium telluride is a compound semiconductor which is fairly difficult to grow into large single crystals. It appears to be insoluble in water. The same cutting and polishing techniques used to prepare silicon and germanium for optical measurements can be used.

Reflection losses for two surfaces amount to 32% at 10 μ. It has good transmission properties, that is, transmission is 38% from about 1 to 10 μ. (See Fig. 4.25.) Because of its desirable transmission characteristics and its good mechanical, chemical, and thermal properties, cadmium telluride is a suitable material for use in irdomes.

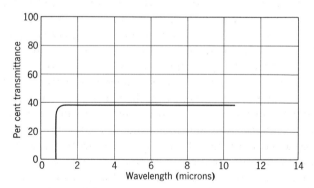

Figure 4.25. Spectral transmission of cadmium telluride. [Adapted from W. D. Lawson et al., *J. Phys. Chem. Solids* **9**, 325 (1959).]

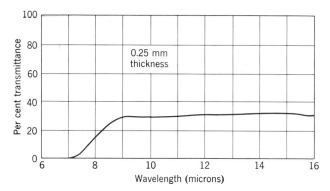

Figure 4.26. Spectral transmission of indium antimonide. (Courtesy of Minneapolis-Honeywell Regulator Company, Minneapolis, Minnesota.)

Indium antimonide (InSb). This material is a soft and brittle inter-metallic compound which, because of its high carrier mobility, has been the subject of much solid state research. Its transmission properties are strongly dependent on impurity concentration but extend approximately from 7 to beyond 16 μ. (See Fig. 4.26.) Sample thicknesses less than 0.010 in. are required for appreciable transmission.

Reflection losses for two surfaces are 53% at 10 μ. It is insoluble in water and is chemically inert. For wavelengths shorter than 7 μ the absorption constant is very high, of the order of 10^4 cm^{-1}.

Germanium (Ge). Very pure single crystals of germanium are available for use as an infrared component material. Germanium is insoluble in water and is fairly inactive chemically.

The transmission properties of germanium are shown in Fig. 4.27. The high reflection loss (53% at 10 μ for two surfaces) can be overcome by SiO (at short wavelengths) or ZnS (at long wavelengths) antireflection coatings. Unless the surface is highly polished, the high index of refraction will cause excessive losses by scattering.

Germanium is a rather brittle material which tends to fracture easily during fabrication. As the temperature increases, the absorption edge moves to longer wavelengths and the transmission at any wavelength beyond it decreases because of the increase in the number of free carriers. The absorption cross section for electrons is only about $\frac{1}{20}$ as great as that for holes in this material. Thus, if one compares free carrier absorption in n-type and p-type material of the same impurity concentration, the n-type material has an absorptivity twenty times smaller. As noted in the remarks pertaining to germanium in Table 4.3, the long wavelength transmission limit for this type of material cannot be specified as precisely

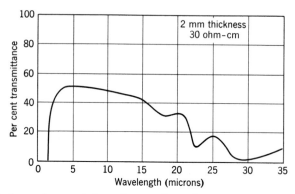

Figure 4.27. Spectral transmission of germanium. [After C. M. Phillipi and N. F. Beardsley (deceased), *WADC Technical Note* 55-194, Wright Air Development Division, Dayton, Ohio (December 1956).]

as the same limit for a dielectric. Very pure *n*-type germanium has good transmission characteristics even at microwave frequencies. The chemical inertness and insolubility of germanium render it a good material for use as a window. It can be soldered to metal.

Sodium chloride (NaCl). Large (1500 cu in.) synthetic single crystals are available from Harshaw Chemical Company, Cleveland, Ohio, and Optovac, Inc., Brookfield, Massachusetts. This material is soft, very soluble in water, and corrodes metals.

The cutoff wavelength of approximately 25 μ is evident in Fig. 4.28 which shows its spectral transmission properties. This material polishes easily. Evaporated films and plastics have been used successfully to

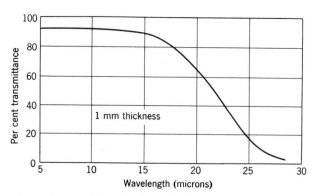

Figure 4.28. Spectral transmission of sodium chloride. (Courtesy of Harshaw Chemical Company, Cleveland, Ohio.)

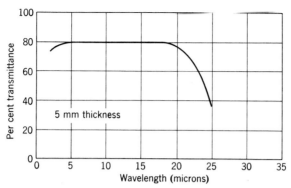

Figure 4.29. Spectral transmission of silver chloride (uncoated). (Courtesy of Harshaw Chemical Company, Cleveland, Ohio.)

protect the surface from water, but the softness of sodium chloride precludes its extensive use as a window or lens material except in environments where it can be carefully protected. It finds wide use in prisms for infrared spectrometers.

Silver chloride (AgCl). Silver chloride is a corrosive, soft, ductile material which is insoluble in water. It is available from Harshaw Chemical Company, Cleveland, Ohio, in the form of synthetic single crystals as large as $3\frac{1}{2}$ in. in diameter and 5 in. long.

Its spectral transmission properties are shown in Fig. 4.29. The reflection loss at $15\ \mu$ for two surfaces is approximately 20%. Unless protected with thin films of silver sulfide, it will darken upon exposure to

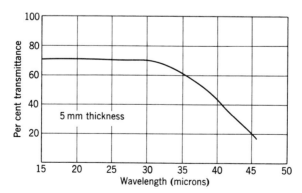

Figure 4.30. Spectral transmission of KRS-5. (Courtesy of Harshaw Chemical Company, Cleveland, Ohio.)

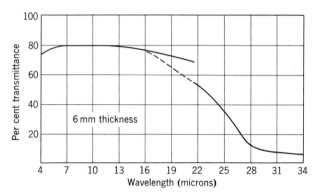

Figure 4.31. Spectral transmission of KRS-6. [Adapted from E. K. Plyler, *J. Research Natl. Bur. Stds.* **41**, 128 (1948).]

ultraviolet. The silver sulfide in turn should be protected by overcoating with polystyrene. Since silver chloride is easily scratched and warped, it has not found extensive use outside the laboratory.

KRS-5 (thallium bromide-thallium iodide)(TlBr-TlI). This material is a synthetic mixed crystal available from Harshaw Chemical Company, Cleveland, Ohio. It is slightly soluble in water and is toxic.

KRS-5 transmits to approximately 45 μ (see Fig. 4.30). Although the reflection losses are high, 30% at 30 μ for two surfaces, coatings can be used to reduce the reflection losses and to facilitate handling without fear of poisoning. As this material has a tendency to cold flow, it is important that it not be left unsupported for long periods of time. For this reason it is not widely used for lenses, windows, or irdomes in infrared systems.

KRS-6 (thallium bromide-thallium chloride)(TlBr-TlCl). This material is a synthetic, mixed crystal which is soft and slightly soluble in water.

The transmission characteristics of KRS-6 are shown in Fig. 4.31. This figure is a composite of measured curves for two wavelength ranges. Approximately 24% reflection loss is encountered at 10 μ for two surfaces. Since KRS-6 covers much the same spectral interval as KRS-5 but is not as transmissive, it is not used to any great extent.

Cesium bromide (CsBr). Synthetic single crystals of cesium bromide are available from Harshaw Chemical Company, Cleveland, Ohio. These crystals are soft and extremely soluble in water.

This material is transparent out to 48 μ as can be seen in Fig. 4.32. At 20 μ the reflection loss is 12% for two surfaces in air. The softness and solubility of cesium bromide greatly restrict its use in infrared systems.

Cesium iodide (CsI). This material is obtainable as a soft, extremely soluble single crystal from Harshaw Chemical Company, Cleveland, Ohio.

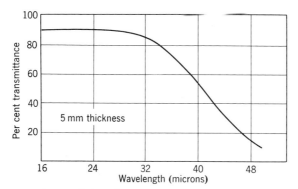

Figure 4.32. Spectral transmission of cesium bromide. (Courtesy of Harshaw Chemical Company, Cleveland, Ohio.)

Its transparency extends out to approximately 60 μ. (See Fig. 4.33.) Reflection loss is 13% at 30 μ for two surfaces. As with cesium bromide, this material finds little use outside the laboratory because of its softness and solubility in water. Perhaps its most important application is as a prism material in spectrometers.

Plastics, Kel-F and Lucite. Although plastics have not been widely used in infrared optical systems, two representative materials are included for the sake of completeness. Kel-F, a polymer of tetrafluorchloroethylene, is a clear plastic which exhibits good transmission in the visible and near infrared. Beyond 2.5 μ considerable structure in its transmission spectrum limits its utility. This material has a tendency to crack and cold flow. It can be bonded to aluminum.

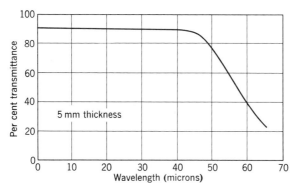

Figure 4.33. Spectral transmission of cesium iodide. (Courtesy of Harshaw Chemical Company, Cleveland, Ohio.)

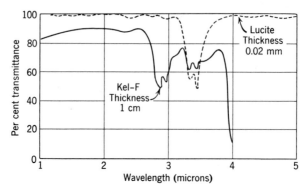

Figure 4.34. Spectral transmission of Kel-F and Lucite. (After Ballard, McCarthy, and Wolfe.[9])

Lucite (polymethylmethacrylate), also sold under the name Plexiglass, is a clear plastic having good transmission in the visible and near infrared, although an absorption band occurs at 3.3 μ. It is soft and easily scratched. (See Fig. 4.34.)

REFERENCES

1. H. H. Blair, *Ohio State University Interim Reports* on Contract No. DA-44-009-eng-658 (September 1951 to June 1956).
2. F. W. Glaze, National Bureau of Standards, Reports 2993, 3173, and 3359 on Contract No. AF 33(616)-53-16 (December 1953 to June 1954).
3. Glaze, Blackburn, and Capps, National Bureau of Standards, *WADC Technical Report 54-457*, on Contract No. AF 33(616)-53-16 (March 1955).
4. B. W. King, Battelle Memorial Institute. *Quarterly Engineering Reports*, on Contract No. AF 33 (616)-3564, for 1956 and 1957.
5. N. J. Dreidl, Bausch and Lomb Optical Company, *WADC Technical Report 55-500*, on Contract No. AF 33 (616)-2769, parts 1, 2, and 3 (July 1957 to October 1958).
6. G. Hass, *J. Opt. Soc. Amer.* **45**, 945 (1955).
7. T. S. Moss, *Optical Properties of Semiconductors*, Academic Press, New York, Butterworths Scientific Publications, London (1959).
8. *Ibid.*, p. 49.
9. Ballard, McCarthy, and Wolfe, *Optical Materials for Infrared Instrumentation—IRIA State of the Art Report*, The University of Michigan, Ann Arbor (January 1959).
10. T. S. Moss, *op. cit.*
11. D. E. Crowley, *Emittance and Reflectance in the Infrared; An Annotated Bibliography*, The University of Michigan, Ann Arbor (April 1959).
12. D. Cabellero, *J. Opt. Soc. Amer.* **37**, 176 (1947).
13. Rountree, Hass, and Cox, Engineer Research and Development Laboratories, *Research Report 1450-RR* (May 1956).

14. J. Strong, *Concepts of Classical Optics*, W. H. Freeman and Co., San Francisco (1958), p. 253.
15. J. Strong, *Procedures in Experimental Physics*, Prentice-Hall, Englewood Cliffs, New Jersey (1953), p. 383.
16. R. A. Sawyer, *Experimental Spectroscopy*, Prentice-Hall, Englewood Cliffs, New Jersey (1951), p. 297.

5
Optical properties
of the atmosphere

5.1 INTRODUCTION

The optical properties of the earth's atmosphere are such as to require consideration of their effects upon the performance of many infrared devices and systems. The purpose of this chapter is to discuss those aspects of the earth's atmosphere which affect the transmission of infrared radiation.

The atmosphere is composed of a mixture of gases in which are suspended a wide variety of particles distributed over a great range in size and which also differ in chemical composition. These gases cause radiation to be absorbed, and the suspended particles scatter the radiation. As a result, the radiation from a source is attenuated and the contrast between the background and the source may be degraded. This can come about in four ways:

1. Radiation from the source may be absorbed by gases in the path.

2. Radiation from the source may be deflected or scattered by particles suspended in the path so that it appears to come from the area which surrounds the source, that is, the background.

3. The gases and particles suspended in the path may themselves radiate. This radiation and its fluctuations may "blur" the scene and reduce the contrast.

4. There may also be scattering "into" the path. Scattering agents in the path may deflect or scatter radiation in such a way that radiation which originated at some distance from the object being viewed is changed in

angle so that it appears to the observer as if it had come from the source or points immediately adjacent to the source.

In this chapter we will discuss only the first two of these effects. The last two effects are more properly related to the topic of radiating backgrounds and so will not be considered. Before discussing the details of absorption and scattering, it will be helpful to describe the earth's atmosphere in detail.

5.2 DESCRIPTION OF THE ATMOSPHERE

The gases present in greatest abundance in the earth's atmosphere are nitrogen, oxygen, water vapor, carbon dioxide, methane, nitrous oxide, carbon monoxide, and ozone. Fortunately, the two gases present at the highest concentrations, N_2 and O_2, are homonuclear. They therefore possess neither a permanent nor an induced electric moment and hence do not exhibit molecular absorption bands.

Over the range of altitudes extending from sea level to approximately 40,000 ft, water vapor and carbon dioxide are by far the most important absorbing molecules. The concentration of H_2O varies between $10^{-3}\%$ and 1% (by volume), depending upon geographical location, altitude, time of year, and local meteorological conditions. Carbon dioxide is much more uniformly distributed; it varies between 0.03 and 0.04% and is greater in an air mass which has been over heavy vegetation than in the atmosphere over the ocean. The distribution is more uniform at higher altitudes where the mixing is more complete. Methane, CH_4, is present at a concentration between 1×10^{-4} and $2 \times 10^{-4}\%$, and is very uniformly distributed in altitude. Nitrous oxide, N_2O, at concentrations of 3×10^{-5} to $4 \times 10^{-5}\%$ and carbon monoxide, CO, with a typical concentration of $2 \times 10^{-5}\%$, have bands which show up if long paths are utilized. Ozone, O_3, is present at concentrations as large as $10^{-3}\%$ at altitudes near 100,000 ft, but it is present at much lower concentrations at other altitudes.

Figure 5.1a shows the molecular absorption spectrum of a typical atmospheric path at low altitudes over the spectral interval extending 1 to 15 μ.[1] Figures 5.1b, c, and d show the molecular spectra of some of the other atmospheric constituents mentioned above. A comparison of Figs. 5.1b, c, and d with Fig. 5.1a illustrates the importance of the H_2O and CO_2 absorption bands. The water vapor bands at 1.1 μ, 1.38 μ, 1.87 μ, 2.7 μ, and 6 μ are easily identifiable. We see CO_2 bands at 2.7 μ, 4.3 μ, and 14.5 μ.

The factors which affect the width and depth of these molecular absorption bands are discussed in the next section. First, however, we shall describe another feature of the atmosphere.

Middleton[2] has said, ". . . any serious limitation of the visual range is due, not to the permanent gases, but to the atmospheric aerosol." The atmosphere contains an assemblage of small, suspended particles. They are not only distributed randomly according to size (their radii may range from 10^{-7} to 10^{-1} cm), but also vary in composition. They may consist of dust from the earth, carbon particles, smoke, water droplets, salt particles, and minute living organisms.

This collection of liquid and solid particles with gas as the suspending

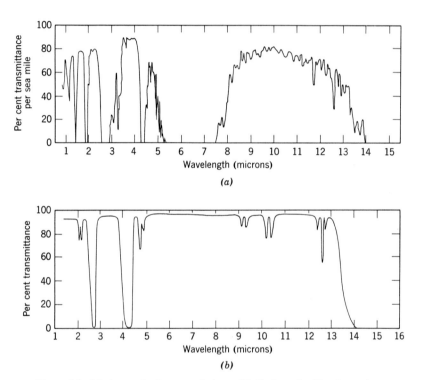

Figure 5.1. (*a*) Atmospheric transmission. (*b*) Carbon dioxide absorption.

medium causes an extinction of infrared radiation that is often far greater than that due to absorption by the vibration of the atoms in a molecule. In Chapter 3 we discussed Rayleigh scattering and derived Eq. 3.90 which showed that the power scattered varied inversely with the fourth power of the wavelength of the radiation. In the visible and near infrared, the effect of Rayleigh scattering is often many orders of magnitude greater than that of molecular absorption. Beyond 1μ, the character of the scattering changes so that it is more correctly described by mathematical expressions different from the Rayleigh scattering expression. This second type is known as Mie scattering in deference to the German meteorologist who first investigated it. Mie scattering is caused by particles considerably larger than individual molecules. Its effect on transmission may be quite large even in the intermediate and far infrared.

Considerable effort has been expended to determine the origin and manner of growth of the water particles present.[3,4] The matter is still not

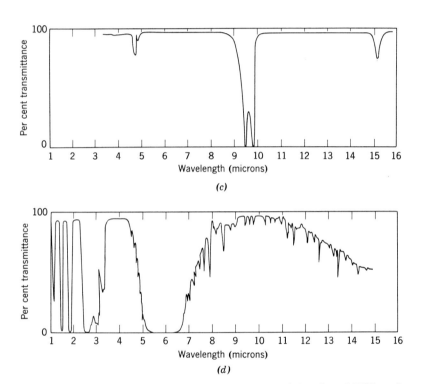

(c) Ozone absorption. (d) Water absorption. (After Howard, Burch, and Williams.[1])

settled, but the argument advanced by Simpson[5] has much to recommend it. Briefly summarized, it takes the following form.

Ozone, produced in the upper atmosphere by the Schumann-Runge solar radiation (hydrogen radiation from the sun of wavelength 1000 to 1500 Å), reacts with nitrogen to form nitrous oxide. A similar reaction probably also occurs between the nitrogen and oxygen of the atmosphere in the presence of electrical discharges. Once formed, the small agglomerates of nitrous oxide provide nuclei for the condensation of water vapor.

Wright,[6] on the other hand, contends that salt nuclei and combustion nuclei provide the centers on which water vapor condenses to form mist, fog, and cloud particles. His point of view is very capably presented in a review article by Mason and Ludlam.[7] Salt or combustion nuclei, such as sulfuric acid, are very hygroscopic and greatly aid the early stages of growth of the droplets. The vapor pressure at the surface of a droplet formed on a soluble nucleus is reduced by an amount proportional to the concentration of the solute. It can be shown that

$$\frac{p_r{'}}{p_\infty} = \exp\left(\frac{2\sigma M}{\rho R T r}\right) - \frac{750\ mC}{(\pi \rho r^3 - m)W}, \tag{5.1}$$

where $p_r{'} \equiv$ the vapor pressure over a droplet of radius r centimeters
        formed on a soluble nucleus,

     $p_\infty \equiv$ the vapor pressure over a plane surface,

     $\sigma \equiv$ the surface tension, in dynes/cm,

     $M \equiv$ the molecular weight of the vapor, in grams/mole,

     $\rho \equiv$ the density of the liquid, in grams/cm^3,

     $R \equiv$ the gas constant, 8.32×10^7 ergs/mole deg K,

     $T \equiv$ the temperature, in deg K,

     $m \equiv$ the mass of the solute, in grams,

     $w \equiv$ the molecular weight of the solute, in grams/mole,

     $C \equiv$ a factor depending upon the nature and concentration of the
        solute. C has the units of grams/mole.

At thermal equilibrium $p_r{'} = p_\infty H/100$, where H is the relative humidity. Using 5.1 and assuming the mass of the solute small compared to the mass of the droplet, the condition for equilibrium is given by

$$\frac{H}{100} = \exp\left(\frac{P}{r}\right) - \frac{Q}{r^3}, \tag{5.2}$$

where $P = 2\sigma M/\rho RT$,
     $Q = 750\ mC/\pi \rho W$.

Wright[8] developed 5.2 to compute the radii of droplets in equilibrium with air at various relative humidities. Figure 5.2 shows his plot of the radii of nuclei versus relative humidity.

The mechanism just described is of great help in explaining the early stages of growth of droplets. Further growth of the particles depends on an extremely complicated set of conditions. Among these are the rate of ascent of the air, its temperature and rate of change of temperature with altitude (lapse rate), the heat transfer coefficient at the surface of the drop, and the rate of diffusion of solute molecules into the new water condensing in the drop. We will not attempt to discuss this complex phenomenon but will refer the interested reader to such treatments as those of Dessens[9] or Mason and Ludlam.[7]

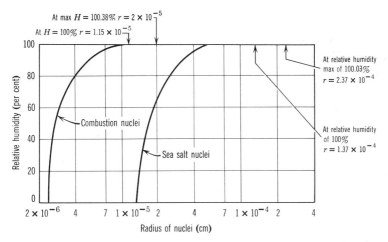

Figure 5.2. Variation in radii with relative humidity. (After H. L. Wright.[8])

Only very general statements can be made about the size and concentration of particles in an arbitrary sample of the atmosphere. This is one of the reasons why calculations of radiation attenuation by the atmosphere are very difficult. The work by Dessens has shown that a sample of clean country air (taken on a day when the visibility was exceptionally good and the humidity low) contains about 100 particles per cm³. Most of these have radii in the range between 0.1 and 1 μ with a small number, perhaps 5% in the range between 1 and 10 μ radius.[9]

Simpson's measurements[5] indicate that about 100,000 particles per cm³ are present in the pale blue haze frequently seen in industrial areas. This haze often extends upward to altitudes of several thousand feet and consists of particles having radii between about 0.03 and 0.2 μ.

Fog particles have radii in the range from 3 to about 60 μ with a pronounced peak in the region of 7 μ.[10] Cloud particles range in radius from about 2 to 30 μ. The peak in the distribution for these particles occurs at a somewhat smaller radius than for fogs. The concentration of droplets in clouds varies between about 50 and 1500 per cm³ whereas the concentrations of particles in fogs are in the range 1 to 50 per cm³.

5.3 MOLECULAR ABSORPTION—THEORY

It would seem, upon a cursory examination of the subject, that since the intensity and spectral position of infrared absorption lines and bands are known, it should be fairly straightforward to calculate the extent to which infrared is transmitted through a given atmospheric path length.

In practice, however, the situation is complicated by several factors, some of which have been alluded to only briefly in this and previous chapters. Even though we neglect scattering, there are still considerations which lead to difficulties in infrared transmittance calculations. These difficulties are:

1. In many wavelength intervals the absorption coefficient is not independent of wavelength.
2. Temperature and total pressure also influence the absorption coefficient.
3. The absorption coefficient is a function of the concentration of the absorbing molecular specie.
4. Absorber concentration, temperature, and total pressure vary with altitude, geographical location, season, local meteorological conditions, etc.

In the following discussion we shall treat some of the theory associated with effects of temperature and pressure on atmospheric transmission. Section 5.4 will discuss the empirical aspects of this subject.

When considering molecular absorption only, it was shown (see 3.69) that the fraction of monochromatic radiation transmitted by a medium is

$$\tau_a(\lambda) = \exp\{-a(\lambda)x\}, \tag{5.3}$$

where a is the absorption coefficient per unit path length,
  x is the path length or range.

In many instances the density of absorbing elements is not constant for the entire path length and to accommodate for this case, Eq. 5.3 must be generalized to

$$\tau_a(\lambda) = \exp\{-a'(\lambda)n_a\}, \tag{5.4}$$

where $a'(\lambda)$ is the absorption coefficient per absorbing molecule,
  n_a is the number of absorbing molecules encountered by a beam of unit cross section in traversing a distance x; it is sometimes called the optical path length $\bar{n}_a x$, where $\bar{n}_a$ is the average density over the path length x.

The average fraction of radiation transmitted in any spectral interval $\Delta\lambda$ centered at λ_p is then

$$\bar{\tau}_a = \frac{1}{\Delta\lambda} \int_{\lambda_p - \Delta\lambda/2}^{\lambda_p + \Delta\lambda/2} \exp\{-a'(\lambda)\bar{n}_a x\}\, d\lambda. \tag{5.5}$$

We have said that n_a is the number of absorbing elements encountered by a beam of unit cross section in going from 0 to x, that is,

$$n_a = \int_0^x n_a(x)\, dx = \bar{n}_a x. \tag{5.6}$$

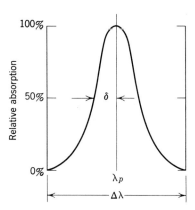

Figure 5.3. Absorption line.

Since $pv = p/\bar{n}_a = kT$, where k is the Boltzmann constant, p is the pressure, and v is the specific volume, the factor p/kT may be substituted for $\bar{n}_a$ in all the following expressions.

If we now wish to compute $\bar{\tau}_a$ we must know how $a'(\lambda)$ varies with wavelength. Let us first consider the case where $\Delta\lambda$ is centered at an absorption line. (See Fig. 5.3.) Lorentz has given the functional relationship between the absorption coefficient and the wavelength as

$$a'(\lambda) = \frac{S/\pi\delta}{[(\lambda - \lambda_p)/\delta]^2 + 1}, \tag{5.7}$$

where S is called the line intensity, that is, $S \equiv \int_{-\infty}^{\infty} a'(\lambda)\, d\lambda$,

δ is one-half the width of the line measured at the points where the absorption has dropped to one-half the maximum; it is called the half width.

Inserting 5.7 into 5.5, we get

$$\bar{\tau}_a = \frac{1}{\Delta\lambda} \int_{\lambda_p - \Delta\lambda/2}^{\lambda_p + \Delta\lambda/2} \exp\left\{ \frac{-\left(\dfrac{S\bar{n}_a}{\pi\delta}\right)x}{\left(\dfrac{\lambda - \lambda_p}{\delta}\right)^2 + 1} \right\} d\lambda. \tag{5.8}$$

By the transformation $(\lambda - \lambda_p)/\delta = \tan\theta$, 5.8 becomes

$$\bar{\tau}_a = \frac{2\delta}{\Delta\lambda} \int_0^{\text{arc tan }(\Delta\lambda/2\delta)} \exp\left\{ -\left(\frac{S\bar{n}_a}{\pi\delta}\right)x \cos^2\theta \right\} \sec^2\theta\, d\theta,$$

$$= \frac{2\delta}{\Delta\lambda} \sum_{m=0}^{\infty} \frac{1}{m!}\left(\frac{-S\bar{n}_a x}{\pi\delta}\right)^m \int_0^{\text{arc tan }(\Delta\lambda/2\delta)} [\cos\theta]^{2(m-1)}\, d\theta. \tag{5.9}$$

To obtain a better physical picture of what these equations represent, consider the special but practical situation in which the spectral interval of interest, $\Delta\lambda$, is much larger than the width of the line, that is, $\Delta\lambda \gg 2\delta$ or $\Delta\lambda/2\delta$ approaches infinity. In such a case 5.9 simplifies by integration by parts to

$$\bar{\tau}_a = 1 - \frac{4S\bar{n}_a}{\pi\Delta\lambda} x \int_0^{\pi/2} \exp\left\{-\left(\frac{S\bar{n}_a}{\pi\delta} x\right)\cos^2\theta\right\}\sin^2\theta\, d\theta,$$

$$= 1 - \frac{S\bar{n}_a}{\Delta\lambda} x \exp\left[\frac{-S\bar{n}_a}{2\pi\delta} x\right]\left\{J_0\left[\frac{jS\bar{n}_a}{2\pi\delta} x\right] - jJ_1\left[\frac{jS\bar{n}_a}{2\pi\delta} x\right]\right\}, \quad (5.10)$$

where the J's are the zero and first order Bessel functions of the first kind.[11] We will now discuss two limiting cases of 5.10. One case, weak line absorption, arises when $S\bar{n}_a x \ll 2\pi\delta$. The other, called strong line absorption, concerns the situation when $S\bar{n}_a x \gg 2\pi\delta$. For weak line absorption, $S\bar{n}_a x/2\pi\delta$ approaches zero and

$$J_0\left(\frac{jS\bar{n}_a x}{2\pi\delta}\right) \to 1,$$

whereas

$$jJ_1\left(\frac{jS\bar{n}_a}{2\pi\delta} x\right) \to -\left(\frac{S\bar{n}_a}{4\pi\delta}\right)x,$$

so that 5.10 becomes

$$\bar{\tau}_a = 1 - \frac{S\bar{n}_a}{\Delta\lambda} x, \quad \text{(weak line).} \quad (5.11)$$

For strong line absorption in which $(S\bar{n}_a/2\pi\delta)x$ is very large, we use the result[11]

$$\lim_{Z\to\infty} J_n(jZ) = \frac{j^n \exp(Z)}{\sqrt{2\pi Z}},$$

to obtain

$$\bar{\tau}_a = 1 - \frac{2\sqrt{S\delta\bar{n}_a x}}{\Delta\lambda}, \quad \text{(strong line).} \quad (5.12)$$

Equations 5.11 and 5.12 relate the concentration of the absorbing molecules to the average transmittance. For weak line absorption the average transmittance varies with $\bar{n}_a$, whereas for strong line absorption $\bar{\tau}_a$ varies as the square root of $\bar{n}_a$.

In general, the absorption spectrum of a single type of absorbing molecule can be predicted by assuming that the absorption coefficient is given by

$$a'(\lambda) = \sum_{m=1}^{N} \frac{S_m/\pi\delta_m}{\left[\dfrac{\lambda - \lambda_{pm}}{\delta_m}\right]^2 + 1}, \quad (5.13)$$

where N is the number of absorption lines in the wavelength interval of interest.

If 5.13 were inserted into 5.5 in order to compute $\bar{\tau}_a$, it is obvious that the result would be exceedingly complex. In some instances the individual absorption lines are sufficiently far apart so that each line can be treated as if it absorbs independently of the other lines. Further simplification is possible since experiments have shown that δ is approximately the same for all lines in a band. In such a case 5.11 becomes for a band composed of N weak lines

$$\bar{\tau}_a = 1 - \frac{\bar{n}_a x}{\Delta\lambda} \sum_{m=1}^{N} S_m. \tag{5.14}$$

Equation 5.12 becomes for a band composed of N strong lines

$$\bar{\tau}_a = 1 - \frac{2\sqrt{\delta\bar{n}_a x}}{\Delta\lambda} \sum_{m=1}^{N} [S_m]^{1/2}. \tag{5.15}$$

When 5.15 can be used to describe measured atmospheric transmission, the absorption is referred to as weak band absorption. In other words, a weak band is comprised of widely spaced strong lines.

As $\bar{n}_a x$ increases, the concept of well-separated absorption lines cannot be applied, since then the individual lines start to overlap. Physically, this is tantamount to saying that for sufficiently large n_a the center portion of the absorption lines is opaque, and increasing n_a produces only a broadening of the lines. In such cases Elder and Strong[12] have found the following analysis to be useful.

Observers of total transmission at constant pressure utilizing low resolution apparatus find that the fraction of radiation transmitted over an entire band varies inversely with $\bar{n}_a$, that is,

$$\mathscr{H}_\tau \propto \bar{n}_a^{-1}, \quad \text{or} \quad d\mathscr{H}_\tau \propto -\frac{d\bar{n}_a}{\bar{n}_a^2}, \tag{5.16}$$

which yields

$$d\bar{\tau}_a \equiv \frac{d\mathscr{H}_\tau}{\mathscr{H}_\tau} = -F\frac{d\bar{n}_a}{\bar{n}_a}, \tag{5.17}$$

where F is a constant.
Integration yields

$$\bar{\tau}_a = C + F\ln\bar{n}_a, \tag{5.18}$$

where C is a constant.

When the transmission varies in this manner, the absorption is called strong band absorption. Further discussion of the experimental aspects of these equations is in Section 5.4.

The next aspect of this problem to be discussed concerns the effect of total pressure and temperature on $\bar{\tau}_a$. The intensity S of any spectral line has been experimentally determined to be approximately independent of pressure. However, δ is found to be a function of pressure and temperature. It has been ascertained through spectroscopic studies that a number of effects determine the width of a spectral absorption line. Of these effects, only the one arising because of the impact of neighboring molecules on the radiating molecule is of importance. This type of broadening is called impact or pressure broadening. Elsasser[13] has pointed out that pressure broadening is the dominant cause of line broadening for infrared absorption by air molecules at low altitudes.

Since pressure broadening results from the collisions which the radiating molecules undergo, we would expect from kinetic theory that the number of impacts per unit area per unit time is inversely proportional to the square root of the absolute temperature (provided the pressure is held constant), and directly proportional to the pressure. However, we will see that the situation is more complicated. The correct relationship between the half width, temperature, and pressure is discussed in the following section. Since variations of total pressure and temperature cannot be determined analytically, their effect on δ can only be given when they have been measured experimentally.

5.4 MOLECULAR ABSORPTION—EMPIRICAL DISCUSSION

Many investigators have found that δ does not vary with total pressure and temperature in the simple way predicted by kinetic theory. Instead it has been shown[14,15,16] that the observed dependence may be described fairly well by

$$\delta = \delta_s[p_T/p_{T_s}]^m \, [T/T_s]^{-q} = Cp_T^m T^{-q}, \qquad (5.19)$$

where m is about 0.5,

q is about 0.25,

δ_s and C are constants,

p_{T_s} is the total pressure at which δ_s is determined,

T_s is the temperature at which δ_s is determined.

In this case 5.15 becomes

$$\bar{\tau}_a = 1 - \frac{2}{\Delta\lambda} \sqrt{Cp_T^m T^{-q} \bar{n}_a x} \sum_{j=1}^{N} \sqrt{S_j} \qquad (5.20)$$

$$= 1 - C'\sqrt{\bar{n}_a p_T^m x},$$

where $\qquad C' = \frac{2\sqrt{CT^{-q}}}{\Delta\lambda} \sum_{j=1}^{N} S_j^{1/2}.$

It was found[17] that for weak bands a better fit to experimental data was obtained if p_T was replaced by $p_T + p_A$ where p_A is the partial pressure of the absorber. Therefore Eq. 5.20 becomes

$$\bar{\tau}_a = 1 - C'\bar{n}_a^{1/2}[p_T + p_A]^{m/2}, \quad \text{(weak band)} \quad (5.21)$$

where values of the exponent $m/2$ vary between 0.5 and 0.25.
For strong band absorption it has been found that

$$\bar{\tau}_a = D \ln \bar{n}_a + E \ln [p_T + p_A] + F, \quad \text{(strong band)} \quad (5.22)$$

fits the experimental data very well, where D, E, and F are experimentally determined constants.

TABLE 5.1*

Window No.	Window Boundaries λ in μ		Window Width $\Delta\nu$ in cps $\times 10^{13}$
I	0.72	0.94	9.7
II	0.94	1.13	5.4
III	1.13	1.38	4.7
IV	1.38	1.90	6.0
V	1.90	2.70	4.7
VI	2.70	4.30	4.1
VII	4.30	6.0	2.0
VIII	6.0	15.0	3.0

* (Adapted from Langer[19])

The work of Elder and Strong[18] has been particularly helpful to those who are interested in calculating the transmission of infrared through the atmosphere because their paper provides a means of relating atmospheric transmission to a parameter which is readily measurable, humidity. They did this by discussing the spectrum in terms of seven "windows" or regions of high transmission between absorption bands and by introducing the idea of factoring the transmission into two terms. One of these terms they called τ_i', the transmittance of the ith window as affected by the selective or molecular absorption, and the other they designated as F_i, the transmittance of the ith window as affected by scattering, a form of attenuation much less strongly dependent upon wavelength. We will call these τ_{ai} and τ_{si} respectively. In the discussion below, we shall follow their treatment quite closely, adding modifications of their work which have been made by Langer,[19] Schuldt,[20] and others.

The spectrum extending from 0.72 to 15.0 μ is divided into eight windows separated by H_2O and CO_2 bands as shown in Table 5.1. Although the frequency of electromagnetic waves is not often stated in cycles per

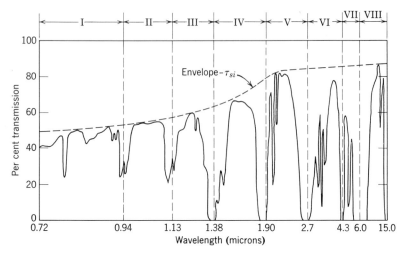

Figure 5.4. Atmospheric scattering and molecular absorption in the interval 0.72 to 15.0 μ. (After Langer.[19])

second if the waves are in the infrared region of the spectrum, we have included a column giving $\Delta\nu$ in cps because it indicates the relationship between the widths of the various windows in terms of energy.

For a source which emits equal amounts of radiant power at all wavelengths, the average infrared transmittance of a path can be written as

$$\bar{\tau}_{(0.72\mu - 15.0\mu)} = \frac{1}{[15.0 - 0.72]} \sum_{i=\text{I}}^{\text{VIII}} \tau_{ai} \cdot \tau_{si} \cdot \Delta\lambda_i, \qquad (5.23)$$

where τ_{ai} is the transmittance of the ith window as affected by selective absorption,

$\qquad \tau_{si}$ is our notation for the transmittance of the ith window as affected by scattering, and

$\qquad \bar{\tau}$ is the transmittance due to the combined effects of absorption and scattering.

Figure 5.4 may serve to clarify Eq. 5.23. The broken line curve labeled "Envelope—τ_{si}" is a smoothed monotonic curve related to the peak transmission of each window and identifies that portion of the attenuation which is due to scattering. Thus,

$$\tau_{ai} = \frac{\text{Area under solid curve in window } i}{\text{Area under broken line curve in window } i},$$

$$\tau_{si} = \frac{\text{Area under broken line curve in window } i}{\text{Area bounded by 0\% line, 100\% line, and boundaries of window } i}.$$

To obtain an approximate expression for $\bar{\tau}$ which can be computed from data which are easily available, we adopt the point of view that variations in $\bar{\tau}$ are produced solely by changes in the water content of the atmosphere: changes in the concentration of molecular H_2O causing changes in τ_{ai} and changes in the size and number of water droplets with humidity affecting τ_{si}. This approximation turns out to be a surprisingly good one because the other gases and particles containing no water are reasonably constant in their effects on transmission. The remainder of this section discusses the factors τ_{ai}, whereas the next two sections discuss τ_{si}, first from a theoretical, then from an empirical, point of view.

Each of the τ_{ai} is dependent upon many factors, the most important of which are

1. The partial pressure of H_2O in the path.
2. The total pressure of all the gases in the path.
3. The temperature of the material in the path.
4. The path length.
5. The nature of the molecular transitions responsible for the absorptions at the boundaries of, and within, each window.

The partial pressure of H_2O is, of course, one of the parameters which determines the number of absorbers encountered by the radiation. Furthermore, partial pressure also has an effect on the pressure broadening of the band.

The total pressure of all gases in the path has a much greater effect on the magnitude of the pressure broadening. Even though neither N_2 nor O_2 has molecular absorption bands itself, frequent collisions with the H_2O molecules damp the oscillations of the water vapor molecules and exert a very marked influence on the band width and hence on the integrated absorption. Since the N_2 and O_2 molecules are present at concentrations approximately one hundred times as great as the H_2O molecules, the "self-broadening" caused by the water molecules can usually be neglected by comparison with the broadening because of collisions with N_2 and O_2 molecules.

The temperature of the gases affects the magnitude and shape of the absorption bands, but this effect is relatively small over the temperature range encountered in the atmosphere. We will neglect all temperature effects in this section.

The path length is also a factor which determines the number of absorbing molecules encountered by the radiation. The method often used to specify the path length is that of stating the number of precipitable millimeters of water in the path. If the humidity is known, the number of precipitable millimeters of water, abbreviated pr. mm H_2O, can be determined by computing the total number of molecules of water in a path of the cross section and length being treated. This number of molecules can be related to a known volume of liquid using Avogadro's number. That volume distributed over the cross section of path used earlier in the calculation determines the thickness of the liquid column which would result if all the water were condensed at one end of the path. Figure 5.5 shows the relationship between path length and precipitable water (for selected values of relative humidity) as a function of temperature.

Howard, Burch, and Williams[21] have shown that the transmission through a path of length l and H_2O partial pressure p is not the same as the transmission through a path of length $l/2$ and H_2O partial pressure $2p$ even though the total number of H_2O molecules available for excitation by the radiation is the same in each case. However, the difference in transmission is small enough so that very useful results can be obtained if we disregard this fact and base our calculations on the absorber concentration as specified by pr. mm of H_2O.

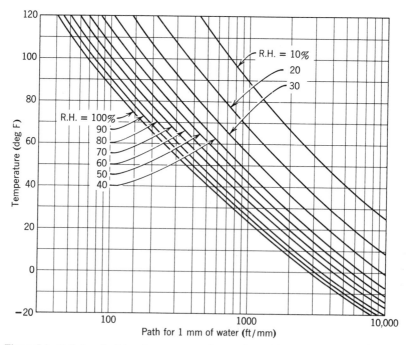

Figure 5.5. Path length (ft/mm) to penetrate 1 mm of precipitable water in atmospheres of various relative humidities (R.H.). (After Langer.[19])

The method which can be used to determine τ_{ai} for some arbitrary concentration of H_2O consists of measuring the total absorption in one of the broad spectral intervals specified earlier. The integral

$$\int_{\lambda_1}^{\lambda_2} \bar{\tau}(\lambda)\, d\lambda, \quad \text{where } (\lambda_2 - \lambda_1) \gg 2\delta,$$

will be independent of the spectral resolution of the monochromator used. In all such measurements the wavelength dependence of the energy falling on the thermocouple of the spectrometer represents what mathematicians call a "Faltung" or convolution of two functions. Changing the slit width changes the shape of the absorption band plotted by the spectrometer, but the integrated absorption is unaffected.

Measurements of this integrated absorption have been made by a number of workers.[12,22,23] The techniques used fall into two broad categories. Some of the data have been obtained by measuring the transmission of long paths outdoors in the real atmosphere.[24,25] Others have measured under very carefully controlled conditions in multiple-pass absorption cells using synthetic atmospheres.[17]

Elder and Strong[12] plotted τ_{ai} versus pr. mm of H_2O for windows I to VI utilizing results obtained in eight different studies of atmospheric absorption. Langer[19] has extended their procedure in several ways. He prepared graphs for windows VII and VIII and also developed two empirical expressions which can be used to compute τ_{ai} in each of the windows for arbitrary values of H_2O concentration.

These expressions are

$$\tau_{ai} = \exp\left(-A_i w^{1/2}\right), \quad (w < w_i) \tag{5.24}$$

and

$$\tau_{ai} = k_i\left(\frac{w_i}{w}\right)^{\beta_i}, \quad (w > w_i) \tag{5.25}$$

where A_i, k_i, and β_i are constants,

$w \equiv$ the water in pr. mm, and

$w_i \equiv$ the value for w which causes the absorption in window i to undergo a transition from the so-called weak-band or square-root absorption to strong band absorption.

Equation 5.24 is useful if $w < w_i$, whereas 5.25 gives a better fit to the measured data if $w > w_i$.

These expressions fit the curves of Elder and Strong and those of Langer quite closely. Table 5.2 lists the values A_i, β_i, k_i, and w_i for each window as given by Langer, together with weighted averages useful over all eight windows.

TABLE 5.2. Constants to be used in Eqs. 5.24 and 5.25*

Window	A_i	k_i	β_i	w_i	Window	A_i	k_i	β_i	w_i
I	0.0305	0.800	0.112	54	V	0.350	0.814	0.1035	0.35
II	0.0363	0.765	0.134	54	VI	0.373	0.827	0.095	0.26
III	0.1303	0.830	0.093	2.0	VII	0.913	0.679	0.194	0.18
IV	0.211	0.802	0.111	1.1	VIII	0.598	0.784	0.122	0.165
					I–VIII	0.211	0.855	0.815	0.60

* After R. M. Langer.[19]

TABLE 5.3. Water vapor transmission coefficients*

w \ Window	I	II	III	IV	V	VI	VII	VIII	I–VIII
0.01	0.997	0.996	0.987	0.979	0.965	0.964	0.915	0.942	0.978
0.02	0.996	0.995	0.982	0.970	0.950	0.949	0.879	0.918	0.969
0.05	0.993	0.992	0.971	0.954	0.925	0.920	0.816	0.875	0.951
0.1	0.990	0.988	0.959	0.935	0.895	0.889	0.749	0.828	0.932
0.2	0.987	0.984	0.940	0.910	0.855	0.846	0.665	0.765	0.907
0.5	0.979	0.975	0.912	0.861	0.784	0.776	0.557	0.685	0.866
1.0	0.970	0.965	0.878	0.810	0.730	0.726	0.487	0.629	0.831
2.0	0.954	0.950	0.830	0.750	0.680	0.680	0.426	0.578	0.793
5.0	0.934	0.922	0.763	0.680	0.618	0.623	0.351	0.517	0.744
10	0.908	0.892	0.715	0.630	0.576	0.584	0.313	0.475	0.706
20	0.874	0.850	0.670	0.582	0.536	0.546	0.274	0.437	0.666
50	0.806	0.774	0.622	0.526	0.488	0.502	0.229	0.390	0.609
100	0.746	0.704	0.576	0.488	0.454	0.469	0.201	0.359	0.562
200	0.688	0.642	0.541	0.452	0.422	0.436	0.175	0.330	0.519
500	0.623	0.568	0.496	0.410	0.384	0.403	0.147	0.296	0.469
1000	0.580	0.517	0.465	0.378	0.357	0.377	0.128	0.272	0.435

* After R. M. Langer.[19]

Table 5.3 contains a useful tabulation of τ_{ai} values for specific values of water vapor concentration in the range 0.01 to 1000 pr. mm. The numbers above the horizontal lines are calculated according to 5.24, those below the lines according to 5.25. The column furthest to the right labeled I-VIII represents a weighted average of the transmission coefficients useful in all the eight windows.

Equations 5.24 and 5.25 together with Tables 5.2 and 5.3 provide information that aid in the calculation of the transmission of infrared through horizontal paths in the atmosphere near sea level. At higher altitudes τ_{ai} can be computed using a relationship suggested by Elder and Strong which takes account of the reduced pressure broadening due to N_2 and O_2 but neglects the self-broadening due to H_2O. Their expression is

$$\tau_{ai}(\text{alt} = h) = \tau_{ai}(\text{sea level}) \left\{ \frac{P(\text{sea level})}{P(\text{alt} = h)} \right\}^{1/8} . \qquad (5.26)$$

However, Larmore[26] states that if Elsasser's model[27] for absorption is used, the exponent in 5.26 should be 0.5. He points out also that the data of

Howard et al.[21] led to a value of 0.3 for the exponent. Modification of that data so as to make it conform to the Elsasser theory would cause the exponent to be used in 5.26 to move nearer to the value 0.5. Since the Elsasser model is somewhat artificial and because the experimental data available place the exponent between 0.12 and 0.35, we will use 0.25 as the exponent.

Assuming a value of 0.25, at 40,000 ft the sea level values of τ_{ai} should be multiplied by approximately 1.5 and at 60,000 ft by about 1.8 in each window. This type of correction must always be applied with caution, especially if the window for which it is being used is bounded by a CO_2 band, since CO_2 absorption has a somewhat different pressure dependence than H_2O.

Calculation of the transmission of slant paths through the atmosphere requires some estimates which come very near to being guesses. The variation of water vapor with altitude varies with meteorological conditions. This causes the gaseous mixture to vary from day to day so that trained judgment is needed to obtain data for the calculation.

Curve	Path Length	Date	Time	Temp.	R.H.	Precipitable Water	Visual Range
A	1000 ft	3-20-56	3 P.M.	37° F	62%	1.1 mm	22 miles
B	3.4 miles	3-20-56	10 P.M.	34.5° F	47%	13.7 mm	16 miles
C	10.1 miles	3-21-56	12 A.M.	40.5° F	48%	52.0 mm	24 miles

Window Definitions

I	0.72 to 0.94 μ	V	1.90 to 2.70 μ
II	0.94 to 1.13 μ	VI	2.70 to 4.30 μ
III	1.13 to 1.38 μ	VII	4.30 to 6.0 μ
IV	1.38 to 1.90 μ	VIII	6.0 to 15.0 μ

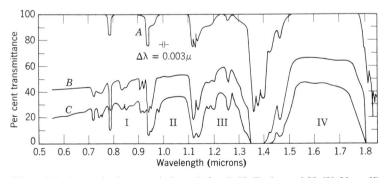

Figure 5.6. Atmospheric transmission. (After J. H. Taylor and H. W. Yates.[25])

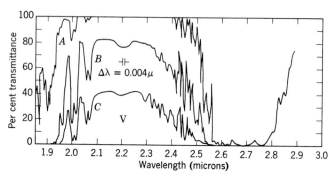

Figure 5.7. Atmospheric transmission. (After J. H. Taylor and H. W. Yates.[25])

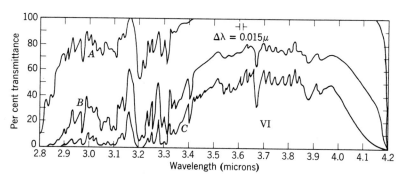

Figure 5.8. Atmospheric transmission. (After J. H. Taylor and H. W. Yates.[25])

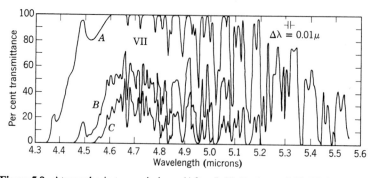

Figure 5.9. Atmospheric transmission. (After J. H. Taylor and H. W. Yates.[25])

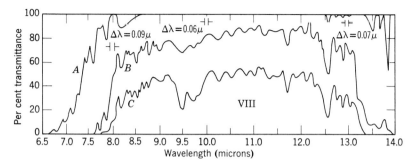

Figure 5.10. Atmospheric transmission. (After J. H. Taylor and H. W. Yates.[25])

We conclude by presenting Figs. 5.6 though 5.10, taken directly from the paper by Taylor and Yates[28] which report the results of a set of high resolution measurements over a long path outdoors. These curves can be enlarged and integrated in spectral regions of interest using a planimeter so that the average τ can be predicted in selected instances.

5.5 SCATTERING BY THE ATMOSPHERE—THEORY

The ratio of the radii of the scattering particles to the wavelength of the scattered radiation can assume values which range over many orders of magnitude. For this reason the interaction of radiation with scattering particles results in many phenomena (for example, the blue of the sky, the appearance of fog, moon halo) which seem totally unrelated. This complexity has given rise to diverse attempts to devise theories which would provide satisfactory explanations for different aspects of essentially the same phenomenon. Of these approaches we will discuss what are considered to be the two most fruitful theoretical treatments of scattering: the approaches based on geometrical optics and on electromagnetic theory. The following discussion will neglect absorption in the scattering elements and will consider only spherical scattering particles.

The total monochromatic power $P_{\text{scat}}(\lambda)$, scattered by a single scattering particle, is

$$P_{\text{scat}}(\lambda) = \int_0^{4\pi} E_{\text{scat}}(\lambda) \, d\omega, \qquad (5.27)$$

where the scattered radiant intensity $E_{\text{scat}}(\lambda)$ is integrated over a surface surrounding the scattering particle. The basic problem which any scattering theory must solve is that of determining the variation of the emittance with wavelength, with angle, and with the characteristics and dimensions of the scattering particle. In general, $E_{\text{scat}}(\lambda)$ is a function of the azimuth as well as the elevation angle. It turns out, however, that for scattering by spheres $E_{\text{scat}}(\lambda)$ is independent of the azimuth angle θ so that 5.27 may be written

$$P_{\text{scat}}(\lambda) = \int_0^{2\pi}\int_0^{\pi} E_{\text{scat}}(\lambda)\sin\phi\, d\phi\, d\theta,$$

$$= 2\pi\int_0^{\pi} E_{\text{scat}}(\lambda)\sin\phi\, d\phi, \tag{5.28}$$

where $\phi = 0$ is the direction of the incident radiation.

Recalling from Eq. 3.81 that

$$E_{\text{scat}}(\lambda) = \gamma_\lambda(\phi)N_s\mathcal{H}_i(\lambda), \tag{5.29}$$

where N_s is the number of particles in an elementary volume,

$\mathcal{H}_i(\lambda)$ is the irradiance, and

$\gamma_\lambda(\phi)$ is a proportionality constant,

we see that the power scattered per particle per unit incident irradiance is

$$\frac{P_{\text{scat}}(\lambda)}{N_s\mathcal{H}_i(\lambda)} = 2\pi\int_0^{\pi}\gamma_\lambda(\phi)\sin\phi\, d\phi, \tag{5.30}$$

$$= \beta', \text{ (see Eq. 3.87)}.$$

If we divide by the cross section of the scattering particle, πb^2, we get a dimensionless ratio $S(\lambda)$, usually called the scattering area ratio, that is,

$$S(\lambda) \equiv \frac{P_{\text{scat}}(\lambda)}{N_s\mathcal{H}_i(\lambda)\pi b^2} = \frac{2}{b^2}\int_0^{\pi}\gamma_\lambda(\phi)\sin\phi\, d\phi \tag{5.31}$$

$$= \frac{\beta'}{\pi b^2},$$

so that the scattering coefficient β (see Eq. 3.66) is for particles of one scattering cross section

$$\beta = S(\lambda)n_s\pi b^2,$$

where n_s is the concentration of scattering elements. For m different types of particles

$$\beta = \pi\sum_{j=1}^{m} n_{sj}S_j(\lambda)b_j^2, \tag{5.32}$$

where n_{sj} is the number of scattering particles of type j, and

S_j is the scattering area ratio for the jth type of particle.

For a large number of particles, where $M(b)\, db$ is the density of particles whose radii are between b and $b + db$, 5.32 becomes

$$\beta = \pi\int_0^{\infty} b^2 S(b)M(b)\, db.$$

We will now compute $S(\lambda)$ using two approaches.

5.5.1 Geometrical Optics Approach

To compute $S(\lambda)$ we must determine the power which is scattered by a particle of known area. This is accomplished by considering $E_{\text{scat}}(\lambda)$ to be composed of radiation which is reflected, refracted, and diffracted into the elemental solid angle.

To compute the radiation reflected from the surface of the scattering sphere, consider Fig. 5.11. The radiation power, dP_i, which is incident on the annular ring, is

$$dP_i(\lambda) = 2\pi b^2 \mathscr{H}_i(\lambda) \sin\theta_i \cos\theta_i\, d\theta_i. \tag{5.33}$$

Since the deflection ϕ of the ray is given by

$$\phi = \pi - 2\theta_i,$$

5.33 becomes

$$dP_i(\lambda) = -\frac{\pi}{2}\, b^2 \mathscr{H}_i(\lambda) \sin\phi\, d\phi. \tag{5.34}$$

Now the radiant power reflected from the annular ring is the reflectivity of the surface times the incident power, that is,

$$dP_r(\lambda) = \rho\, dP_i(\lambda),$$

$$= \frac{\rho\pi}{2}\, b^2 \mathscr{H}_i(\lambda) \sin\phi\, d\phi. \tag{5.35}$$

From Eqs. 3.55 and 3.57 for unpolarized radiation,

$$\rho = \frac{1}{2}\{r_\perp{}^2 + r_\parallel{}^2\} = \frac{1}{2}\left\{1 + \frac{\cos^2[\theta_i + \theta_r]}{\cos^2[\theta_i - \theta_r]}\right\}\frac{\sin^2[\theta_i - \theta_r]}{\sin^2[\theta_i + \theta_r]},$$

where θ_r is shown in Fig. 5.12.

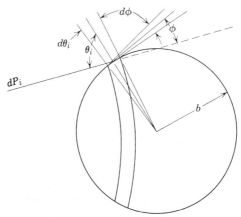

Figure 5.11. Geometrical arrangement.

$$\phi = \pi - 2\theta_i$$

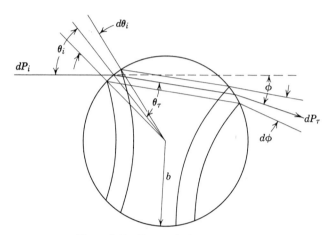

Figure 5.12. Geometrical arrangement.

From Snell's law

$$\frac{\sin \theta_i}{\sin \theta_r} = \frac{n_p}{n_m},$$

where n_m is the index of refraction of the medium, and n_p is the refractive index for the material of which the spherical particle is composed.

However, the radiation reflected from the annular ring is also the surface radiant intensity E_r multiplied by the solid angle subtended by the annular ring, that is,

$$dP_r(\lambda) = E_r(\lambda)2\pi \sin \phi \, d\phi, \qquad (5.36)$$

so from 5.35 and 5.36

$$E_r(\lambda) = \frac{\rho}{4} b^2 \mathscr{H}_i(\lambda). \qquad (5.37)$$

To compute the refracted radiant intensity E_r, consider Fig. 5.12. The power incident on the annular ring is

$$dP_i(\lambda) = \mathscr{H}_i(\lambda)2\pi b^2 \sin \theta_i \cos \theta_i \, d\theta_i,$$

$$= \mathscr{H}_i(\lambda)\pi b^2 \sin 2\theta_i \, d\theta_i. \qquad (5.38)$$

The power transmitted through the two surfaces of the sphere is

$$dP_r(\lambda) = dP_i(\lambda)[1 - \rho]^2. \qquad (5.39)$$

But this is also equal to the radiant intensity $E_r(\lambda)$ from another annular ring into which the radiation is refracted multiplied by the solid angle it subtends at the center of the sphere, that is,

$$dP_r(\lambda) = E_r(\lambda)2\pi \sin \phi \, d\phi.$$

$E_r(\lambda)$ can be shown to be given by

$$E_r(\lambda) = [1 - \rho]^2 \frac{\mathscr{H}_i(\lambda) b^2 \sin 2\theta_i \, d\theta_i}{2 \sin \phi \, d\phi} .$$ (5.40)

Now

$$\phi = 2(\theta_i - \theta_r),$$

as can be seen in Fig. 5.12. By utilizing Snell's law and some algebraic manipulation it can be shown that

$$\frac{d\theta_i}{d\phi} = \frac{\sin \theta_i \cos \theta_r}{2 \sin (\phi/2)} .$$

The power diffracted per unit solid angle is computed in a manner quite similar to the diffraction arising when radiation passes through a circular aperture (see 3.97), that is

$$E_d(\lambda) = \frac{b^2 \mathscr{H}_i(\lambda) J_1^2 \left[\dfrac{2\pi b \sin \phi}{\lambda} \right] \cos^4 \phi/2}{\sin^2 \phi} ,$$ (5.41)

where, as in 3.97, J_1 is the first order Bessel function of the first kind. It turns out that the total scattered power cannot be computed merely by adding E_r, E_r, and E_d. This arises because the radiation from the three phenomena interfere due to the fact that the radiation scattered in each case is not completely incoherent. Bricard[29] has taken this into account, and has found that for a large number of scattering particles randomly distributed in size, the average power scattered per unit solid angle per scattering particle is

$$E_{scat}(\lambda) = E_r(\lambda) + E_r(\lambda) + E_d(\lambda)$$
$$+ \frac{2 \cos^2 (\phi/2)}{\sin \phi} J_1 \left[\frac{2\pi b}{\lambda} \sin \phi \right] \sqrt{\rho} b^2 \mathscr{H}_i(\lambda) \sin \left[\frac{4\pi b}{\lambda} \sin \phi \right].$$ (5.42)

The scattering area ratio can then be calculated from Eqs. 5.28, 5.29, 5.31, 5.37, 5.40, 5.41, and 5.42.

$$S(\lambda) = \frac{1}{N_s} \int_0^\pi \left\{ \frac{\rho}{2} + [1 - \rho]^2 \frac{\sin 2\theta_i}{\sin \phi} \left(\frac{d\theta_i}{d\phi} \right) + 2 J_1^2 \left[\frac{2\pi b}{\lambda} \right] \cos^4 \frac{\phi}{2} \right.$$
$$\left. + \frac{4 \cos^2 \phi/2}{\sin \phi} \rho^{1/2} J_1 \left[\frac{2\pi b}{\lambda} \sin \phi \right] \sin \left[\frac{4\pi b}{\lambda} \sin \phi \right] \right\} \sin \phi \, d\phi.$$ (5.43)

This integral is difficult to evaluate in closed form and as a result numerical methods are generally employed. If (b/λ) is greater than about 3, the last term may be neglected. If (b/λ) is less than 1.5, the above theory is inadequate and one must employ the Mie or electromagnetic theory of scattering.

In the foregoing discussion, absorption was neglected. Zanotelli[30] has shown that in cases where

$$1 - \exp(-ax) \approx ax,$$

the incident power absorbed in the sphere is

$$P_a(\lambda) = \frac{4\pi\alpha b^3}{3n_p} \mathcal{H}_i(\lambda), \tag{5.44}$$

where α is the absorptivity and n_p is the index of refraction of the particle. For water $n_p = \frac{4}{3}$.

For 10 μ radius water droplets the absorbed power is smaller than the scattered power by a factor of about 10^5, and hence the assumption of negligible absorption is valid. For more absorbing particles, for example, carbon and wood, we must rely on the Mie theory.

It turns out that $S(\lambda)$ given in 5.43 is fairly independent of wavelength for large values of b/λ. This is evident to us by the "whiteness" of fog, which is composed of relatively large droplets. Scattering of this type is referred to as nonselective scattering.

5.5.2 Electromagnetic Theory Approach—Mie's Theory

Mie[31] has considered the scattering problem in terms of the wave picture of radiation in which the electromagnetic waves are subject to certain boundary conditions. Stratton and Houghton[32] applied this theory to the problem of atmospheric transmission. Mie has derived expressions for the electric and magnetic vectors describing the radiation per unit solid angle scattered by a sphere. To do this he started with the basic Maxwell equations. Since the scattered fields represent outgoing spherical waves, it is judicious to state Maxwell's equations, the incident radiation, and the scattered radiation in terms of spherical coordinates. It can be shown[33] that if the incident linearly polarized radiation is described by

$$\mathbf{E}_i = \mathbf{u}_x E_{ix0} \exp\{2\pi j\nu t - \Gamma z\}, \tag{5.45}$$

$$\mathbf{H}_i = \mathbf{u}_y H_{iy0} \exp\{2\pi j\nu t - \Gamma z\}, \tag{5.46}$$

where $\mathbf{u}_x$ and $\mathbf{u}_y$ are unit vectors in the x and y directions, then $\mathbf{E}_i$ and $\mathbf{H}_i$ can be represented in spherical coordinates by

$$\mathbf{E}_i = \text{curl}\,[\mathbf{r}v_i] + \frac{1}{\Gamma}\text{curl curl}\,[\mathbf{r}w_i], \tag{5.47}$$

$$\mathbf{H}_i = m\left\{-\text{curl}\,[\mathbf{r}w_i] + \frac{1}{\Gamma}\text{curl curl}\,[\mathbf{r}v_i]\right\}. \tag{5.48}$$

Here $m \equiv j\Gamma\lambda/2\pi \approx jn_m$ (for dielectrics) and the scalars v_i and w_i are given by

$$w_i = \exp\{2\pi j\nu t\} \cos\theta \sum_{l=1}^{\infty} \frac{(-1)^l(2l+1)}{l(l+1)} P_l'(\cos\phi) g_l\left(\frac{2\pi r}{\lambda}\right), \quad (5.49)$$

and

$$v_i = \exp\{2\pi j\nu t\} \sin\theta \sum_{l=1}^{\infty} \frac{(-1)^l(2l-1)}{l(l+1)} P_l'(\cos\phi) g_l\left(\frac{2\pi r}{\lambda}\right), \quad (5.50)$$

where $P_l'(\cos\phi)$ is the associated Legendre polynomial, and

$g_l\left(\dfrac{2\pi r}{\lambda}\right)$ is the spherical Bessel function derived from $J_{l+\frac{1}{2}}$, the

$(l + \frac{1}{2})$ order Bessel function of the first kind, that is,

$$g_l \equiv \left[\frac{\lambda}{4r} J_{l+\frac{1}{2}}\left(\frac{2\pi r}{\lambda}\right)\right]^{\frac{1}{2}}.$$

The functional form used to represent the incident wave prescribes the form of the scattered wave. By considering the boundary conditions between the medium (assumed to have an index of refraction n_m equal to unity) and the sphere (index of refraction given by n_p), as well as the conditions to be satisfied at infinity, the scattered wave can be specified by functions w_s and v_s which are similar to those for the incident wave, that is,

$$w_s = \exp\{j\omega t\} \cos\theta \sum_{l=1}^{\infty} c_l \frac{(-1)^{l+1}(2l+1)}{l(l+1)} P_l'(\cos\phi) h_l^{(2)}\left[\frac{2\pi r}{\lambda}\right], \quad (5.51)$$

$$v_s = \exp\{j\omega t\} \sin\theta \sum_{l=1}^{\infty} d_l \frac{(-1)^{l+1}(2l+1)}{l(l+1)} P_l'(\cos\phi) h_l^{(2)}\left[\frac{2\pi r}{\lambda}\right], \quad (5.52)$$

where $\quad h_l^{(2)}\left[\dfrac{2\pi r}{\lambda}\right] \equiv \sqrt{\dfrac{\lambda}{4r}} H_{l+\frac{1}{2}}^{(2)}\left(\dfrac{2\pi r}{\lambda}\right),$

$H_{l+\frac{1}{2}}^{(2)}\left(\dfrac{2\pi r}{\lambda}\right)$ is a Bessel function of the second kind,

and the c_l's and d_l's are constants determined by the boundary conditions and the magnitude of the incident vectors as well as by Γ. For observations far removed from the scattering particles, that is, $r \gg b$, these expressions can be simplified and yield electric field and magnetic intensity vector components.

$$E_\phi = H_\theta = -\frac{j\lambda}{2\pi r} \exp\left\{-j\frac{2\pi r}{\lambda} + j\omega t\right\} \times$$

$$\cos\theta \sum_{l=1}^{\infty} \frac{2l+1}{l(l+1)} \left\{d_l \frac{P_l'(\cos\phi)}{\sin\phi} + c_l \frac{d}{d\phi}[P_l'(\cos\phi)]\right\}, \quad (5.53)$$

and

$$-E_\theta = H_\phi = \frac{j\lambda}{2\pi r} \exp\left\{-j\,\frac{2\pi r}{\lambda} + j\omega t\right\} \times$$

$$\sin\theta \sum_{l=1}^{\infty} \frac{2l+1}{l(l+1)} \left\{ c_l \frac{P_l'(\cos\phi)}{\sin\phi} + d_l \frac{d}{d\phi}\,[P_l'(\cos\phi)] \right\}. \quad (5.54)$$

The radial components of **E** and **H** have been neglected since they decrease very rapidly with increasing r. These equations represent fields which are outgoing, spherical waves. Their amplitude and polarization depend upon direction.

TABLE 5.4. $\gamma_\lambda(\phi)$ from the Mie theory, for water droplets, in units of $(\lambda^2/4\pi^2)\mathrm{cm}^2$ *

ϕ \ $2\pi b/\lambda$	0.5	1.0	1.5	2.0	2.5	3.0	4.0	5.0	6.0
0°	0.00071	0.05260	0.6465	3.937	14.20	41.69	197.7	585.8	1253.0
10	0.00069	0.05152	0.6288	3.790	13.45	38.68	173.7	478.0	927.1
20	0.00066	0.04846	0.5790	3.382	11.42	30.85	116.6	251.8	349.0
30	0.00061	0.04383	0.5054	2.800	8.705	21.00	57.71	76.70	56.58
40	0.00055	0.03827	0.4196	2.155	5.938	12.02	19.48	13.168	27.61
50	0.00048	0.03248	0.3331	1.546	3.626	5.646	4.333	9.441	28.16
60	0.00042	0.02706	0.2551	1.038	1.984	1.1549	2.064	10.234	9.396
70	0.00037	0.02253	0.19087	0.654	0.9808	0.6463	2.744	4.872	3.863
80	0.00033	0.01915	0.14213	0.3867	0.4519	0.3214	2.375	2.635	5.914
90	0.00032	0.01700	0.10765	0.21354	0.2149	0.3936	1.230	1.909	3.573
100	0.00032	0.01598	0.08472	0.10831	0.13202	0.4806	0.4746	2.390	1.5505
110	0.00034	0.01588	0.07025	0.04918	0.12225	0.4630	0.4771	1.354	3.052
120	0.00037	0.01645	0.06138	0.02028	0.14755	0.35880	0.8279	0.5854	2.730
130	0.00042	0.01741	0.05605	0.01096	0.19170	0.23317	1.0006	1.300	0.8624
140	0.00046	0.01851	0.05284	0.01370	0.2462	0.14452	0.8518	2.222	2.555
150	0.00050	0.01956	0.05088	0.02262	0.3034	0.1188	0.6302	2.065	4.647
160	0.00054	0.02042	0.04970	0.03288	0.3540	0.1446	0.6083	1.593	3.456
170	0.00056	0.02096	0.04906	0.04068	0.3888	0.1855	0.7196	1.829	2.880
180	0.00057	0.02115	0.04887	0.04358	0.4013	0.2042	0.8709	2.155	3.507

* Adapted from LaMer and Sinclair[35]

With these equations we can calculate the average value of the power of the scattered radiation per solid angle as a function of ϕ since it is the vector product of **E** and **H**. In this way Mie was able to compute R_ω, the radiance, as an infinite series in ϕ.

For water, $[n_p = 1.33]$, $\gamma_\lambda(\phi)$ is given as a function of $2\pi b/\lambda$ and ϕ in Table 5.4. Figure 5.13 shows the relationship between $S(\lambda)$ and $2\pi b/\lambda$ for water droplets. The calculations for Fig. 5.13 were made by Houghton and Chalker.[34]

Since $S(\lambda)$ can be seen to be functionally related to (b/λ) we shall from now on refer to $S(\lambda)$ as $S(b/\lambda)$. The numerical calculation of R_ω may be rather tedious since these series converge slowly when (b/λ) is large. If this is the case, the approximation afforded by the geometrical optics approach may prove more satisfactory.

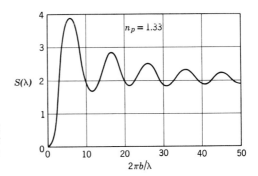

Figure 5.13. Plot of scattering area ratio, S, versus $2\pi b/\lambda$ for $n = 1.33$. (After H. G. Houghton and W. R. Chalker.[34])

5.6 SCATTERING BY THE ATMOSPHERE— EMPIRICAL DISCUSSION

The discussions of scattering in the preceding sections have treated the problem in terms of scattering particles which are:

1. Much smaller than $\lambda \rightarrow$ Rayleigh scattering.
2. Comparable in size to $\lambda \rightarrow$ Mie scattering.
3. Much larger than $\lambda \rightarrow$ nonselective scattering.

It is evident from the discussions given concerning the various types of scattering that the type of treatment involved in Rayleigh or molecular scattering considers the more fundamental aspects of the relationship between scattered radiation and the composition and properties of matter. The appealing basic nature of the reasoning and the elegant simplicity of Rayleigh scattering theory has prompted a number of attempts to find a relationship of the same form which is applicable to and which describes Mie scattering. Now the transmitted intensity in the ith window as affected by scattering can be written as (see Eqs. 3.66, 3.84, and 3.85),

$$\tau_{si} = \exp\left\{-\beta(\lambda_i)x\right\}, \tag{5.55}$$

where λ_i is the midpoint of the ith window. From Eq. 5.32, $\beta(\lambda_i)$ was found to be

$$\beta(\lambda_i) = \pi \sum_{j=1}^{m} n_{sj} b_j^{2} S_j(b/\lambda_i). \tag{5.32}$$

Since $S(b/\lambda)$ is a function of the ratio (b/λ), then $\beta(\lambda)$ may be written in the form

$$\beta(\lambda) = A\lambda^{-q}, \tag{5.56}$$

where A and q are constants determined essentially by the size and distribution of the scattering particles. An expression of the form of 5.56 is quite similar to that obtained for Rayleigh scattering.

Cognizant that gases in the atmosphere as well as suspended particles scatter radiation, Granath and Hulburt[36] and Anderson[37] have generalized 5.56 to

$$\beta(\lambda) = A_1\lambda^{-q} + A_2\lambda^{-4}.$$

Although this represents a more general relationship, the precision of measurements under ordinary conditions render Eq. 5.56 entirely adequate.

The task remaining is to determine the constants A and q. In order to determine q experimentally, we utilize 5.56 by taking the logarithm of both sides, that is,

$$\ln \beta(\lambda) = \ln A - q \ln \lambda. \tag{5.57}$$

The slope of $\ln \beta$ versus $\ln \lambda$ is $-q$. According to Middleton,[2] when seeing conditions are exceptionally good, $q = 1.6$. On days when seeing conditions are "average," $q = 1.3$. If the atmosphere contains enough haze so that the visual range (to be discussed below) is less than 6 km, a good value for q suggested by Löhle[38] based on work by Wolff[39] is

$$q = 0.585V^{\frac{1}{3}}, \tag{5.58}$$

where V is the visual range in kilometers.

Unfortunately, it is very difficult to determine the constant A from experimental data plotted according to 5.57. Instead, it is more convenient to determine A by recourse to the concept of visual range. The visual range is a commonly used parameter which provides a measure of the extent to which the transmission properties of the atmosphere cause the visual contrast at 0.55 μ to be reduced. It is reported daily by the United States Weather Bureau in many localities. The apparent contrast C_x of a source when viewed at a distance x is defined by

$$C_x \equiv \frac{R_{\omega sx} - R_{\omega bx}}{R_{\omega bx}},$$

where $R_{\omega sx}$ and $R_{\omega bx}$ are the apparent radiances of the source and its background when viewed from a distance x. The distance at which the ratio

$$C_x/C_0 = \frac{\dfrac{R_{\omega sx} - R_{\omega bx}}{R_{\omega bx}}}{\dfrac{R_{\omega s0} - R_{\omega b0}}{R_{\omega b0}}}, \tag{5.59}$$

where 0 refers to the intrinsic $[x = 0]$ condition, is reduced to 2 % is defined as the visual range, that is,

$$C_{x=V}/C_0 = 0.02. \tag{5.60}$$

Standard practice is to evaluate the visual range for $\lambda = 0.55 \mu$.

If we consider the case where the source radiance is much greater than that of the background for any viewing distance, that is, $R_{\omega s} \gg R_{\omega b}$ and that the background radiance is constant, that is, $R_{\omega b0} = R_{\omega bx}$, then for the visual range 5.59 becomes

$$C_{x=V}/C_0 = \frac{R_{\omega sV}}{R_{\omega s0}} = 0.02, \tag{5.61}$$

or

$$\ln\left(\frac{R_{\omega sV}}{R_{\omega s0}}\right) = -3.91.$$

When considering scattering the only mechanism of importance, a condition which is true at $0.55\,\mu$ where absorption is negligible, the transmission coefficient for the visual range is

$$\tau_{0.55\mu} = \left(\frac{R_{\omega sV}}{R_{\omega s0}}\right)_{0.55\mu} = \exp\left(-\beta V\right), \tag{5.62}$$

or

$$\ln\left(\frac{R_{\omega sV}}{R_{\omega s0}}\right)_{0.55\mu} = -\beta V. \tag{5.63}$$

Combining 5.63 with 5.61 we obtain

$$\beta = \frac{3.91}{V}, \quad \text{at} \quad \lambda = 0.55\,\mu. \tag{5.64}$$

From 5.56 we see that

$$A = \frac{3.91}{V}(0.55)^q. \tag{5.65}$$

Substituting 5.65 into 5.55 we obtain

$$\tau_{si} = \exp\left\{-\frac{3.91}{V}\left[\frac{\lambda_i}{0.55}\right]^{-q}x\right\}. \tag{5.66}$$

This relationship enables us to compute the transmission at the midpoint of the ith window for any value of x if V is known. Of course, the transmission of this window will also be affected by absorption, which must be computed separately.

In conclusion let us consider a practical example to illustrate the use of the foregoing information to compute atmospheric transmission. Consider the transmission of infrared radiation for a known source distribution in the spectral interval from 1.13 to 6.0 μ on a day when the visual range is 5 km, the relative humidity is 75 %, and the air temperature is 60°F over a path length of 5000 ft (1.525 km). The first step is to compute the number of pr. mm of H_2O in the path length. Figure 5.5 shows that

1.0 pr. mm of H_2O would be encountered in approximately 350 ft or 14.3 pr. mm are encountered in 5000 ft. We see that the following is true using Table 5.2, and Eqs. 5.24 and 5.25.

Window	Relationship between w and w_i	Use Equation	τ_{ai}
III	$w > w_i$	5.25	$\tau_{a\text{III}} = 0.70$
IV	$w > w_i$	5.25	$\tau_{a\text{IV}} = 0.60$
V	$w > w_i$	5.25	$\tau_{a\text{V}} = 0.55$
VI	$w > w_i$	5.25	$\tau_{a\text{VI}} = 0.56$
VII	$w > w_i$	5.25	$\tau_{a\text{VII}} = 0.29$

Since V, the visual range, is 5 km, q is found to be 1 from 5.58. Therefore we may calculate the following from Eq. 5.66.

Window	λ_i	τ_{si}
III	1.26	0.60
IV	1.64	0.67
V	2.30	0.75
VI	3.50	0.81
VII	5.15	0.88

The product $\tau_{ai}\tau_{si}$ for each window is therefore:

Window	$\tau_{ai}\tau_{si}$
III	0.42
IV	0.40
V	0.41
VI	0.45
VII	0.25

To compute the power transmitted in the ith wavelength interval, the power emitted by the source in that interval should be multiplied by $\tau_{ai}\tau_{si}$.

REFERENCES

1. J. N. Howard, D. E. Burch, and D. Williams, *AF Cambridge Research Center Tech. Report* 55-213 (November 1955).
2. W. E. K. Middleton, *Vision Through the Atmosphere*, University of Toronto Press (1952), p. 20.
3. H. L. Wright, *Proc. Phys. Soc.* **48**, 675 (1936).

4. H. Dessens, *Quart J. Royal Meteorol. Soc.* **75**, 23 (1949).
5. G. C. Simpson, *Quart. J. Royal Meteorol. Soc.* **67**, 99 (1941).
6. H. L. Wright, *op. cit.*
7. B. J. Mason and F. H. Ludlam, *Reports on Progress in Physics* **14**, 147 (1951).
8. H. L. Wright, *Quart. J. Royal Meteorol. Soc.* **65**, 411 (1939).
9. H. Dessens, *Ann. de Geophys.* **2**, 343 (1946).
10. H. G. Houghton and W. H. Radford, *Mass. Inst. of Tech. Papers in Phys. Ocean. and Meteorol.* **6**, No. 4, Cambridge (1938).
11. E. T. Whittaker and G. N. Watson, *A Course of Modern Analysis*, Cambridge University Press, London (1952).
12. T. Elder and J. Strong, *J. Franklin Institute* **255**, 189 (1953).
13. W. M. Elsasser, *Harvard Meteorological Studies* No. 6, Blue Hill Meteorological Observatory, Harvard University (1942), p. 28.
14. G. Falckenberg, *Met. Zeits.* **45**, 334 (1928).
15. F. A. Brooks, *Mass. Inst. of Tech. Papers in Phys. Ocean. and Meteorol.* **8**, No. 2 Cambridge (1941).
16. F. Schnaidt, *Gerl. Beitz z. Geophys.* **54**, 203 (1939).
17. J. N. Howard, D. E. Burch, and D. Williams, *J. Opt. Soc. Amer.* **46**, 237 (1956).
18. T. Elder and J. Strong, *op. cit.*
19. R. M. Langer, *Report on Sig. Corps Contract* No. DA-36-039-SC-72351 (May 1957).
20. S. B. Schuldt, *Report on Army Ord. Contract* No. DA-11-022-501-ORD-2964 (1959).
21. J. N. Howard, D. E. Burch, and D. Williams, *AF Cambridge Research Center Tech. Report* 55-213 (November 1955), p. 54.
22. J. D. Strong, *OSRD Report* 5986 (Nov. 30, 1945) (Harvard).
23. H. Fischer, *ASAF Tech. Report* F-TR-2104-ND (Oct. 1946).
24. H. A. Gebbie et al., *Proc. Roy. Soc.* **A206**, 87 (1951).
25. J. H. Taylor and H. W. Yates, *Naval Research Laboratory Report No. 4759*, published also as PB121199 (May 11, 1956).
26. L. Larmore, *Proc. IRIS*, **1**, No. 1, 14 (June 1956). (Paper unclassified).
27. W. M. Elsasser, *op. cit.*, p. 36.
28. J. H. Taylor and H. W. Yates, *op. cit.*
29. J. Bricard, *La Meteorol.* **15**, 83 (1939).
30. G. Zanotelli, *Atti Accad. Ital. Rend. Cl. Sci. Mat. Nat.* **2**, 42 (1940).
31. G. Mie, *Ann. der Phys.* **25**, 377 (1908).
32. J. A. Stratton and H. G. Houghton, *Phys. Rev.* **38**, 159 (1931).
33. H. C. Van de Hulst, *Light Scattering by Small Particles*, J. Wiley and Sons, New York (1957), p. 121.
34. H. G. Houghton and W. R. Chalker, *J. Opt. Soc. Amer.* **39**, 955 (1949).
35. V. K. LaMer and D. Sinclair, *OSRD Report* No. 1857 (Dept. of Commerce, Office of Publication Board, No. PB 944), Washington, D.C.
36. L. P. Granath and E. O. Hulburt, *Phys. Rev.* **34**, 140 (1929).
37. S. H. Anderson, *Aviation* **28**, 930 (1930).
38. F. Löhle, *Phys. Zeits.* **45**, 199 (1944).
39. M. Wolff, *Das Licht* **8**, 105, 128 (1938).

6
The physics of
semiconductors

6.1 INTRODUCTION

Semiconductors have a special prominence in infrared detection systems. As has been seen, they are useful in infrared optical systems, both as lenses and as filters. Semiconductor diodes and transistors are used as elements in infrared electronic systems. Their most important application in infrared technology is found in infrared detectors, where semiconductors are more widely used than any other material. In order that we may arrive at an understanding of the detection mechanisms to be developed in Chapter 9, this chapter shall be devoted to a study of the physics of semiconductors. All discussion of photoeffects shall be deferred to Chapter 8. The reader interested in a more complete discussion of semiconductors is referred to the many excellent books available.[1]

6.2 THE PERIODIC LATTICE

Solids can be divided according to their electrical properties into three classes: metals, insulators, and semiconductors. In terms of their electrical resistivities it is generally considered that metals have resistivities less than 10^{-3} ohm cm, dielectrics more than 10^{12} ohm cm, and semiconductors occupy a range intermediate between metals and dielectrics. This artificial separation, of limited use only, is by no means to be considered definitive. Another criterion by which solids can be divided depends on whether their atoms are arranged in a regular array, termed a periodic lattice, or in a disorganized manner. The former are termed crystalline solids, the latter amorphous solids. Some materials will exhibit both phases, for example, selenium has both amorphous and crystalline forms.

Semiconductors, metals, and some dielectrics have a periodic lattice; thus they are crystalline solids. The existence of the periodic lattice is basic to their electrical properties. This periodicity, that is, the regular array of atoms spaced in a repetitive structure, can exist in microscopic localized crystalline regions separated by discontinuities in the periodicity, termed grain boundaries, or can exist in large, macroscopic regions. The former, an array of crystallites, is referred to as a "polycrystalline solid," the latter a "single crystal." Obviously, the polycrystalline solid is composed of an aggregate of small single crystal regions.

The electrical and physical properties of semiconductors are influenced by their crystallinity—the degree to which they approach being single crystals. The widespread use of semiconductor materials in devices today has been largely due to the development of the ability to produce large single crystals of high purity. The method generally used for transistors and diodes is the Czochralski[2] method, in which a "seed" crystal, a small single crystal of the material being prepared, is dipped into a container of molten semiconductor closely controlled to a temperature just above the melting point. The seed is rotated and simultaneously withdrawn from the melt at a speed of a few inches per hour. As it is pulled from the melt, a single crystal builds up at the base of the seed, forming at the solid-liquid interface. By continuing this process until most of the melt has solidified on the seed, large single crystals several inches long can be prepared. Although certain semiconductors cannot be prepared in single crystal form in this manner, the ones which concern us most can be.

The regular array of atoms arranged in a crystal lattice can have only a limited choice of symmetry. A simple illustration of this in two dimensions can be seen in the arrangement of tiles in a floor. These may be three-sided, four-sided, or six-sided, but not, for example, five-sided. It is not possible to fit pentagons together in a regular array; there are always gaps in the array into which the pentagons will not fit. Even the three-sided regular figure is not independent of the six-sided, for a hexagon is composed of six equilateral triangles. Most crystals possess either cubic or hexagonal symmetry. These in turn may be broken down into subgroups possessing certain symmetry properties. The study of these is beyond the scope of this book. The interested reader is referred to Kittel[3] for an introductory discussion.

Let us consider germanium, symbol Ge, an elemental semiconductor. Referring to Fig. 6.1, part of the periodic table, we see that Ge is a group IVA element.

IIA	IB	IIB	IIIA	IVA	VA	VIA	VIIA
12 Mg			13 Al	14 Si	15 P	16 S	17 Cl
20 Ca	29 Cu	30 Zn	31 Ga	32 Ge	33 As	34 Se	35 Br
	47 Ag	48 Cd	49 In	50 Sn	51 Sb	52 Te	53 I
	79 Au	80 Hg	81 Tl	82 Pb	83 Bi		

Figure 6.1. A portion of the periodic table.

As such, it is characterized by having an outermost shell only half filled with a normal complement of eight electrons. Since these four electrons are not as tightly bound to the nucleus as those in the inner lying shells, they can be shared more readily by the parent atom with other atoms of the lattice.

Consider now the periodic arrangement of Ge atoms in a block of solid Ge. A unique arrangement of the atoms exists so that the potential energy associated with the arrangement is a minimum. This is found in crystalline Ge. Each of the four valence electrons is shared with each of the four

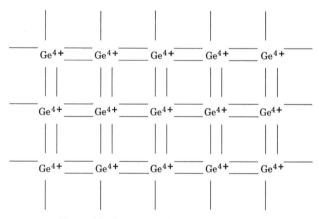

Figure 6.2. Representation of the Ge lattice.

$$Zn^{2+} \quad S^{2-} \quad Zn^{2+} \quad S^{2-} \quad Zn^{2+}$$
$$S^{2-} \quad Zn^{2+} \quad S^{2-} \quad Zn^{2+} \quad S^{2-}$$
$$Zn^{2+} \quad S^{2-} \quad Zn^{2+} \quad S^{2-} \quad Zn^{2+}$$
$$S^{2-} \quad Zn^{2+} \quad S^{2-} \quad Zn^{2+} \quad S^{2-}$$

Figure 6.3. Representation of ZnS lattice.

nearest neighbors of the Ge atom in an arrangement known as covalent bonding. A two-dimensional representation of this type of bonding is shown in Fig. 6.2.

In this representation the Ge^{4+} symbols represent the atomic nuclei surrounded by three closed shells containing 28 electrons. The paired lines represent two valence electrons being shared by the nuclei, one from each of the nuclei being connected. The pair represents the bond between the two nuclei and is shared equally. It is this electron bond which maintains the nuclei at their proper spacing at positions termed lattice sites.

The other major type of lattice bonding is found in certain compound semiconductors. In this type, which occurs in compounds formed from elements arising at the opposite sides of the periodic table, the valence electrons from the one type of atom are transferred to the other. The bond formed in this manner is known as ionic or polar. Zinc sulfide, ZnS, has such a lattice. (See Fig. 6.3.)

Zinc is a IIB element, sulfur is VIA. Zinc has therefore two valence electrons; sulfur has six. The arrangement of the atoms in order to provide the lowest potential energy is found when zinc gives its two electrons to the nearest sulfur atom. Zinc, having lost two valence electrons, is doubly ionized positively, and has its outermost electron shell closed. Sulfur, having added two electrons to its outer shell, is doubly ionized negatively and has its outermost shell closed. This ionic, or electrostatic, bond is stronger than the covalent one, requiring more energy to break it.

Most materials show a type of bond which is intermediate between the covalent and the ionic. The degree to which the bond approaches the ionic is known as the ionicity of the bond. In general, the bond character progresses from covalent to ionic from the bottom to the top and from the center to the sides of the periodic table.

6.3 ENERGY BANDS IN A PERFECT CRYSTAL

Having discussed the forces which bind atoms in a periodic lattice, we shall now consider the energies which an electron is allowed to have in the lattice. At first glance we may wonder at the significance of the question, for it may seem that an electron can acquire any energy, all energies being permitted. This is not the case. Electrons in the crystal are permitted to have certain energies and not others. The explanation for this is in the realm of quantum mechanics, which is outside the scope of this book. It is sufficient to say that just as electrons in an atom are forced to have only certain discrete energies, so also electrons in a crystal are allowed only discrete energies. The discrete levels of shell electrons in an atom are broadened into energy bands by the proximity of the other atoms in the periodic lattice. Within each band there exists a multiplicity of discrete energy levels. The Pauli exclusion principle states that only two electrons having opposite spins can occupy the same level, that is, have the same energy. The allowed energy bands of interest in electronic conduction phenomena are two: the valence band and the conduction band. They are separated by a region of forbidden energies referred to as the forbidden band. This band structure is represented schematically in Fig. 6.4.

To give an order of magnitude approximation to the energies, the width of the forbidden band in semiconductors ranges from about 1/100 of an electron volt (1 ev = 1.6 × 10⁻¹⁹ joule) to a few electron volts. The distance between levels is of the order of 10⁻¹⁴ ev. Thus the levels form practically a continuum. The forbidden band widths of a number of semiconductors are listed in Table 6.1.

Electrons having energies in the conduction band are referred to as "electrons in the conduction band," "conduction band electrons," or "free electrons." Their energies are sufficiently great so that they are not attached to any atom. Since the uniform character of the lattice does not obstruct their movement, they wander freely through the crystal. The number of free electrons and their ease of movement account for the magnitude of the electrical conductivity.

Let us consider the valence band. Assume that in some portion of the lattice an electron is missing from one of the bonds. (See Fig. 6.5.)

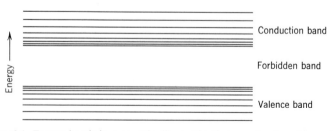

Figure 6.4. Energy bands in a crystal. (Separation between levels not to scale.)

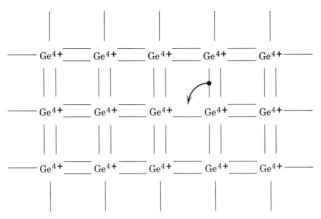

Figure 6.5. Hole motion.

The absence of an electron is termed a "hole." It is possible for an electron in a bond on an adjacent atom to move into the bond from which the electron is missing, as illustrated in Fig. 6.5. The original bond is

TABLE 6.1. Width of forbidden band in pure semiconductors at room temperature

Semiconductor	Band Gap (ev)	Semiconductor	Band Gap (ev)
Ge	0.67	GaAs	1.45
Se	2	GaSb	0.80
Si	1.12	InAs	0.35
Sn (gray)	0.08	InP	1.25
Te	0.33	InSb	0.18
AlSb	1.6	Mg_2Sn	0.33
CdS	2.4	PbS	0.37
ZnS	3.6	PbSe	0.27
CdSe	1.8	PbTe	0.30
CdTe	1.5	HgTe	0.02

now complete, but the hole appears in the adjacent bond. This bond in turn can be filled by an electron from a bond on a neighboring atom, and so on. This motion of electrons in one direction can be represented by a motion of holes in the opposite direction. Although this process appears to be rather jumpy or chaotic, the nature of the periodic lattice is such that hole flow can be characterized as a smooth movement. Holes not in motion are termed "bound holes." Holes in motion are "free holes" and have energies in the valence band. Free electrons and holes, contributing to the conductivity of the semiconductor, are called "current carriers" or simply "carriers." In the presence of an electric field, holes flow in the direction of the field, whereas electrons flow against the field. Since the flow of holes is opposite in direction to the electron flow, the hole has a positive charge, in contrast to the negative charge of the electron. Thus the holes also contribute to the conductivity. In addition, there is an effective mass associated with free electrons, that is, electrons with energies in the conduction band and also an effective mass for free holes, that is, holes with energies in the valence band. The effective masses of both are in general less than that of the rest mass of an electron, $m = 9.108 \times 10^{-31}$ kg, with free holes usually being heavier than free electrons. We shall see in a later portion of this chapter that the effective masses influence the electrical properties of the semiconductor.

At any temperature above absolute zero the atoms comprising the lattice, bound to each other by the electron bonds, are undergoing rapid vibrational motion about their lattice sites. The energy of these vibrations is sufficient to free some electrons and holes, the number of which is dependent upon the forbidden energy gap and upon the temperature. In pure semiconductors this process, known as thermal excitation, produces simultaneously a free electron and a free hole, termed an electron-hole pair. The numbers of free electrons and holes determine to a large extent the electrical conductivity. We shall see that materials with wide forbidden bands which require large energies for excitation, have fewer free carriers (electrons and holes), thus lower conductivity, than materials having a smaller band gap. Pursuing this observation, we realize that dielectric materials, with very low conductivity, must have wide forbidden gaps. Metals, having very high conductivity, must have small gaps. As a matter of fact, most metals have no forbidden band, their conduction bands touching the valence bands. Semiconductors, having forbidden bands less than those of dielectrics, have electrical conductivities greater than dielectrics, but smaller than metals. Thus we see that there is no sharp distinction from an energy point of view between metals, semiconductors, and dielectrics, and the classification of them in terms of the electrical resistivities given at the beginning of this chapter is an arbitrary one.

6.4 IMPERFECTIONS IN THE LATTICE

Thus far the discussion has considered the most simple form of semiconductor known as "intrinsic." It is possible to influence the properties of a semiconductor in a controllable manner by adding impurities. Let us once again consider the Ge lattice. Suppose that at some particular lattice site a Ge atom has been replaced by an atom of arsenic (As). (See Fig. 6.6.)

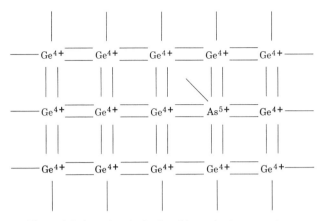

Figure 6.6. Arsenic substitutional impurity in germanium.

This replacement is known as a substitutional impurity, since the As atom has been substituted for a Ge atom at a lattice site. Arsenic, in the fifth column of the periodic table, has five valence electrons. Four of these electrons are shared with the four neighboring Ge atoms. The fifth, finding no Ge atom on which to bond, is loosely held to the parent atom. The energy required to free this electron is far less than that required to free an electron in a bond. Thus, the parent atom can donate a free electron to the crystal, if the necessary energy is available. Such an impurity is called a "donor." The site is said to be ionized if the electron has been removed from the donor. A semiconductor in which the conductivity is controlled by donors is known as "n-type" (that is, negative-type). The energy levels associated with donors are located in the forbidden band, in many cases just below the bottom of the conduction band. If the donor level is sufficiently close to the bottom of the conduction band, that is, if it is "shallow," so that at some temperature most of the free electrons have come from donor centers, the semiconductor is in an "extrinsic" condition.

Consider now the effect of substituting for a Ge atom an atom having a valence less than that of Ge, for example, boron (B). (See Fig. 6.7.) In this case B, having a valence of three, can share only three electrons with Ge. The absence of an electron in the fourth bond constitutes a hole bound to the B site. This unfilled bond can be filled by an electron moving from a neighboring bond into it in a process again known as ionization. This event is seen to produce a free hole. The energy required for this is less than that needed to free an electron from a Ge site. Thus the energy levels associated with these sites lie in the forbidden band, generally just

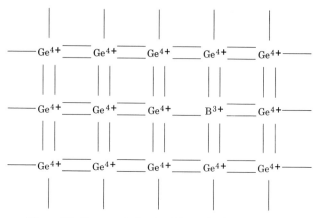

Figure 6.7. Boron substitutional impurity in germanium.

above the valence band. Since such sites contribute to the electrical properties of the semiconductor by accepting an electron from the valence band, these sites are called "acceptors." Material in which the conductivity is controlled by the acceptors is known as "*p*-type" (that is, positive-type). A semiconductor having acceptors in levels shallow enough to be ionized and in numbers sufficient to control the conductivity is also said to be extrinsic, in contrast to the intrinsic semiconductor in which most of the carriers are formed by excitation across the forbidden band. Figure 6.8 shows the energy level scheme of a semiconductor containing both donors and acceptors. Possible excitation processes are indicated by arrows. It should be noted that in extrinsic excitation a free electron-hole pair is not formed. Rather, excitation at a donor center produces a free electron and a bound hole, that is, a hole bound to the donor site. Excitation at an acceptor site produces a free hole and an electron bound to the acceptor site.

The influence of these impurity atoms upon the electrical properties of a semiconductor is remarkable. In concentrations of only one impurity atom per million normal lattice or host atoms, the impurity can completely

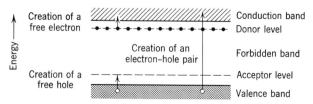

Figure 6.8. Energy bands and excitation processes in an extrinsic semiconductor.

dominate the electrical properties of the lattice. The reason lies in the dependence of the free carrier concentration upon the energy required for excitation. We shall see that it is an exponential function, indicating that only a very small fraction of the normal lattice sites may be ionized at any one temperature, say room temperature. Imagine a lattice containing one part per million impurity concentration. The levels associated with these

TABLE 6.2. Impurity ionization levels in Ge and Si*

Periodic Table Column	Element	Ge Donor or Acceptor	$\mathscr{E}_i$ (ev)	Si Donor or Acceptor	$\mathscr{E}_i$ (ev)
I	Li	D	0.0093^c	D	0.033^c
	Cu	A	0.041^v	A	0.49^v
		A	0.33^v	D	0.24^v
	Au	D	0.053^v	A	0.54^c
		A	0.15^v	D	0.33^v
		A	0.20^c		
		A	0.04^c		
II	Zn	A	0.033^v	A	0.55^c
III	Cd	A	0.06^v	A	0.30^v
	B	A	0.0104^v	A	0.045^v
	Al	A	0.0102^v	A	0.057^v
	Ga	A	0.0108^v	A	0.065^v
	In	A	0.0112^v	A	0.16^v
	Tl	A	0.014^v		
V	P	D	0.0120^c	D	0.044^c
	As	D	0.0127^c	D	0.049^c
	Sb	D	0.0097^c	D	0.039^c
	Bi	D	0.012^c		
VII	Mn	A	0.16^v	D	0.53^c
		A	0.35^c		
VIII	Fe	A	0.27^c	D	0.55^c
		A	0.34^v	D	0.40^v
	Co	A	0.25^v		
		A	0.31^c		
	Ni	A	0.22^v		
		A	0.30^c		
	Pt	A	0.04^v		
		A	0.20^c		

* Adapted from Schultz and Morton[4] and Bube[5]
v: Measured from valence band edge.
c: Measured from conduction band edge.

impurities may lie close enough to the band edge to be completely ionized at room temperature. On the other hand, the gap width could be large enough so that less than one atom of the host lattice out of every million is ionized. Thus the impurity sites, although low in concentration, can contribute a greater number of carriers to the semiconductor than does the lattice.

We have discussed extrinsic semiconductors containing substitutional impurities. Another important class of impurities enters into the host lattice as interstitial atoms, the impurity atoms not replacing host atoms but fitting in between the lattice sites. It is possible also to have vacant lattice sites, which can act either as donors or as acceptors. Impurities, both substitutional and interstitial, vacancies and other structural defects, such as lattice dislocations, all fall into the general category of lattice imperfections.

Table 6.2 lists the ionization energies of donors and acceptors in germanium and silicon. Much of the information is taken from Schultz and Morton[4] and from Bube.[5]

6.5 FERMI-DIRAC STATISTICS FOR AN INTRINSIC SEMICONDUCTOR

In order to derive some of the fundamental relationships concerning the electrical properties of semiconductors, we must first determine the free carrier concentrations in a semiconductor at thermal equilibrium. Since the fraction of electrons which are free depends upon the energy they possess, it is necessary to investigate their energy distribution. To do this we have recourse to the Fermi-Dirac, or simply Fermi, statistics, which are applicable to electrons and holes (see Section 2.3.2). The Fermi statistics are of a type known as quantum statistics, and must be used when considering the aggregate properties of electrons. Under certain conditions, valid for most semiconductors, we can simplify matters by using

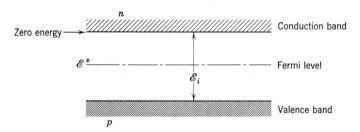

Figure 6.9. Energy levels in an intrinsic semiconductor.

Maxwell-Boltzmann, or classical statistics. We begin by considering again the energy level diagram of an intrinsic semiconductor, Fig. 6.9.

Let the density or concentration of free electrons, that is, the number per unit volume, be n. Let the density of free holes be p. Assume the material is in thermal equilibrium. We shall assign our reference zero of energy to be at the bottom of the conduction band.

As was stated in Chapter 2, the probability that an energy level $\mathscr{E}$ (a level of energy $\mathscr{E}$) is occupied is given by $f(\mathscr{E})$, the Fermi function.

$$f(\mathscr{E}) = \frac{1}{1 + \exp\left(\dfrac{\mathscr{E} - \mathscr{E}^*}{kT}\right)} . \tag{6.1}$$

In this expression k is Boltzmann's constant, 1.38×10^{-23} joule/deg K and T is the absolute temperature. At an energy $\mathscr{E} = \mathscr{E}^*$, the value of $f(\mathscr{E})$ is 0.5. Thus $\mathscr{E}^*$ is the energy of a particular level, termed the Fermi level, at which the probability of occupation is 0.5. By this is meant that if it were possible to have sites in the lattice having an ionization energy for donors of $\mathscr{E}^*$, half of the sites would be occupied and half would be empty.

If the exponential term in the denominator is much greater than unity, that is, if $(\mathscr{E} - \mathscr{E}^*) \gg kT$, then $f(\mathscr{E})$ can be approximated by

$$f(\mathscr{E}) \approx \exp -\left(\frac{\mathscr{E} - \mathscr{E}^*}{kT}\right) . \tag{6.2}$$

This is known as the Maxwell-Boltzmann, or simply Boltzmann, approximation.

The following is a discussion of the number of available energy levels or states. Near the bottom of the conduction band the density of states $g(\mathscr{E})$, that is, the number of energy levels per unit volume per unit energy at which an electron can reside, is given by

$$g(\mathscr{E}) = 4\pi\left(\frac{2m_e}{h^2}\right)^{3/2} \mathscr{E}^{1/2} . \tag{6.3}$$

In this expression h is Planck's constant, 6.63×10^{-34} joule sec and m_e is the effective mass of an electron. As has already been pointed out, the value of m_e is in general less than that of the rest mass m of an electron, 9.11×10^{-31} kg.

The density $n(\mathscr{E})$ of electrons in the conduction band at a level $\mathscr{E}$ is the product of the density of energy levels at $\mathscr{E}$ with the occupational probability at $\mathscr{E}$.

$$n(\mathscr{E}) = g(\mathscr{E})f(\mathscr{E}) = 4\pi\left(\frac{2m_e}{h^2}\right)^{3/2} \frac{\mathscr{E}^{1/2}}{1 + \exp\left(\dfrac{\mathscr{E} - \mathscr{E}^*}{kT}\right)} . \tag{6.4}$$

The density n of electrons present at all levels in the conduction band is the integral over all energies of $n(\mathscr{E})$.

$$n = \int_0^\infty n(\mathscr{E})\, d\mathscr{E} = 4\pi \left(\frac{2m_e}{h^2}\right)^{3/2} \int_0^\infty \frac{\mathscr{E}^{1/2}\, d\mathscr{E}}{1 + \exp\left(\dfrac{\mathscr{E} - \mathscr{E}^*}{kT}\right)}. \quad (6.5)$$

The reduced dimensionless variables η and η^* are introduced as

$$\eta \equiv \frac{\mathscr{E}}{kT}, \qquad \eta^* \equiv \frac{\mathscr{E}^*}{kT},$$

and a quantity N_c is defined as

$$N_c = 2\left(\frac{2\pi m_e kT}{h^2}\right)^{3/2}.$$

N_c is called the effective density of states in the conduction band, and represents approximately the density in a slice kT wide at the bottom of the band. For our purposes we shall consider only the classical case, in which $\eta^* \leqslant -2$, valid for most semiconductors of reasonable purity. The reader is referred to the textbook by Smith[6] for a complete treatment. In the classical case, the approximate solution of 6.5 is

$$n = N_c \exp \eta^*. \quad (6.6)$$

The concentration of free electrons is seen to depend linearly upon the effective density of states and exponentially upon the position of the Fermi level. As the Fermi level is raised, that is, approaches the bottom of the conduction band, the free electron density increases in an exponential fashion.

The density of free holes in the valence band can be computed in a manner analogous to that used for electrons in the conduction band. The probability that a level in the valence band contains a hole, that is, does not contain an electron, is $f_p(\mathscr{E})$ where

$$f_p(\mathscr{E}) = 1 - f(\mathscr{E}). \quad (6.7)$$

From Eq. 6.1 we see that

$$f_p(\mathscr{E}) = \frac{1}{1 + \exp\left(\dfrac{\mathscr{E}^* - \mathscr{E}}{kT}\right)}. \quad (6.8)$$

The density of states $g_p(\mathscr{E})$ in the valence band is given in an expression similar to Eq. 6.3 by

$$g_p(\mathscr{E}) = 4\pi \left(\frac{2m_h}{h^2}\right)^{3/2} (-\mathscr{E} - \mathscr{E}_i)^{1/2}, \qquad \text{for } \mathscr{E} < -\mathscr{E}_i, \quad (6.9)$$

where m_h is the effective mass of a free hole and $\mathscr{E}_i$ is the intrinsic excitation energy, that is, the width of the forbidden band. The density of free holes p is given by

$$p = \int_{-\infty}^{-\mathscr{E}_i} g_v(\mathscr{E}) f_v(\mathscr{E}) \, d\mathscr{E} = 4\pi \left(\frac{2m_h}{h^2}\right)^{3/2} \int_{-\infty}^{-\mathscr{E}_i} \frac{(-\mathscr{E} - \mathscr{E}_i)^{1/2} \, d\mathscr{E}}{1 + \exp\left(\dfrac{\mathscr{E}^* - \mathscr{E}_i}{kT}\right)}. \quad (6.10)$$

We utilize the reduced variables η^* and $\eta_i \equiv \mathscr{E}_i/kT$, and define a quantity N_v as

$$N_v \equiv 2\left(\frac{2\pi m_h kT}{h^2}\right)^{3/2}.$$

In a manner similar to that for N_c in the conduction band, N_v is called the effective density of states in the valence band, and represents approximately the density in a slice kT wide at the top of the valence band. Once again we shall consider classical statistics only. In this case, upon integration, the density of holes is found to be

$$p = N_v \exp\left(-\eta^* - \eta_i\right). \quad (6.11)$$

In intrinsic semiconductors the density of free electrons equals that of free holes, $n = p$. This density is termed n_i, the intrinsic concentration. We find for intrinsic materials by equating Eq. 6.6 with 6.11 that

$$\left(\frac{m_e}{m_h}\right)^{3/2} \exp \eta^* = \exp\left(-\eta^* - \eta_i\right), \quad (6.12)$$

or the location of the Fermi level is given by

$$\eta^* = -\frac{\eta_i}{2} - \frac{3}{4}\ln\frac{m_e}{m_h}. \quad (6.13)$$

If $m_e = m_h$, then $\eta^* = -\eta_i/2$. For intrinsic semiconductors and for the case in which the effective mass of the electron equals that of the hole, the Fermi level lies in the center of the forbidden band, that is, halfway between the top of the valence band and the bottom of the conduction band.

Consider the product of the free electron concentration and the free hole concentration.

$$np = N_c N_v \exp\left(-\eta_i\right). \quad (6.14)$$

In intrinsic material, with $n = p = n_i$, we introduce the expression for η^*, 6.13, into that for $n = n_i$, Eq. 6.6,

$$n_i = N_c \exp\left(\frac{-\eta_i}{2} - \frac{3}{4}\ln\frac{m_e}{m_h}\right). \quad (6.15)$$

In terms of N_c and N_v, Eq. 6.15 is

$$n_i^2 = N_c N_v \exp(-\eta_i). \tag{6.16}$$

Upon comparing Eq. 6.14 with 6.16 we see that

$$np = n_i^2. \tag{6.17}$$

Since in deriving Eq. 6.14 we merely multiplied n with p, and did not specify that the material be intrinsic, we see that the product np is equal to the square of n_i in extrinsic material as well as intrinsic. This is a very useful relationship, pointing out that if the material has many donor centers, giving rise to a large free electron concentration, the free hole concentration is reduced, and vice versa. The intrinsic concentrations at room temperature of several semiconductors are listed in Table 6.3. Here cgs units are used in keeping with normal practice in semiconductor physics.

TABLE 6.3 Parameters of selected intrinsic semiconductors at room temperature

Material	n_i (cm^{-3})	μ_e (cm^2/volt sec)	μ_h (cm^2/volt sec)	τ (μsec)
Si	1.5×10^{10}	1700	350	1000
Ge	2.4×10^{13}	3800	1800	1000
PbS	2.9×10^{15}	800	500	100
PbSe	2×10^{17}	1200	600	2
PbTe	6×10^{16}	2000	800	30
InAs	2×10^{15}	40,000	600	5
InSb	1.8×10^{16}	60,000	600	0.2
Te	1×10^{16}	1700	500	30

6.6 FERMI-DIRAC STATISTICS FOR AN EXTRINSIC SEMICONDUCTOR

Up to this point, discussion has centered about intrinsic conditions. Attention must be given also to extrinsic conditions. Consider the case of an impurity semiconductor containing N_d donors per cm^3. The donors all lie at an energy $\mathscr{E}_d$ below the conduction band. The electron density n is still given by Eq. 6.6. For the sake of simplicity it is assumed that no acceptor levels are present and that the valence band is sufficiently far from the conduction band so that all the electrons come from the donor levels. The energy level diagram is shown in Fig. 6.10.

Figure 6.10. Energy level diagram for an *n*-type semiconductor.

The density of nonionized donors $(N_d)_e$, that is, the density of electrons bound to donor centers, is determined from the product of the occupational probability $f(\mathcal{E})$, evaluated at $\mathcal{E} = -\mathcal{E}_d$ and the density of states $g(\mathcal{E})$ evaluated at $\mathcal{E} = -\mathcal{E}_d$. The latter is N_d. Thus

$$(N_d)_e = N_d f(-\mathcal{E}_d) = \frac{N_d}{1 + \frac{1}{2}\exp(-\eta_d - \eta^*)}, \qquad (6.18)$$

where $\eta_d \equiv \mathcal{E}_d/kT$. The factor $\frac{1}{2}$ arises from the spin degeneracy.

Since the concentration of ionized donors $(N_d)_h$ plus the concentration of nonionized donors equals the total donor concentration, we have

$$(N_d)_h = N_d - (N_d)_e = \frac{N_d}{1 + 2\exp(\eta_d + \eta^*)}. \qquad (6.19)$$

But if the valence band is sufficiently far from the conduction band so that little intrinsic excitation takes place, the concentration of free electrons equals the concentration of ionized donors.

$$n = N_c \exp \eta^* = \frac{N_d}{1 + 2\exp(\eta_d + \eta^*)}. \qquad (6.20)$$

Eliminating η^* from the two expressions for n in Eq. 6.20 gives

$$n^2 = \frac{1}{2}(N_d - n)N_c \exp(-\eta_d). \qquad (6.21)$$

At sufficiently low temperatures $n \ll N_d$ and this simplifies to

$$n = \frac{1}{\sqrt{2}}(N_d N_c)^{1/2} \exp\left(\frac{-\eta_d}{2}\right). \qquad (6.22)$$

Thus the electron concentration under these conditions depends exponentially upon the donor ionization energy. In addition, the Fermi level lies midway between the donor level and the bottom of the conduction band.

We can derive completely analogous expressions for the classical, one acceptor level semiconductor. Defining the acceptor concentration to be N_a, located at an energy level $\mathcal{E}_a$ above the top of the valence band, or a

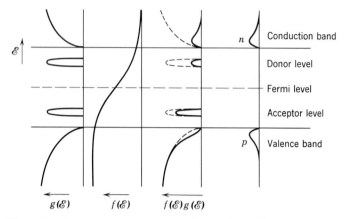

Figure 6.11. Dependence upon energy of $g(\mathscr{E})$, $f(\mathscr{E})$, $f(\mathscr{E})g(\mathscr{E})$, n, and p.

reduced energy η_a, we then obtain for the case in which $p \ll N_a$ an expression similar to Eq. 6.22.

$$p = \frac{1}{\sqrt{2}}(N_a N_v)^{1/2} \exp\left(\frac{-\eta_a - \eta_i}{2}\right). \tag{6.23}$$

In a p-type semiconductor at low temperatures, for which $p \ll N_a$, the Fermi level lies midway between the top of the valence band and the acceptor level. To complete the section on Fermi-Dirac statistics consider Fig. 6.11, which shows graphically the functions $f(\mathscr{E})$, $g(\mathscr{E})$, $f(\mathscr{E})g(\mathscr{E})$, and the free electron and hole concentrations, n and p.

6.7 ELECTRICAL CONDUCTIVITY AND THE HALL EFFECT

We have derived expressions describing the density of free electrons and holes in intrinsic and extrinsic semiconductors at thermal equilibrium. Our concern now is with the motion of these carriers in the presence of an electric field. Let us consider for the sake of simplicity that only electrons are present. These electrons perform random movements through the lattice similar to Brownian motion. Each will move in a straight line until it meets with some imperfection in the lattice or with a vibrating lattice atom. At this point it will move off in another direction until it again meets an imperfection or lattice vibration and has its path deflected. Motion of this type produces no net flow of electrons in any direction.

Consider a situation where the density of electrons at some point suddenly becomes much greater than the average density throughout the lattice. It will be shown that such a situation can be caused by injection at a rectifying contact or by photon absorption in a localized region. In this event the electrons will immediately move out from the point of concentration in an attempt to distribute themselves uniformly throughout the lattice. This phenomenon is termed diffusion.

Another form of motion takes place in the presence of an electric field. Consider an n-type semiconductor having an electric field impressed upon it. The electrons will move toward the positive terminal, that is, opposite to the direction of the field. This type of motion is termed drift. The electrons moving in the field will still collide with points in the lattice, being deflected from their initial paths, but the drift in the field superimposed upon the random motion will cause the electrons to have a net motion toward the positive terminal. (See Fig. 6.12.)

As has been seen, holes have a sign of electric charge opposite to electrons. Thus the net motion of holes will be in the direction of the electric field.

At low fields semiconductors obey Ohm's law. At high fields two phenomena, velocity saturation and avalanche breakdown, cause deviations from Ohm's law. In the former, the carriers reach a terminal velocity at high fields. Increasing the field strength no longer increases the velocity. In the latter, carriers having very high velocities colliding with lattice sites transfer sufficient energy to cause excitation of additional carriers. For the cases of interest to us, Ohm's law will be obeyed and we will not consider deviations.

Ohm's law states that the current density vector **J** is proportional to the electric field vector **E**, with the proportionality constant being the electrical conductivity σ.

$$\mathbf{J} = \sigma \mathbf{E}. \qquad (6.24)$$

In the one-dimensional case for an isotropic crystal the current density J_x in the x-direction is proportional to the field E_x in the x-direction.

$$J_x = \sigma E_x. \qquad (6.25)$$

Figure 6.12. Motion of electrons in an electric field.

However, we can also express the current density as the carrier density multiplied by the average velocity v_x and the electronic charge q. For an n-type semiconductor this is

$$J_x = nqv_x. \tag{6.26}$$

It can be shown that the velocity v_x is linearly related to the field E_x. The constant of proportionality is called the electron mobility μ_e.

$$v_x = \mu_e E_x. \tag{6.27}$$

Since μ_e is the ratio of velocity to field, the cgs dimensions of it are cm/sec per volt/cm or cm²/volt sec. The room temperature mobilities of several intrinsic semiconductors are listed in Table 6.3. Combining Eqs. 6.25, 6.26, and 6.27, we arrive at

$$\sigma = nq\mu_e. \tag{6.28}$$

We see that the conductivity, in units of (ohm cm)$^{-1}$, is equal to the product of electron concentration, electronic charge, and electron mobility. In a p-type semiconductor the related expression is

$$\sigma = pq\mu_h, \tag{6.29}$$

where μ_h is the hole mobility. In the general case, current is carried by both electrons and holes, resulting in

$$\sigma = nq\mu_e + pq\mu_h. \tag{6.30}$$

Equation 6.30 can also be written in terms of the mobility ratio b defined as $b \equiv \mu_e/\mu_h$, as

$$\sigma = q\mu_h(nb + p). \tag{6.31}$$

From Eqs. 6.24 and 6.30 we see that the electron current density vector $\mathbf{J}_{e,E}$ and hole current density vector $\mathbf{J}_{h,E}$ due to drift in the electric field $\mathbf{E}$ are given by

$$\mathbf{J}_{e,E} = nq\mu_e\mathbf{E}, \tag{6.32}$$

and

$$\mathbf{J}_{h,E} = pq\mu_h\mathbf{E}. \tag{6.33}$$

As seen in Eq. 6.27 the velocity is linearly related to the field. This seems surprising, for one might expect the rate of change of velocity, that is, the acceleration, to be proportional to the field, according to Newton's second law. Although complete explanation of this is beyond the scope of this book, an indication of it can be given.

We have seen that the electrons suffer collisions with the points in the lattice which limit the mean free path between collisions. Thus motion in the presence of a field is a series of starts and stops. During any small path between collisions the electron is accelerated. During the collision it is decelerated. As a consequence we can ascribe to it an average velocity, averaged over a long time including many collisions. This average velocity is the drift velocity.

The term applied to the process limiting the mobility is scattering. The question arises as to what the major sources of scattering are, that is, what phenomena limit the mobility. Electron collisions with other electrons can be shown to be of little importance. The two major scattering mechanisms are those of ionized impurity scattering and of lattice scattering, although neutral impurity and dislocation scattering also occur. The last two will not be treated here.

Consider first ionized impurity scattering. In a lattice containing impurity atoms a certain fraction, depending, as we have seen, on the temperature and energy level, will be ionized. For donor centers this means that they have lost their extra electron and have a net positive charge. For acceptors it means that they have acquired an extra electron which completes the formerly incomplete bond and thus have a net negative charge.

An electron wandering in the vicinity of an ionized impurity will be deflected by the charge attraction or repulsion as the case may be, and will thus have the component of its velocity in the direction of the field changed. The derivation of the expression for the mobility μ_I for ionized impurity scattering is due to Conwell and Weisskopf.[7] A more exact treatment is due to Brooks.[8] The latter expression is

$$\mu_I = \frac{2^{7/2} K_e^2 (kT)^{3/2}}{\pi^{3/2} m_e^{1/2} q^3 N_I} \frac{1}{\ln(1 + \delta) - \delta/(1 + \delta)}, \qquad (6.34)$$

where

$$\delta = \frac{6}{\pi} \frac{K_e m_e k^2 T^2}{n \hbar^2 q^2}.$$

In this expression K_e is the dielectric constant, $\hbar$ is Planck's constant divided by 2π, and N_I is the density of ionized impurities. We see that the mobility increases with the approximate $\frac{3}{2}$ power of the temperature, and is inversely proportional to the concentration of ionized impurities.

For lattice scattering, the electrons collide with atoms of the host lattice which are undergoing vibration at their lattice sites. The lattice vibrations, which are quantized, are termed phonons. The phonons also deflect the electrons. The derivation of the expression for the mobility μ_L limited by lattice scattering is due to Bardeen and Shockley,[9] among others.

$$\mu_L = \frac{(8\pi)^{1/2} \hbar^4 c_{ll}}{3 \varepsilon_{1n}^2 m_e^{5/2} (kT)^{3/2}}, \qquad (6.35)$$

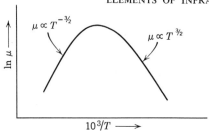

Figure 6.13. Temperature dependence of mobility.

where c_{ll} is the average longitudinal elastic constant and ε_{1n} is the shift of the edge of the conduction band per unit dilation. We note that the mobility for lattice scattering is inversely proportional to the $\frac{3}{2}$ power of the temperature. It can be shown[10] that the mobility in the presence of both scattering mechanisms is found by taking the sum of the reciprocals. Thus we have

$$\frac{1}{\mu} = \frac{1}{\mu_L} + \frac{1}{\mu_I}. \tag{6.36}$$

Examining this, together with the temperature dependence of μ_I and μ_L, we see that at low temperatures the mobility is limited by ionized impurities and at high temperatures by lattice scattering. The sum of these two terms results in a mobility having ideally a temperature dependence that at some intermediate temperature goes through a maximum. (See Fig. 6.13.)

In Fig. 6.13 $10^3/T$ is plotted on the abscissa and ln μ on the ordinate. Plotting the logarithm of a parameter versus $10^3/T$ is a common type of representation for such parameters as carrier concentrations, conductivity, and as we shall see, Hall coefficient.

Let us return now to the subject of electrical conductivity. In an n-type semiconductor at very low temperatures, very few free electrons are present, and they come from donor levels which are only partially ionized. At some higher temperature the donors will be completely ionized. As the temperature is raised above this point, the electron concentration remains substantially constant, since intrinsic excitation will be small compared to the electrons present from the donors. However, at some higher temperature the exponential increase in electrons due to intrinsic excitation will take over and the concentrations will again rise rapidly. This is depicted in Fig. 6.14.

Consider now the electrical resistivity ρ, the reciprocal of the conductivity. In n-type material ρ is given by

$$\rho = \frac{1}{nq\mu_e}, \tag{6.37}$$

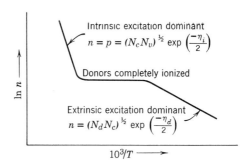

Figure 6.14. Temperature dependence of electron concentration.

whereas in p-type material it is

$$\rho = \frac{1}{pq\mu_h}. \tag{6.38}$$

We can obtain the temperature dependence of ρ by taking the inverse of the product of μ and n given in Figs. 6.13 and 6.14. This is shown in Fig. 6.15.

In almost all of the region shown, ρ decreases as T increases. This negative temperature dependence of resistivity distinguishes semiconductors from metals which have a positive temperature dependence of resistivity.

At sufficiently high temperatures in the intrinsic range, to the left of point A in Fig. 6.15, the resistivity is exponentially dependent upon the reciprocal of the absolute temperature. The carrier concentration n_i in this region, given by Eq. 6.16, is seen to be of the form

$$n_i \propto T^{3/2} \exp\left(-\mathscr{E}_i/2kT\right). \tag{6.39}$$

If the mobility is dominated by lattice scattering, it will be proportional to the reciprocal of the $\frac{3}{2}$ power of the temperature. The product $n_i\mu$ will be

$$n_i\mu \propto \exp\left(-\mathscr{E}_i/2kT\right). \tag{6.40}$$

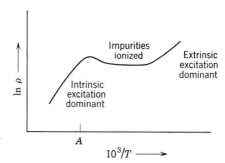

Figure 6.15. Temperature dependence of resistivity.

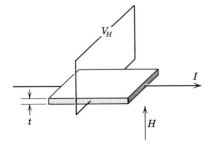

Figure 6.16. The Hall effect.

If dominated by ionized impurity scattering, the product $n_i\mu$ will be

$$n_i\mu \propto T^3 \exp\left(-\mathscr{E}_i/2kT\right). \qquad (6.41)$$

At sufficiently high temperatures the T^3 factor will disappear as the mobility goes over to being limited by lattice scattering. In any case, at sufficiently high temperatures the exponential term dominates over the T^3 term, and the product $n_i\mu$ becomes exponentially dependent upon temperature. Since this is related to the reciprocal of the resistivity, we see that at these temperatures the resistivity is exponentially dependent upon temperature according to

$$\rho \propto \exp\left(\mathscr{E}_i/2kT\right). \qquad (6.42)$$

A plot of $\ln \rho$ versus $1/T$ appears in this region as a straight line having a slope of $\mathscr{E}_i/2k$, shown to the left of point A in Fig. 6.15. Thus we see that a measurement of ρ as a function of T provides a means of determining the width of the forbidden band.

Let us examine next an experimental method to determine carrier concentration. Although measuring electrical resistivity will give us the product of mobility and concentration, we would like to know the value of each independently. The experiment which can be performed to determine carrier concentration utilizes the Hall effect, one of a group of "galvanomagnetic" phenomena, involving interactions of electric and magnetic fields with carriers. When the Hall effect is measured together with a resistivity measurement, the mobility can be determined.

Consider a slab of a semiconductor placed in a transverse magnetic field, Fig. 6.16. A longitudinal current I, consisting in general of both electrons and holes, is deflected by a magnetic field H in a direction normal to the plane of the current and field. Although the electrons and holes are charges of opposite sign, they are deflected in the same direction, since they are initially moving in opposite directions. This deflection causes a net charge to build up on one side of the semiconductor, giving rise to a transverse electric field, that is, an electric field normal to the plane of the current and magnetic field vectors. The voltage due to the transverse field is called the Hall voltage and can be measured by an external circuit. The magnitude of the Hall voltage is

$$V_H = \frac{R_H I H}{t} , \qquad (6.43)$$

where t is the sample thickness in the direction of the magnetic field and R_H is a proportionality constant called the Hall constant. In the general case in which both electrons and holes are present, it can be shown[11] that the Hall constant is given by

$$R_H = -\frac{3\pi}{8q} \frac{nb^2 - p}{(nb + p)^2} , \qquad (6.44)$$

where b has been defined to be the mobility ratio. If $nb^2 \gg p$, Eq. 6.44 reduces to

$$R_H = -\frac{3\pi}{8qn} . \qquad (6.45)$$

This is the Hall coefficient for n-type material. If $nb^2 \ll p$, Eq. 6.44 reduces to

$$R_H = \frac{3\pi}{8qp} , \qquad (6.46)$$

which is the expression for p-type material. The sign of the Hall coefficient is an indication of the material type, whereas the magnitude indicates the carrier concentration. Examination of Eq. 6.44 shows that R_H can be either negative or positive, depending upon the relative values of nb^2 and p. In a semiconductor having more donors than acceptors, R_H must always be negative, since b is never less than unity. The value of b depends greatly upon the semiconductor, being about two in germanium, and one hundred in the compound semiconductor indium antimonide. In general, b varies with both temperature and purity.

In an n-type or p-type semiconductor the value of the electron and hole mobilities can be determined from knowledge of the resistivity and the Hall coefficient. From Eqs. 6.37 and 6.45 we have

$$\mu_e = -\frac{8}{3\pi} \frac{R_H}{\rho} , \qquad (6.47)$$

the corresponding expression for μ_h being

$$\mu_h = \frac{8}{3\pi} \frac{R_H}{\rho} . \qquad (6.48)$$

Thus measurement of the Hall coefficient and resistivity enables us to determine both the carrier concentration and the mobility.

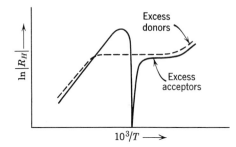

Figure 6.17. Hall coefficient as a function of temperature.

In an n-type semiconductor, in which R_H is always negative, the temperature dependence of R_H is similar to the temperature dependence of the inverse of n. (See Fig. 6.11.) Thus R_H appears as the dotted curve labeled "Excess donors" in Fig. 6.17.

On the other hand, if there is an excess of acceptors, then at very low temperatures R_H will be positive, since p will be greater than nb^2. However, as the temperature rises both n and p will increase due to intrinsic excitation. Since b is greater than unity, eventually the term nb^2 will predominate and the sign will change. At very high temperatures the Hall curve will have the form of that for excess donors. The curve for excess acceptors is so labeled in Fig. 6.17. Only the magnitude of R_H is plotted.

The general shape of the Hall curves can be understood in the following manner. Consider first the one labeled "Excess donors." A simple model of a semiconductor having these characteristics is one with a single donor level a slight distance below the bottom of the conduction band.

At absolute zero n and p are zero; therefore R_H is at negative infinity. As the temperature rises a few degrees, electrons are excited from donor centers into the conduction band. $|R_H|$ given by Eq. 6.45 decreases as n increases. Above some low temperature the donor centers are exhausted and as the temperature rises, R_H remains constant, shown as the flat portion of the curve. Finally, intrinsic excitation occurs and n increases. At the same time, p increases, but since nb^2 is always greater than p, the sign of R_H never changes.

Consider now the curve labeled "Excess acceptors." An example of this is a semiconductor having a single acceptor level just above the top of the valence band. At absolute zero no acceptors are ionized and R_H is infinite. As the temperature rises, some acceptors are ionized and R_H decreases according to Eq. 6.46, flattening out as before when exhaustion of the acceptors occurs. As intrinsic excitation becomes appreciable, the term nb^2 in Eq. 6.44 approaches p and the difference approaches zero. Above this temperature R_H becomes negative and $|R_H|$ increases. Finally the denominator of Eq. 6.44 predominates over the numerator, at some higher temperature, and $|R_H|$ decreases as predicted by Eq. 6.45.

6.8   RECOMBINATION AND LIFETIME

We have seen that at thermal equilibrium the concentrations of free electrons and holes are completely determined by the temperature, width of the forbidden band, and carrier masses according to Eqs. 6.6, 6.11, and 6.17. It is possible to upset this equilibrium and increase the concentrations by means of carrier injection. We shall see in Section 6.10 that this may be accomplished electrically by applying forward bias to a rectifying junction. It may be also accomplished optically by allowing photons of energy greater than the forbidden gap width in intrinsic semiconductors, or greater than the impurity ionization energy in extrinsic materials, to be absorbed by the semiconductor. In either case free carriers are produced over and above the concentrations obtained at thermal equilibrium. If the carrier injection is suddenly halted, the concentrations begin to decrease, ultimately reaching the value characteristic of the semiconductor in thermal equilibrium. This process is known as recombination.

At first glance we might assume that the recombination process is a direct action between free electrons and free holes. A free electron wandering through the lattice could happen upon an atom with an unfilled valence bond, that is, a hole, and drop into it. Since it must lose energy to do so, a photon of energy equal to that lost by the electron is given off. This method of recombination, termed direct radiative, is the inverse of the generation process in which a photon is annihilated and a free pair formed. Direct recombination is not the major recombination process usually found in semiconductors, although in certain materials, for example, indium antimonide, it does appear to be of importance. The radiation given off in the process is termed recombination radiation. The application of this process as a source of infrared energy has been mentioned in Chapter 3.

Another possible recombination mechanism, similar to direct, involves the emission of several quantized units of lattice energy rather than a single photon. The electron in the act of dropping into the valence band gives off its excess energy in a number of discrete bits of lattice vibrational energy. These bits of quantized vibrational energy are termed phonons. It is also possible that the energy release upon recombination of an electron and a hole can excite a third carrier to some higher energy level. This is termed Auger recombination, (pronounced "oh-zhā").

The major recombination mechanism in most semiconductors is an indirect process by way of certain lattice imperfections termed recombination centers. These centers have the ability to capture electrons or holes and hold them until their opposite number appears to complete the annihilation. In this process the minority carrier, that is, an electron in a p-type semiconductor, a hole in an n-type, is captured by the recombination center, which therefore loses its original neutrality and acquires an extra charge corresponding to the captured carrier. The majority carrier, drawn by the coulomb force to this center, recombines with the minority carrier and the center reverts to its original neutral state. This process may be repeated over and over, with carriers recombining in large numbers at the center. The energy level characterizing a recombination center usually lies deep within the forbidden band.

Another center of importance is known as a trapping center or trap. The trap has a high affinity for one type of carrier but not the other. Thus, for example, the minority carriers, happening upon the center, become trapped, or bound to the center. However, the ability to capture the majority carrier, expressed mathematically in terms of the capture cross section, is low. Therefore, it is likely that the trapped minority carrier will be thermally re-excited before the majority carrier is captured. During the time the carrier is in the trap it is not available for conduction, thus does not contribute to the electrical conductivity. The energy levels associated with traps lie within the forbidden band, sufficiently far from the band edge so that they can trap carriers and hold them without their being immediately released, but close enough so that the probability of release is greater than that of capture of a carrier of opposite sign.

The average length of time spent by carriers while in the free state is termed the lifetime. Two forms of lifetime may be differentiated. In the diffusion process, which we shall see is basic to phenomena taking place at potential barriers, the minority carrier lifetime is of importance. Since the existence of minority carriers is necessary to devices based upon diffusion, recombination of injected minority carriers is the factor limiting the performance of the device. On the other hand, in conductivity phenomena, the lifetime of importance is the time during which either carrier is free. Since both contribute to the current, if one carrier recombines, the other one remaining free still contributes to the conductivity. This lifetime is the one of importance in devices such as intrinsic photoconductors, which we shall see depend upon a radiation-induced change in conductivity of the semiconductor.

Let us consider now the expression relating the excess carrier concentration to the lifetime. Suppose we irradiate a crystal with photons of wavelength sufficiently short to cause excitation. The carrier density, which in the absence of radiation is given by n_0, increases by an amount Δn_0, termed the excess carrier concentration. If we remove the radiation, the excess carrier concentration begins to decrease due to recombination. At first there will be many sites at which the carriers can recombine, so the instantaneous recombination rate will be high. As these recombination centers fill up, fewer and fewer will remain, and it will become harder for the remaining carriers to find unoccupied sites. Thus the recombination rate will decrease with time. We find the concentration decays in an exponential manner given by

$$n - n_0 = \Delta n_0 \exp\left(-t/\tau\right), \qquad (6.49)$$

where n is the instantaneous value of the concentration and τ is the lifetime, the time during which the concentration drops to $1/e$ of its original value.

The equilibrium concentration of thermally excited carriers is maintained by a process in which carriers are being continuously generated and are continuously recombining. If equilibrium is disturbed, the rate of change of carrier concentration is determined by taking the derivative of Eq. 6.49.

$$\frac{dn}{dt} = -\frac{\Delta n_0}{\tau} \exp\left(-t/\tau\right). \qquad (6.50)$$

Upon introducing Eq. 6.49 into 6.50 we find

$$\frac{dn}{dt} = \frac{n_0}{\tau} - \frac{n}{\tau}. \qquad (6.51)$$

The first term on the right of Eq. 6.51 is the rate of generation; the second is the rate of recombination. At equilibrium $dn/dt = 0$, generation rate equals recombination rate, and $n = n_0$.

Equation 6.51 is the expression for an indirect recombination process. The corresponding expression for direct recombination can be shown to be

$$\frac{dn}{dt} = \frac{-(np - n_0 p_0)}{\tau(n_0 + p_0)}. \qquad (6.52)$$

The recombination processes which we have considered are characteristic of the bulk semiconductor, that is, processes occurring deep within the material, far from the surface. The bulk lifetimes of several intrinsic semiconductors at room temperature are listed in Table 6.3. Surface effects also exist which limit the lifetime. The surface of a crystal is often a region of gross crystalline defects, formed in the act of preparing the crystal. The action of abrasive powders used to polish the surface (termed "lapping") deforms the surface and produces large numbers of recombination centers. The lifetime of carriers which recombine at a crystal surface is characterized in terms of a "surface recombination velocity," describing how rapidly carriers diffusing from the bulk to the surface recombine.

The surface recombination velocity s is defined as the recombination rate per unit area per unit excess carrier concentration just below the surface. Surface recombination may also be expressed in terms of a lifetime τ_s for surface recombination. The overall sample lifetime, including the bulk lifetime τ_v and surface lifetime τ_s, can be shown to be[12]

$$\frac{1}{\tau} = \frac{1}{\tau_s} + \frac{1}{\tau_v}. \tag{6.53}$$

Shockley[13] has shown that for long rods of rectangular cross section having transverse dimensions $2B$ and $2C$ the value for τ_s in the extremes of very low and very high recombination velocities is given by

$$\text{For } s \to 0, \qquad \frac{1}{\tau_s} = s(1/B + 1/C), \tag{6.54}$$

$$\text{For } s \to \infty, \qquad \frac{1}{\tau_s} = \frac{\pi^2 D}{4}(1/B^2 + 1/C^2), \tag{6.55}$$

where D is the diffusion constant of the minority carrier discussed in the next section. In the first case volume effects dominate the lifetime. In the second case surface effects are dominant.

It is important to the action of many types of semiconductor devices that the lifetime be long. Thus it is desirable that the surface recombination velocity be low. The factors responsible for recombination at the surface are not clearly understood. Certain etches, for example, will produce surfaces with a low value of recombination velocity, whereas others will produce surfaces with a high value. The study of surface effects is of great importance to semiconductor device technology.

6.9 MOTION OF CARRIERS IN THE PRESENCE OF ELECTRIC AND MAGNETIC FIELDS

We are concerned in this section with developing the equations of motion of carriers in isotropic crystals in electric and magnetic fields. By isotropic is meant that the various semiconductor properties, such as electrical conductivity, are independent of direction in the crystal. These equations form the foundation upon which the theories of the photoconductive effect, the photoelectromagnetic effect, and the photovoltaic effect rest. We shall see in Chapter 8 that these effects are the ones most widely used in infrared photon detectors. The theories of these effects are developed in detail in Chapter 9.

In Section 6.7 two types of carrier motion were considered, diffusion and drift. The drift of carriers in an electric field, holes moving in the

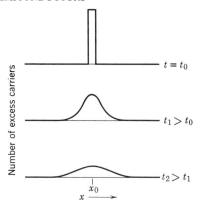

Figure 6.18. Carrier diffusion.

direction of the field and electrons against it, was seen to be the electric current carried by a semiconductor under the influence of an electrical bias. Diffusion will now be discussed in more detail. Consider a uniform, homogeneous, isotropic semiconductor at thermal equilibrium with its surroundings, the distribution in energy of thermally excited carriers being described by the Fermi statistics. Imagine a plane separating the material into two identical regions. The random motion of the carriers will cause some to cross the plane from, say, left to right. At the same time others will cross from right to left. Since the two motions are opposite in direction but equal in magnitude, there will be no average net flow in either direction. Therefore, no net electrical currents exist.

Suppose the concentration at some point in the crystal is now increased suddenly, say by means of photon excitation in some well-defined region. The excess carriers generated by the radiation will tend to move away from their common center just as additional air molecules introduced into a room near one corner would move away from the corner to fill the room. This process, known as diffusion, is illustrated in Fig. 6.18.

As the carriers move away from their common center, the distribution of the carriers throughout the material will approach uniformity. As the distribution approaches uniformity, the rate of approach to uniformity will diminish. During the diffusion process the density of excess carriers will also be diminishing by means of recombination. Thus we have a picture of excess carriers simultaneously moving away from the point of excitation and recombining in an exponential manner. We shall see that just as recombination may be described in terms of a characteristic time, the lifetime, so the diffusion process may be described in terms of a characteristic length, the diffusion length, which is a measure of the average distance a carrier will travel before recombining.

In order that a more quantitative expression of the diffusion process may be developed, a simple picture of the process shall be examined. Imagine a section of crystal of unit cross-sectional area. Suppose that excess carriers are generated within the section under consideration. Assume that all diffuse a mean distance l in the x-direction in a mean time $\bar{t}$ which is short compared to the lifetime. Therefore recombination shall be assumed to be negligible.

The initial distribution of excess carriers will be modified by diffusion. Let the density of excess carriers at a point $x = 0$ in the center of the section be n. At a distance l to the right of this, that is, in the $+x$-direction, the density will differ from n. The difference may be expressed in terms of the gradient of the excess charge density. At this point the density $n(l)$ will be

$$n(l) = n + \frac{dn}{dx}\, l, \tag{6.56}$$

and the average excess density $\bar{n}$ in the slice will be

$$\bar{n} = n + \frac{1}{2}\frac{dn}{dx}\, l. \tag{6.57}$$

During the time $\bar{t}$ one-half of the carriers will move to the left and one half to the right. The average number $\bar{N}_1$ per unit area in the region to the right of $x = 0$ and moving to the left, crossing the surface at $x = 0$ during the time $\bar{t}$, will be one-half those in a volume of a unit cross-sectional area and length l.

$$\bar{N}_1 = \tfrac{1}{2}l\left(n + \frac{1}{2}\frac{dn}{dx}\, l\right). \tag{6.58}$$

The number in the region to the left of $x = 0$ and moving to the right will be $\bar{N}_2$, where

$$\bar{N}_2 = \tfrac{1}{2}l\left(n - \frac{1}{2}\frac{dn}{dx}\, l\right). \tag{6.59}$$

The net flow across the surface, given by $\bar{N}_1 - \bar{N}_2$, will be

$$\bar{N}_1 - \bar{N}_2 = \frac{1}{2}\frac{dn}{dx}\, l^2. \tag{6.60}$$

The net flow per unit time will be

$$\frac{\bar{N}_1 - \bar{N}_2}{\bar{t}} = \frac{1}{2}\frac{l^2}{\bar{t}}\frac{dn}{dx}. \tag{6.61}$$

The net flow of current per unit area, that is, the current density, will be

$$J = q\left(\frac{1}{2}\frac{l^2}{\bar{t}}\right)\frac{dn}{dx}. \tag{6.62}$$

Examining Eq. 6.62 we see that the net current flow of excess carriers is proportional to the gradient of the excess charge density. The proportionality is normally expressed in terms of the diffusion constant D, where

$$D \equiv \frac{1}{2} \frac{\bar{l}^2}{\bar{t}}. \tag{6.63}$$

Since the charges move in the direction of the gradient, we describe both the electron current density vector $\mathbf{J}_{e,D}$ and the hole current density vector $\mathbf{J}_{h,D}$ as

$$\mathbf{J}_{e,D} = qD_e\nabla n, \tag{6.64}$$

$$\mathbf{J}_{h,D} = -qD_h\nabla p, \tag{6.65}$$

where D_e is the diffusion constant for electrons,
D_h is that for holes,
∇n and ∇p are the gradients of the excess charge densities.

Since the diffusion of the carriers is away from the direction in which they increase, the hole current is negative. Because the electrons have a negative charge, the product of the negative sign due to direction of diffusion and that due to charge makes the electron current positive.

Consider now the equations describing the flow of current in the presence of both diffusion and drift. The diffusion currents are described by Eqs. 6.64 and 6.65. The drift currents are described by Eqs. 6.32 and 6.33. Thus the electron current density vector $\mathbf{J}_e$ and hole current density vector $\mathbf{J}_h$ for both diffusion and drift are given by

$$\mathbf{J}_e = \mathbf{J}_{e,E} + \mathbf{J}_{e,D} = nq\mu_e\mathbf{E} + qD_e\nabla n, \tag{6.66}$$

$$\mathbf{J}_h = \mathbf{J}_{h,E} + \mathbf{J}_{h,D} = pq\mu_h\mathbf{E} - qD_h\nabla p. \tag{6.67}$$

The holes are positive charges moving in the direction of the electric field, whereas the electrons are negative charges moving against the field. Thus both drift currents are positive.

Consider now the condition of thermal equilibrium in which the electric field $\mathbf{E}$ arises not due to an external voltage source but because of some potential barrier within the material, such as at a p-n junction. The subject of junctions is more thoroughly discussed in the next section, but at present it is necessary only to know that internal electric fields may arise. The principle of detailed balancing, which arises from thermodynamic considerations,[14] requires that for every electron crossing an area in one direction per unit time another with the same energy must cross in the opposite direction. Applying this to hole current, for example, this states that under these conditions $\mathbf{J}_h$ equals zero. Thus

$$pq\mu_h\mathbf{E} - qD_h\nabla p = 0. \tag{6.68}$$

But p is determined by Eq. 6.11. Taking the gradient of p and inserting it into Eq. 6.68, we find

$$pq\mu_h\mathbf{E} - q D_h p\left[\frac{1}{kT}\mathbf{\nabla}(-\mathscr{E}^* - \mathscr{E}_i)\right] = 0. \qquad (6.69)$$

However, $1/q\,\mathbf{\nabla}(\mathscr{E}^* + \mathscr{E}_i)$ is the gradient of the internal potential difference, that is, the negative of the electric field $\mathbf{E}$, which we are considering. Introducing this into Eq. 6.69 gives us

$$q p \mathbf{E}\left(\mu_h - \frac{q}{kT}D_h\right) = 0. \qquad (6.70)$$

This equality is true only if the term in the parentheses is zero. Thus

$$D_h = \frac{kT\mu_h}{q}. \qquad (6.71)$$

A similar derivation for electrons results in

$$D_e = \frac{kT\mu_e}{q}. \qquad (6.72)$$

This relationship involving the diffusion constant and the mobility is known as the Einstein relationship.

We shall now have recourse to the equation of continuity, which describes the charge flow, or current, in a semiconductor. This same equation occurs in many fields, for example, heat transfer and hydrodynamics. It may be described as a relationship involving the rate of change of material, in this case electric charge, within a closed region. The rate of increase of charge density in the region is due to the difference between the net rate of generation g and the net rate of recombination r in the volume, minus the net loss through flow across the boundary. In terms of hole flow this is expressed as

$$\frac{\partial p}{\partial t} = (g - r) - \frac{1}{q}\mathbf{\nabla}\cdot\mathbf{J}_h, \qquad (6.73)$$

and for electrons it is

$$\frac{\partial n}{\partial t} = (g - r) + \frac{1}{q}\mathbf{\nabla}\cdot\mathbf{J}_e. \qquad (6.74)$$

Consider the equation of continuity for holes, Eq. 6.73. Imagine that holes are flowing from a region in which they are generated to one in which they recombine at a rate expressed as p/τ_h where τ_h is the hole lifetime. In the steady state $\partial p/\partial t$ is zero and the generation rate in the region of recombination is zero.

Combining the equation of continuity with that for diffusion, substituting Eq. 6.65 into 6.73, we have

$$r = \frac{p}{\tau_h} = D_h \nabla \cdot \nabla p = D_h \nabla^2 p.$$ (6.75)

The solution of Eq. 6.75 in one dimension is of the form

$$p = p_0 \exp\left(-x/L_h\right).$$ (6.76)

Substituting Eq. 6.76 into 6.75 we find

$$L_h^2 = D_h \tau_h = \frac{kT\mu_h\tau_h}{q}.$$ (6.77)

The quantity L_h is known as the diffusion length for holes. It is a characteristic distance for the diffusion process which describes how far holes will diffuse before their density is reduced to $1/e$ of its initial value. Similarly, for electrons

$$L_e^2 = D_e \tau_e = \frac{kT\mu_e\tau_e}{q}.$$ (6.78)

We see that semiconductors in which the carrier mobilities are high and the lifetimes long have large values of diffusion length. This parameter is one of great importance in semiconductor devices which rely upon carrier injection in one form or another. The carriers will diffuse on the average only the diffusion length, and will therefore influence their surroundings only over this length.

The final topic in this section concerns the motion of carriers in the presence of magnetic fields. An electric charge crossing lines of magnetic flux is deflected in a direction normal to the plane of the velocity vector and the field vector. The direction along this normal is determined by the sign of the electric charge. The contribution to the total electron current density vector due to this Lorentz force can be shown to be described by an expression of the form

$$\mathbf{J}_{e,B} = -\mu_e \mathbf{J}_e \times \mathbf{B},$$ (6.79)

whereas the contribution to the hole current density vector is

$$\mathbf{J}_{h,B} = \mu_h \mathbf{J}_h \times \mathbf{B},$$ (6.80)

where $\mathbf{B}$ is the magnetic induction vector.

Combining these expressions with the ones for diffusion and drift, Eqs. 6.66 and 6.67, we arrive at the expression for the electron and hole current density vectors describing diffusion, drift, and the deflection of carriers by a magnetic field.

$$\mathbf{J}_e = \mathbf{J}_{e,E} + \mathbf{J}_{e,D} + \mathbf{J}_{e,B} = nq\mu_e\mathbf{E} + qD_e\nabla n - \mu_e\mathbf{J}_e \times \mathbf{B}.$$ (6.81)

$$\mathbf{J}_h = \mathbf{J}_{h,E} + \mathbf{J}_{h,D} + \mathbf{J}_{h,B} = pq\mu_h\mathbf{E} - qD_h\nabla p + \mu_h\mathbf{J}_h \times \mathbf{B}.$$ (6.82)

6.10   RECTIFYING JUNCTIONS

Having developed the subjects of diffusion and drift, we can now consider action taking place at potential barriers in semiconductors. Such action forms the basis of many semiconductor effects including rectification, transistor action, and photovoltaic action. We shall develop here the theory of rectification and in Chapter 9 the theory of the photovoltaic effect. Readers interested in transistor action are referred to Shockley.[15]

Rectification in solids is caused by the presence of potential barriers, arising either within the solid or at the surface of the solid. Applying an electric field in one direction causes a large current to flow. Applying it in the opposite causes a small current to flow. A large number of carriers are available for flow in one direction; a small number are available for flow in the other. The current-voltage representation typical of a rectifier is shown in Fig. 6.19.

The region lying to the right of the current axis, characterized by high current-low voltage, is known as the forward region. The bias here is said to be applied in the forward direction. The region lying to the left, characterized by low current-high voltage, is known as the reverse region. The bias here is said to be applied in the reverse direction.

Rectification in solids can take place at a contact between a metal and a semiconductor, or at an internal semiconductor junction between an n-type region and a p-type region, known as a p-n junction. Metal semiconductor contacts are used in point contact diodes, but are not widely used as infrared detectors. However, the action of radiation on p-n junctions in semiconductors give rise to the widely used photovoltaic effect at p-n junctions. To lay the basis for the study of the photovoltaic effect in Chapter 9, consider the theory of rectification at p-n junctions.

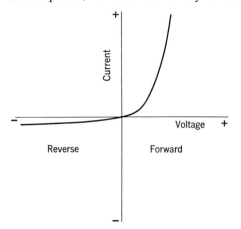

Figure 6.19. Rectification characteristic.

Several methods exist for the preparation of *p-n* junctions. The grown junction is prepared by doping the crystal melt with one type of impurity, say a donor impurity, and pulling a crystal from the melt. After the crystal is partly pulled, an excess of the other type, in this case, an acceptor impurity, is added. The impurity type changes at the point in the crystal at which the second impurity was added. If a slice of the crystal containing the junction is cut, it will be *n*-type on one side, *p*-type on the other. (See Fig. 6.20.) It should be noted that the second impurity must be added in sufficient quantity to more than compensate for the first impurity.

Other preparation methods include alloying, rate growth, and diffusion techniques. The alloy junction is prepared by fusing a low melting point metal, for example, indium, on the surface of an already doped crystal, for example, *n*-type germanium. The indium when melted dissolves some germanium, producing a thin *p*-type layer in contact with the *n*-type bulk crystal. The rate growth method utilizes the fact that the solubility of certain doping agents in semiconductors is a function of the crystal growth rate. By varying the pulling speed during growth, the amount of dissolved impurity of either type can be varied, forming a junction at some point. In the diffusion process a piece of semiconductor of one type is placed in a sealed evacuated container together with a piece of a metal which will produce substitutional impurities of the opposite type. The container is heated to a temperature sufficient to give rise to a large value of the vapor pressure of the metal. The metallic vapors diffuse into the semiconductor, forming a "skin" of material opposite in type to that of the interior. The junction is revealed by slicing into the crystal.

Let us now consider the electrical properties of *p-n* junctions. Imagine that the junction is formed by placing into intimate contact a piece of *n*-type material with a piece of *p*-type material. At the instant before contact the energy bands of each will be as shown in Fig. 6.21*a*.

There are more electrons in the upper energy states of the conduction band in the *n*-type material than in the *p*-type and more holes in the lower energy states of the valence band in the *p*-type material than in the *n*-type. When the pieces are joined, some of the electrons from the *n*-type material will diffuse into the *p*-type material, settling in empty acceptor levels immediately adjacent to the junction. The holes will behave

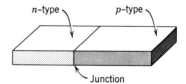

Figure 6.20. A *p-n* junction.

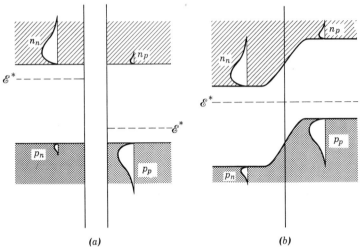

(a)   (b)

Figure 6.21. Energy bands at a *p-n* junction. (*a*) Before contact. (*b*) After contact.

similarly, moving from the *p*-type material into the *n*-type. There they recombine with electrons, causing the donors adjacent to the junction to be completely ionized. Since both pieces were electrically neutral originally, each has now acquired a net charge, the *n*-type being charged positively and the *p*-type negatively. The existence of this charge dipole layer gives rise to an electric field directed across the junction, confined principally to the very narrow region on either side of the junction in which the space charge is to be found.

The action of the charge dipole layer is to depress the conduction band on the *n*-type side and raise the valence band on the *p*-type side, forming a potential barrier which impedes the flow of electrons in the conduction band and holes in the valence band. Only those electrons having energies sufficiently great to occupy levels above the top of the barrier may cross it. (Not considered here are the very narrow junctions employed in the tunnel diode, in which an electron can penetrate the barrier by a quantum mechanical process known as "tunneling.") It can be shown[16] that the action of the dipole layer is just sufficient to cause the Fermi levels to become aligned. (See Fig. 6.21*b*.) In a semiconductor in thermal equilibrium with its surroundings the Fermi levels throughout the material are aligned.

Consider now the action of an external electric field upon the semiconductor. Figure 6.22*a* illustrates the *p-n* junction in thermal equilibrium. When a voltage is applied to the semiconductor, a portion of it, in many cases, almost all of it, will appear across the potential barrier. If the direction of the applied field is such that positive polarity is applied to the

p-type material, the Fermi level will be displaced in the direction shown in Fig. 6.22b.

Examination of Fig. 6.22b shows that the potential barrier hindering the flow of current across the junction has been reduced, allowing carriers to diffuse across the junction. This process, known as carrier injection, is basic to the operation of a number of semiconductor devices. Since lowering of the barrier by forward bias causes electrons to flow into the p-type region and holes to flow into the n-type, we see that it is the minority carriers which are injected. In order to preserve space charge neutrality, that is, so that the p-type region does not become charged negatively due to the added electrons, and the n-type does not become charged positively due to the added holes, majority carriers flow into the semiconductor from the electrodes to which the bias voltage is applied. It is assumed here that these electrodes are ohmic, that is, nonrectifying. Thus the action of applying a forward bias, causing minority carrier injection, increases both the majority and minority carrier concentrations. When this happens, the expressions developed for the free carrier concentrations at thermal equilibrium, Eqs. 6.6 and 6.11, no longer hold. Since there are now more free carriers present, the current flow due to the applied voltage is relatively high. This is the region of forward bias of the diode characteristic, a region of high current and low voltage.

If now the bias is reversed, positive polarity being applied to the n-type material and negative to the p-type, the potential barrier will be raised. This impedes the flow of current across the barrier since there are fewer carriers now above the top of the barrier than before the barrier was raised. Thus the current flow due to the applied voltage is relatively low. This is the region of reverse bias of the diode characteristic, a region of low current and high voltage.

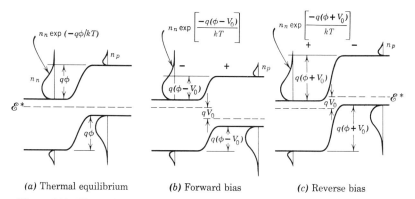

(a) Thermal equilibrium   (b) Forward bias   (c) Reverse bias

Figure 6.22. Illustrating the effects of forward and reverse bias upon the junction.

Having thus examined qualitatively the electrical behavior, known as the diode characteristic, of the *p-n* junction, we now consider the quantitative expression of it. We have seen that for the classical semiconductor the Fermi function may be approximated by the Maxwell-Boltzmann expression, Eq. 6.2.

$$f(\mathscr{E}) \approx \exp -\left(\frac{\mathscr{E} - \mathscr{E}^*}{kT}\right). \tag{6.2}$$

Remember that $\mathscr{E}$ is the energy at some level measured with respect to the bottom of the conduction band in the region of the material being considered. Since $\mathscr{E} - \mathscr{E}^*$ is greater on the *p*-side than on the *n*-side, the probability of occupation of a level at energy $\mathscr{E}$ on the *n*-side is greater than on the *p*-side. In thermal equilibrium the density of electrons on the *n*-side having energies above the barrier ϕ is approximately exponentially dependent on ϕ and approximately equal to the density of electrons in the *p*-region. (See Fig. 6.22a.) The latter is not strictly true but is a good approximation. Thus we have

$$n_p = n_n \exp\left(-q\phi/kT\right), \tag{6.83}$$

where n_p is the concentration of electrons in the *p*-region, and n_n is the concentration in the *n*-region. If we now apply a bias $+V_0$ in the forward direction, the barrier is reduced and the density of electrons Z above the new barrier of height $q(\phi - V_0)$, becomes

$$Z = n_n \exp\left[-\frac{q(\phi - V_0)}{kT}\right]. \tag{6.84}$$

With forward bias applied a greater number of electrons now exists on the *n*-side above the barrier than on the *p*-side. The excess will diffuse across the barrier into the *p*-side for a distance equal to the minority carrier (electron) diffusion length in the *p*-region. Remembering the diffusion current density to be related to the gradient of the excess charge density, we express the magnitude of the current density as

$$J_e = q D_e \frac{Z - n_p}{L_e}. \tag{6.85}$$

Combining Eqs. 6.83, 6.84, and 6.85, we have

$$J_e = \frac{q D_e n_p}{L_e} \left[\exp\left(qV_0/kT\right) - 1\right]. \tag{6.86}$$

In a similar manner the hole current density is given by

$$J_h = \frac{q D_h p_n}{L_h} \left[\exp\left(qV_0/kT\right) - 1\right], \tag{6.87}$$

where p_n is the concentration of holes in the *n*-type region.

The total current across the junction is

$$J = J_e + J_h = q\left(\frac{D_h p_n}{L_h} + \frac{D_e n_p}{L_e}\right) [\exp(qV_0/kT) - 1]. \quad (6.88)$$

Equation 6.88 is the diode equation for a p-n junction. It can be seen that the current rises exponentially with voltage in the forward direction when $qV_0/kT > 1$. In the reverse direction V_0 is negative and the current saturates to a value J_{sat} given by

$$J_{sat} = q\left(\frac{D_h p_n}{L_h} + \frac{D_e n_p}{L_e}\right). \quad (6.89)$$

This is shown in Fig. 6.23.

Expressing the diffusion lengths in terms of the diffusion constants according to Eqs. 6.77 and 6.78, we can rewrite Eq. 6.88 as

$$J = q\left\{p_n\left[\frac{D_h}{\tau_h}\right]^{1/2} + n_p\left[\frac{D_e}{\tau_e}\right]^{1/2}\right\}\{\exp[qV_0/kT] - 1\}. \quad (6.90)$$

We see from examination of Eq. 6.90 that the current is dependent upon the minority carrier lifetimes τ_h and τ_e, the minority carrier concentrations p_n and n_p, and the diffusion constants D_e and D_h. To have small reverse currents, that is, small values of J_{sat}, which is required for a good diode characteristic, it is necessary that the lifetimes be long, the minority carrier concentrations be low, and the diffusion constants be small. Since the product of the minority carrier concentration with the majority carrier concentration in the same region is a constant, low minority concentrations are attained by having high majority concentrations. In addition, since the diffusion constants are proportional to the square roots of the mobilities, the mobilities should be low.

We should note that we have assumed a small applied voltage. If the applied voltage is large, the density of electrons above the barrier does not

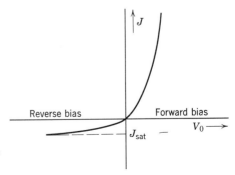

Figure 6.23. The diode characteristic of a p-n junction.

undergo the exponential change expressed in Eq. 6.84. In addition, the carrier movement will be a combination of diffusion and drift at high fields. We have also assumed implicitly that the junction has zero width, that is to say, we have assumed that no carriers recombine within the junction. If, on the other hand, the potential barrier is wide, the potential changing only slowly with distance, the carriers may be generated and recombine within the barrier. In the limit of this case the carriers "see" no barrier and thus rectification does not exist.

REFERENCES

1. See, for example; W. Shockley, *Electrons and Holes in Semiconductors*, D. van Nostrand Co., Princeton, New Jersey (1950); C. Kittel, *Introduction to Solid State Physics*, John Wiley and Sons, New York (1953); A. Dekker, *Solid State Physics*, Prentice-Hall, Englewood Cliffs, New Jersey (1957); A. van der Ziel, *Solid State Physical Electronics*, Prentice-Hall, Englewood Cliffs, New Jersey (1957); R. A. Smith, *Semiconductors*, University Press, Cambridge (1959).
2. See, for example, J. N. Shive, *The Properties, Physics, and Design of Semiconductor Devices*, D. van Nostrand Co., Princeton (1959), p. 458.
3. C. Kittel, *op. cit.*, Chap. 1.
4. M. L. Schultz and G. A. Morton, *Proc. Inst. Radio Engrs.* **43**, 1819 (1955).
5. R. H. Bube, *Photoconductivity of Solids*, John Wiley and Sons, New York (1960), p. 136.
6. R. A. Smith, *op. cit.*, Chap. 4.
7. E. Conwell and V. F. Weisskopf, *Phys. Rev.* **69**, 258A (1946); **77**, 388 (1950).
8. H. Brooks, *Phys. Rev.* **83**, 879 (1951).
9. J. Bardeen and W. Shockley, *Phys. Rev.* **80**, 72 (1950).
10. W. Shockley, *op. cit.*, p. 287.
11. *Ibid.*, p. 217.
12. C. Kittel, *op. cit.*, p. 367.
13. W. Shockley, *op. cit.*, p. 323.
14. See, for example; R. C. Tolman, *The Principles of Statistical Mechanics*, Oxford University Press, London (1938), p. 165.
15. W. Shockley, *op. cit.*, Chap. 4.
16. C. Kittel, *op. cit.*, p. 590.

7
Sources of noise

7.1 INTRODUCTION

The ultimate detection capability of any detection system is limited by fluctuations of a random nature. These fluctuations, termed "noise," have been studied in considerable detail in the last quarter century. An understanding of such fluctuation phenomena is basic to any attempt to realize the ultimate performance of a detection system. This chapter discusses the various mechanisms giving rise to noise, and develops the relationships describing the amplitude and frequency spectrum of the noise. The concepts developed here will be used in Chapters 9 and 10 to describe the noise limitations of infrared detectors.

The noise appearing in an infrared system may arise from a variety of sources, which may be placed in three categories, noise in detectors, in amplifiers, and from the radiating background. As seen in Chapter 8 most detectors utilize semiconductor effects to transduce the radiant signal into an electrical signal. Fluctuations in the concentration and motion of the current carriers in the semiconductor give rise to fluctuations in the electrical output of the detector. The ability to detect the presence or absence of a signal rests upon the ability to discriminate against the noise. Amplifiers may also make use of semiconductor devices, or they may use electron tubes which, too, are sources of noise. The system designer must endeavor by careful circuit design and component selection to reduce the noise generated in the amplifier to a level below that of the detector.

The third noise phenomenon, background or photon noise, is due to the quasi-random arrival of photons from the surroundings of the detector. It is of interest because it is the factor which ultimately limits the performance of infrared detectors. Two aspects of it need to be considered: generation by a radiating source and detection by an infrared detector.

235

The derivation of the spectral content of photon noise emitted by a radiating body appears in Section 2.5. The application of this to a determination of the ultimate performance of an infrared detector depends upon knowledge of the characteristics of the detector. Derivation of the photon noise limited performance of detectors is in Section 9.4.

A secondary noise phenomenon which we will not consider here is that termed scanning noise, which arises from nonuniform radiance of a background being scanned or the nonuniform sensitivity of the surface of a detector over which a small image is moved.

7.2 THERMAL NOISE—NYQUIST THEOREM

There is a form of electrical noise due to the random motions of the charge carriers known as thermal noise or Johnson noise.[1] It appears in any resistive material, even a pure metal. Other forms of noise, loosely grouped under the term excess noise, are found in certain types of resistive materials. Granular carbon resistors, for example, exhibit a form of excess noise called current noise. The current noise level may be sufficiently high at low frequencies to mask the thermal noise. On the other hand, wire-wound resistors do not exhibit excess noise, only thermal noise. The measured thermal noise power from a given resistor can be changed by two means only, namely, by reducing the resistor temperature and by reducing the measurement bandwidth. The value of the resistance determines the magnitude of the noise voltage, but not of the power.

Our understanding of the spectral distribution of thermal noise is due to Nyquist.[2] Although other derivations exist,[3] we shall follow the original one of Nyquist. Consider two resistors R_1 and R_2 connected together, both at a temperature T. They are ideal, that is, noninductive and noncapacitive. Each develops a noise power and transfers part of it to the other. The second law of thermodynamics requires that on the average one cannot acquire energy at the expense of the other, since they are at the same temperature. Thus at equilibrium each receives as much energy as it gives. If now an ideal filter of bandwidth Δf is interposed between the resistors, the average net acquisition of energy of each must still be zero, again according to the second law. Since Δf is arbitrary, the noise power in any bandwidth is independent of the resistance value. Let the frequency dependence of the open circuit noise voltages be denoted as $v_{N_1}(f)$ and $v_{N_2}(f)$. The mean square noise voltages in the bandwidth Δf are $\overline{[v_{N_1}(f)\,\Delta f]^2}$ and $\overline{[v_{N_2}(f)\,\Delta f]^2}$. The noise power P_N dissipated in R_2 by the emf in R_1 is given by

$$P_{N_2} = \overline{[v_{N_1}(f)\,\Delta f]^2}\,\frac{R_2}{(R_1 + R_2)^2},\qquad(7.1)$$

whereas the power dissipated in R_1 by the emf in R_2 is

$$P_{N_1} = \overline{[v_{N_2}(f)\,\Delta f]^2}\, \frac{R_1}{(R_1 + R_2)^2}. \qquad (7.2)$$

At thermal equilibrium, from the second law we find the power dissipations must be equal. Therefore

$$\overline{[v_{N_1}(f)\,\Delta f]^2}R_2 = \overline{[v_{N_2}(f)\,\Delta f]^2}R_1. \qquad (7.3)$$

In order to determine the functions $\overline{[v_N(f)\,\Delta f]^2}$ we have recourse to the original argument of Nyquist. The derivation follows a pattern similar to that used in deriving the Planck radiation law in Section 2.3.2. We multiply the energy per degree of freedom by the number of degrees of freedom to obtain the total energy. However, where the Planck radiation law requires a three-dimensional treatment, the Nyquist derivation of the thermal noise expression requires a one-dimensional treatment. Suppose that R_1 and R_2 are equal, having resistance R_1 and are connected by an ideal lossless transmission line of characteristic impedance R. (See Fig. 7.1.) Thus the resistors are matched to each other and to the transmission line, which is the condition for maximum power transfer. All the power transferred by each resistor is absorbed by the other. The theorem of equipartition of energy[4] states that the energy per degree of freedom is $\frac{1}{2}kT$, where k is Boltzman's constant and T is the absolute temperature. The electromagnetic energy, progressing along the transmission line, possesses two degrees of freedom, corresponding to the two planes of polarization, per mode of vibration. Thus the energy per mode of vibration is given by kT. The number of modes of vibration per unit bandwidth is given by $2L/v$, where L is the line length and v is the propagation velocity. At equilibrium the energy stored in the line in a bandwidth Δf will be

$$\frac{2L}{v}\,\Delta f kT. \qquad (7.4)$$

Half of this energy is carried in each direction. In each second the energy transferred into each resistor is given by

$$\left(\frac{1}{2}\frac{v}{L}\right)\left(\frac{2L}{v}\Delta f kT\right) = kT\Delta f. \qquad (7.5)$$

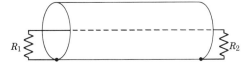

Figure 7.1. Two resistors connected by an ideal transmission line.

This represents the average power transferred from one resistor to the other. The average power is also given by Eqs. 7.1 and 7.2. Setting $R_1 = R_2$ and combining with Eq. 7.5, we find

$$P_N = \frac{\overline{[v_N(f)\Delta f]^2}}{4R} = kT\Delta f. \tag{7.6}$$

Equation 7.6 represents the noise power delivered under matched conditions. The open circuit noise power, four times that delivered under matched conditions, is $4kT\,\Delta f$. The mean square noise voltage in the bandwidth Δf is, from Eq. 7.6,

$$\overline{v_N{}^2} \equiv \overline{[v_N(f)\,\Delta f]^2} = 4kTR\,\Delta f, \tag{7.7}$$

where we have represented $\overline{[v_N(f)\,\Delta f]^2}$ as $\overline{v_N{}^2}$. Since the term on the right is a function of bandwidth but not of frequency, we see that the noise voltage is independent of frequency. Noise of this type is known as "white" noise. We can express the thermal noise voltage v_N in the bandwidth Δf as simply

$$v_N \equiv \left(\overline{v_N{}^2}\right)^{1/2} = \left\{\overline{[v_N(f)\,\Delta f]^2}\right\}^{1/2} = (4kTR\,\Delta f)^{1/2}. \tag{7.8}$$

The noise current i_N in the bandwidth Δf, determined by dividing the open circuit noise power by the noise voltage, is

$$i_N \equiv \left(\overline{i_N{}^2}\right)^{1/2} = \left\{\overline{[i_N(f)\,\Delta f]^2}\right\}^{1/2} \equiv \left(\frac{4kT\,\Delta f}{R}\right)^{1/2}. \tag{7.9}$$

Equations 7.6, 7.8, and 7.9 show clearly that the thermal noise power in a resistor is independent of resistance, but the noise current and voltage are not.

The value of kT at room temperature ($295°$K) is 4.07×10^{-21} watt sec. The noise power in a 1 cps bandwidth at room temperature is therefore 4.07×10^{-21} watts. The thermal noise voltage per cycle in a 1 ohm resistor at room temperature is 1.28×10^{-10} volts. Thus the expression for the thermal noise voltage in a resistance R and a bandwidth Δf at room temperature is

$$v_N = 1.28 \times 10^{-10}(R\,\Delta f)^{1/2} \text{ volts}. \tag{7.10}$$

In a network involving more than one resistor, each at a different temperature, the noise voltage is considered as arising from each independently. Two resistors, R_1 and R_2, in series at temperatures T_1 and T_2 have a mean square noise voltage output of

$$\overline{v_N{}^2} = 4k\,\Delta f[R_1T_1 + R_2T_2]. \tag{7.11}$$

Equation 7.11 can be rewritten in terms of an equivalent noise temperature T_n, the temperature at which R_1 and R_2 would have the same mean square noise voltage as given by Eq. 7.11.

$$\overline{v_N{}^2} = 4kT_n(R_1 + R_2)\,\Delta f. \tag{7.12}$$

Therefore, from Eqs. 7.11 and 7.12

$$T_n = T_1 \frac{R_1}{R_1 + R_2} + T_2 \frac{R_2}{R_1 + R_2}. \tag{7.13}$$

For two resistors R_1 and R_2 in parallel, at T_1 and T_2, the mean square noise currents add as

$$\overline{i_N{}^2} = 4k\left(\frac{T_1}{R_1} + \frac{T_2}{R_2}\right)\Delta f, \tag{7.14}$$

which can be rewritten in terms of an equivalent noise temperature T_n as

$$\overline{i_N{}^2} = 4kT_n\left(\frac{1}{R_1} + \frac{1}{R_2}\right)\Delta f. \tag{7.15}$$

In this case T_n is found from Eqs. 7.14 and 7.15 to be

$$T_n = T_1 \frac{1/R_1}{1/R_1 + 1/R_2} + T_2 \frac{1/R_2}{1/R_1 + 1/R_2}. \tag{7.16}$$

Thus the mean square noise voltage $\overline{v_N{}^2}$ for two resistors R_1 and R_2 in parallel, at T_1 and T_2 is given by

$$\overline{v_N{}^2} = \overline{i_N{}^2}\,\overline{\overline{R}}{}^2 = 4kT_n\overline{\overline{R}}\,\Delta f = 4k\,\frac{R_1R_2}{(R_1 + R_2)^2}(R_1T_2 + R_2T_1)\,\Delta f, \tag{7.17}$$

where $\overline{\overline{R}}$, the parallel resistance, is given by

$$\overline{\overline{R}} = \frac{R_1R_2}{R_1 + R_2}. \tag{7.18}$$

It should be noted that the expressions derived above for thermal noise become invalid at frequencies of the order of 10^{13} cps–10^{14} cps. This is because the energy per degree of freedom, determined to be kT on the basis of classical theory, is more correctly given according to quantum theory by the Planck expression

$$\frac{hf}{\exp\left(\dfrac{hf}{kT}\right) - 1}, \tag{7.19}$$

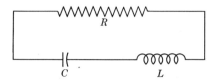

Figure 7.2. *RLC* circuit.

where h is Planck's constant. The noise power P_N is exactly given by

$$P_N = \frac{hf\,\Delta f}{\exp\left(\dfrac{hf}{kT}\right) - 1},$$ (7.20)

and the noise voltage v_N is

$$v_N = \left(\frac{4Rhf\,\Delta f}{\exp\left(\dfrac{hf}{kT}\right) - 1}\right)^{\frac{1}{2}}.$$ (7.21)

However, in applications involving infrared detectors and their associated amplifiers, the electrical frequencies involved are sufficiently low to justify the use of classical theory, and the Eqs. 7.6 through 7.17 are valid.

In a network consisting of capacitors, inductors, and resistors, the noise generated is due only to the thermal noise in the resistances, including lead resistance and the winding resistance of the inductors. Neglecting the small ohmic contributions from leads and windings, the presence of inductors and capacitors does not change the noise voltage level in the circuit. This may at first appear strange, since these elements contribute to establishing the impedance level of the circuit. However, Callen and Welton[5] have shown that noise fluctuations are associated with processes in which energy is dissipated. Since an ideal capacitor or inductor is nondissipative, we would not expect them to contribute noise. We can also establish by a direct approach that capacitors and inductors do not determine the noise level. Consider the *RLC* series circuit in Fig. 7.2. Let us examine the thermal noise appearing across the capacitor in the circuit at resonance. We view the resistor as a source of thermal noise to the reactance elements, which contribute to the circuit impedance, and which we shall initially assume to be noiseless. The noise voltage generated at the terminals of the resistor is given by Eq. 7.8 as

$$v_N = (4kTR\,\Delta f)^{\frac{1}{2}}.$$ (7.8)

Since at resonance the circuit impedance is purely resistive, the noise current flowing in the circuit will be

$$i_N = \frac{v_N}{R}.$$ (7.22)

The noise voltage v_N' appearing at the terminals of the capacitor will be

$$v_N' = i_N \chi_c = \frac{i_N}{\omega C} = \frac{(4kTR\,\Delta f)^{\frac{1}{2}}}{\omega RC}. \qquad (7.23)$$

where χ_c is the capacitive reactance.

Let us now approach the problem by an alternate method, considering the noise in the resistive component of the impedance Z_R. At resonance Z_R is simply

$$Z_R = \frac{\chi_c^2}{R} = \frac{1}{\omega^2 C^2 R}. \qquad (7.24)$$

The thermal noise v_N'' generated in Z_R is then

$$v_N'' = (4kTZ_R\,\Delta f)^{\frac{1}{2}} = \left(\frac{4kT\,\Delta f}{\omega^2 C^2 R}\right)^{\frac{1}{2}} = \frac{(4kTR\,\Delta f)^{\frac{1}{2}}}{\omega CR}. \qquad (7.25)$$

Examination of Eqs. 7.23 and 7.25 shows the noise voltages v_N' and v_N'' obtained are identical although two different points of view have been used to compute them. Since the first approach assumed the reactance elements to be noiseless, we can always consider the thermal noise in a circuit as arising from the resistive elements alone.

7.3   NOISE IN ELECTRON TUBES

Noise in electron tubes arises from several sources. The one considered first, basic to the thermionic emission process, is shot noise. Shot noise owes its existence to two factors: the discreteness of the electronic charge and the random emission of electrons by a heated cathode. The emission of an electron from the cathode, and the passage of it to the anode, is an isolated event, uncorrelated to the emission and passage of the other electrons. Because of the random character of the process, the amount of current varies from instant to instant about an average value. These fluctuations, first analyzed correctly by Schottky,[6] are termed "shot" noise. We shall see that shot noise is frequency independent, that is, "white," is greater in multielement tubes than in diodes, and is less in space charge limited emission than in temperature limited emission.

A second source of noise is the flicker effect, found principally in tubes having oxide cathodes. It is greatest at low frequencies, where it may be considerably greater than shot noise, but drops off as the frequency is increased until it falls below shot noise. It is a manifestation of "patchy" areas on the cathode, areas having different work functions which fluctuate in value with time.

In addition, three other sources of noise deserve mention. The first, partition noise, is found in multielement tubes and is a result of a random distribution of current between the various positive grids. The second is noise due to ionization of residual gas in tubes, the positive ions produced contributing noise not only by means of their generation and passage through the tube, much as shot noise is generated, but also because their presence lowers the space charge barrier surrounding the cathode. Since the space charge acts to suppress shot noise, lowering it increases the noise. The third, microphonism, results from changes in interelectrode capacitances induced by mechanical vibration.

7.3.1 Shot Noise in Saturated and Exponential Diodes

In order to arrive at an analytic expression for shot noise, we shall make use of Fourier analysis. The use of Fourier series to describe noise spectra would seem to be open to question, since noise is a nonrepetitive phenomenon, involving frequencies which apparently may range from zero to infinity. The noise power over all frequencies therefore appears to be infinite if the expressions derived by classical methods are used. However, quantum mechanical treatment establishes that the total noise power is finite. In practice also the noise generated is amplified and displayed by instruments having a finite bandwidth. The use of Fourier series to derive low frequency approximations therefore appears justified in large part by the measuring techniques.

The current-voltage characteristic relevant to thermionic emission may be divided into three regions: the retarding field or exponential region, the space charge limited region, and the saturated or temperature limited region. (See Fig. 7.3.)

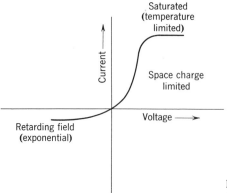

Figure 7.3. Current-voltage characteristic of a diode.

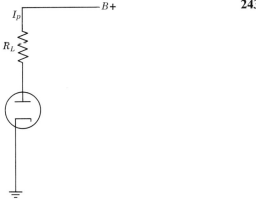

Figure 7.4. Saturated diode and plate resistor.

We shall consider first shot noise in a saturated diode. In saturated emission all electrons emitted from the cathode reach the anode. The anode current, controlled by the cathode temperature, is independent of the anode voltage. Thus fluctuations in the cathode emission reach the anode unchanged.

Consider the saturated diode and plate resistor shown in Fig. 7.4. An electron leaving the cathode at zero velocity is accelerated by the electric field. For a constant voltage on the plate and an interelectrode distance d the current i at the instant when the velocity of the electron is v is given by

$$i = \frac{qv}{d}, \tag{7.26}$$

where q is the electronic charge. Since the acceleration is constant, the average velocity is one-half the final velocity v_f when the electron just reaches the anode. Because the average velocity is given by the distance divided by the transit time τ, we find

$$d = \frac{v_f \tau}{2}. \tag{7.27}$$

But since

$$v = v_f \frac{t}{\tau}, \tag{7.28}$$

we have

$$i = \frac{2qt}{\tau^2}. \tag{7.29}$$

We wish now to know the magnitude of the noise current flowing through the resistor R_L due to shot noise in the diode. The average current I_p is determined by n_{e0}, the average number of electrons emitted per unit time by the cathode which reach the anode. Since the electronic charge is q, we have

$$I_p = n_{e0}q. \tag{7.30}$$

If there were no noise, the value of I_p obtained would be independent of the measurement interval. Since there is noise, the value of I_p obtained for very short intervals fluctuates considerably about the average obtained over a long interval. Probability theory[7] tells us that the mean square deviation of a large number of random events is equal to the average value of that number. Applying this to our example, we find

$$\overline{(n_e - n_{e0})^2} = n_{e0}, \tag{7.31}$$

where n_e is the instantaneous value of the rate of electron emission.

Since we measure current, not count electrons, it is necessary to rewrite Eq. 7.31 in terms of current. We apply Campbell's theorem[8] describing the effect $F(t)$ caused in the output circuit by the arrival of the electrons at the plate. The theorem indicates that the average current $\bar{I}(t)$ is given by

$$\bar{I}(t) = n_{e0} \int_{-\infty}^{\infty} F(t)\, dt, \tag{7.32}$$

and the mean square deviation $\overline{i_N^2}$ is given by

$$\overline{i_N^2} \equiv \overline{[I(t) - \bar{I}(t)]^2} = n_{e0} \int_{-\infty}^{\infty} F^2(t)\, dt, \tag{7.33}$$

where $I(t)$ is the instantaneous value of the current.

We now introduce the Fourier transforms which will enable us to relate events in the time domain to events in the frequency domain.

$$F(t) = \frac{1}{(2\pi)^{1/2}} \int_{-\infty}^{\infty} G(\omega) \exp(j\omega t)\, d\omega. \tag{7.34}$$

$$G(\omega) = \frac{1}{(2\pi)^{1/2}} \int_{-\infty}^{\infty} F(t) \exp(-j\omega t)\, dt. \tag{7.35}$$

The Fourier transform of $F^2(t)$, obtained through use of Parseval's theorem,[9] is

$$\int_{-\infty}^{\infty} F^2(t)\, dt = \int_{-\infty}^{\infty} G(\omega)G^*(\omega)\, d\omega, \tag{7.36}$$

where $G^*(\omega)$ is the complex conjugate of $G(\omega)$. Since the product $G(\omega)G^*(\omega)$ will be an even function in our example, we can write

$$\overline{i_N^2} = \frac{2I_p}{q} \int_0^{\infty} G(\omega)G^*(\omega)\, d\omega. \tag{7.37}$$

The function $F(t)$ in the example we are considering is the instantaneous current due to the passage of an electron from cathode to anode. Thus $F(t)$ is simply i of Eq. 7.29. Substituting this into Eq. 7.35 gives

$$G(\omega) = \frac{1}{(2\pi)^{1/2}} \frac{2q}{\tau^2} \int_0^\tau t \exp(-j\omega t)\, dt. \tag{7.38}$$

We now introduce a new variable, $\theta \equiv \omega\tau$, called the transit angle. The value of $G(\omega)$ upon integration is

$$G(\omega) = \left(\frac{2}{\pi}\right)^{1/2} \frac{q}{\theta^2} \left[(\theta \sin\theta + \cos\theta - 1) - j(\sin\theta - \theta\cos\theta) \right]. \tag{7.39}$$

The product $G(\omega)G^*(\omega)$ is

$$G(\omega)G^*(\omega) = \frac{2q^2}{\pi} \frac{1}{\theta^4} \left[\theta^2 + 2(1 - \cos\theta - \theta\sin\theta) \right]. \tag{7.40}$$

If $\theta \approx 0$, we have the example of saturated emission at low frequencies, that is, not limited by transit time effects. This will be true for audio amplifiers, which are used in almost all infrared systems work. In this case Eq. 7.40 reduces to

$$G(\omega)G^*(\omega)\Big|_{\theta \approx 0} = \frac{q^2}{2\pi}. \tag{7.41}$$

Integrating Eq. 7.37 over the bandwidth Δf gives

$$\overline{i_N^2} = \frac{qI_p}{\pi} \Delta\omega = 2qI_p \Delta f. \tag{7.42}$$

Equation 7.42 is the analytic expression for the mean square shot noise current in saturated diodes. As was the case for thermal noise, the spectrum is white, depending on the bandwidth but not the frequency. Referring to Fig. 7.4, the noise voltage v_N developed across the resistor R_L due to shot noise in a diode operated in the saturated region is seen to be

$$v_N = (2qI_p \Delta f)^{1/2} R_L. \tag{7.43}$$

The preceding arguments were based upon a diode operating under saturated conditions. The arguments are equally valid for operation in the retarding field or exponential region. Although some of the electrons emitted from the cathode do not reach the anode, the arrival of those with sufficient energy still constitutes random events. Therefore expressions 7.42 and 7.43 are valid in this region.

We shall now express the noise in terms of the diode noise resistance. In the retarding field region the plate current is given by

$$I_p = I_s \exp(qV_p/kT_c), \tag{7.44}$$

where I_s is the saturation current, T_c is the cathode temperature, and V_p is the anode voltage. In this region V_p is negative. The conductance is

$$g \equiv \frac{dI_p}{dV_p} = \frac{qI_p}{kT_c},$$ (7.45)

where T_c is the cathode temperature. From Eqs. 7.42 and 7.45, therefore, the mean square noise current is given by

$$\overline{i_N^2} = 2kT_cg \, \Delta f.$$ (7.46)

Thus the mean square shot noise current in the retarding field region is proportional to the conductance.

The equivalent noise resistance R_N is that resistance at the input which will give rise to thermal noise in the output in an amount equal to the tube noise. If the equivalent noise emf is $(\overline{e_N^2})^{1/2}$, we have

$$\overline{e_N^2} = \frac{\overline{i_N^2}}{g^2}.$$ (7.47)

In terms of R_N, $\overline{e_N^2}$ is expressed as

$$\overline{e_N^2} = 4kTR_N \, \Delta f,$$ (7.48)

or

$$\overline{i_N^2} = 4kTR_Ng^2 \, \Delta f.$$ (7.49)

Equation 7.49 represents the mean square shot noise current in an electron tube operated at any point of its current-voltage characteristic. Applying it to the exponential diode, and combining Eq. 7.49 with 7.46, we see that

$$R_N = \frac{T_c}{2Tg}.$$ (7.50)

Assuming a cathode temperature, T_c, of 1000°K and an ambient temperature, T, of 295°K, we find the equivalent noise resistance of the exponential diode to be

$$R_N = \frac{1.7}{g}.$$ (7.51)

7.3.2 Shot Noise in Space Charge Limited Diodes

The operation of a diode biased at low voltages in the forward direction is governed by a space charge or electron cloud in front of the cathode. The current-voltage characteristic in this region is the so-called $\frac{3}{2}$ power law, the current density being proportional to the $\frac{3}{2}$ power of the voltage, first derived by Child[10] in 1911. The space charge acting as a virtual cathode in front of the cathode constitutes a potential hill which the emitted electrons must climb in order to reach the anode. Those emitted with velocities sufficiently great will surmount the hill. The presence of a group of electrons having velocity just sufficient to pass the barrier will momentarily raise the barrier to the following electrons, tending to turn more of them back toward the cathode. Thus a noise burst at one instant tends to reduce the noise at the next. The over-all effect is to reduce the shot noise by a factor which depends upon tube design, and can be as high as ten. We call this factor Γ^2, the space charge reduction factor. In terms of Γ^2 we write the shot noise expression, applying it in this instance to space charge limited diodes, as

$$\overline{i_N{}^2} = 2qI_p\Gamma^2\,\Delta f. \tag{7.52}$$

The analytic expression for Γ^2 involves a derivation considered too detailed for the purposes of this book. The interested reader is referred to the works of North[11] and Rack.[12] A condensed version is found in Beck.[13]

The value of Γ^2 derived by van der Ziel[14] is

$$\Gamma^2 = \frac{2kT_c\theta g}{qI_p}, \tag{7.53}$$

where θ is a dimensionless parameter having a value of 0.644. Thus

$$\overline{i_N{}^2} = 4kT_c\theta g\,\Delta f. \tag{7.54}$$

Combining Eqs. 7.53 and 7.52 with 7.49, we find for the equivalent noise resistance of the space charge limited diode

$$R_N = \frac{\theta T_c}{gT}. \tag{7.55}$$

At a cathode temperature of $1000°K$ we find

$$R_N = \frac{2.2}{g}. \tag{7.56}$$

7.3.3 Shot Noise in Triodes

The analysis of shot noise in triodes is simplified if we assume no grid current. In this case we once again consider the three regions of the tube characteristic: the saturated region, the retarding region, and the space charge limited region. In the saturated region the situation is the same as for the diode and the mean square noise current is given by Eq. 7.42.

$$\overline{i_N{}^2} = 2qI_p\,\Delta f. \tag{7.42}$$

In the exponential region of the triode the transconductance g_m is given by

$$g_m = \frac{\partial I_p}{\partial V_g} = \frac{q I_p \kappa}{k T_c},$$ (7.57)

where V_g is the grid voltage and κ has a value usually between 0.5 and 0.7. The mean square noise current is then given by

$$\overline{i_N^2} = 2q I_p \Delta f = \frac{2 k T_c g_m \Delta f}{\kappa}.$$ (7.58)

By analogy to Eq. 7.49 we can express $\overline{i_N^2}$ as

$$\overline{i_N^2} = 4 k T R_N g_m^2 \Delta f.$$ (7.59)

Assuming a value of $\kappa = 0.6$, with $T_c = 1000°$K, we find for the triode in the exponential region

$$R_N = \frac{T_c}{2 T g_m \kappa} = \frac{2.8}{g_m}.$$ (7.60)

Comparing Eq. 7.60 with 7.51 we see that the triode operated in the exponential region is more noisy than the diode in the same region.

In the space charge region of the triode we once again express the mean square noise current as we did for the diode, Eq. 7.52.

$$\overline{i_N^2} = 2q I_p \Gamma^2 \Delta f.$$ (7.52)

In this case van der Ziel[15] shows that Γ^2 is given by

$$\Gamma^2 = \frac{2 k T_c g_m \theta}{q I_p \sigma}.$$ (7.61)

Once again $\theta = 0.644$. The value of σ is approximately 0.9. Combining Eqs. 7.52 and 7.61 with 7.59, we find

$$R_N = \frac{\theta T_c}{g_m T \sigma},$$ (7.62)

which has a value at $T_c = 1000°$K of

$$R_N = \frac{2.5}{g_m}.$$ (7.63)

The derivations above concerning shot noise in triodes were based upon an assumption of no grid current. In practice this is generally not true. However, as long as the impedance of the input circuit is kept small with respect to the grid resistance, as is usually the case, the noise due to grid current will be small compared to the other noise.

7.3.4 Partition Noise

In multigrid tubes such as tetrodes and pentodes, an additional type of noise, termed partition noise, is found. This additional noise is due to the random division of current between the grids, tending to reduce the smoothing action of the space charge. Harris[16] has shown the equivalent noise resistance for a pentode to be

$$R_N = \frac{2.5}{g_m} \frac{I_p}{I_s} \left(1 + 8 \frac{I_s}{g_m} \right), \tag{7.64}$$

where I_s is the current to the screen grid. The factor $2.5/g_m$ represents the noise resistance of a triode. The remaining factors multiply the equivalent noise resistance by a factor ranging between three and five. Wherever possible, the use of multigrid tubes in an amplifier input stage should be avoided.

7.3.5 Flicker Noise

A dominant source of noise at low frequencies is flicker noise, also called $1/f$ noise, the mechanism of which has not been completely clarified. It is much greater in oxide cathode tubes than metal filament tubes, and is characterized by a power spectrum which varies approximately inversely with frequency. The first theory of the effect is due to Schottky,[17] who attributed it to variation in emission from large patches of the cathode surface. The empirical expression fitting the measured data is of the form

$$\overline{i_N^2} = \frac{A I_p^{\alpha} \Delta f}{f^{\beta}}, \tag{7.65}$$

where $\alpha \approx 2$, $\beta \approx 1$, and A is a proportionality factor. Equation 7.65 predicts infinite power at zero frequency. Although this cannot be true, the $1/f$ dependence has been verified at frequencies much less than 1 cps. In order to minimize this noise, which is generally the dominant noise in an amplifier at low frequencies, it is best to choose operating frequencies above the flicker noise region.

7.3.6 Microphonics

A source of low frequency noise in electron tubes arises from vibration of the electrodes when the tube is subjected to mechanical shock. This effect, termed microphonism or microphonics, is due to changes in the interelectrode capacity caused by displacing the electrodes from their normal position. The frequency dependence of this noise varies with the tube construction, different configurations having different resonant frequencies. Since the infrared system designer is usually concerned with

TABLE 7.1. Equivalent noise resistances of receiving tubes*

Tube	g_m, μmhos	R_N, ohms
Triode Amplifiers		
6AC7	11,250	220
6AK5	6,670	385
6C4	2,200	1,140
6F4	5,800	430
6J5**	2,600	960
6J4	12,000	210
6J6**	5,300	470
6SC7**	1,325	1,890
6SL7**	1,600	1,560
6SN7**	2,600	960
7F8**	5,650	440
9002	2,200	1,140
Sharp Cutoff Pentodes		
1L4	1,025	4,300
6AC7	9,000	720
6AG5	5,000	1,640
6AJ5	2,750	2,650
6AK5	5,000	1,880
6AS6	3,500	4,170
6SH7	4,900	2,850
6SJ7	1,650	5,840
9001	1,400	6,600
Remote Cutoff Pentodes		
1T4	750	20,000
6AB7	5,000	2,440
6SG7	4,700	4,000
6SK7	2,000	10,500
9003	1,800	13,000

* After Valley and Wallman[20] and *Radiotron Designers Handbook.*[21]
** One unit of a dual triode tube.

low noise audio amplifiers often required to operate under adverse shock and vibration conditions, elimination of microphonics is important. Selected low microphonic tubes should be used in amplifier input stages. Potting the amplifiers in a casting resin, with the tubes coated with a rubberlike material, serves to reduce microphonics, as does shock mounting. Discussions of microphonics are given by Rockwood and Ferris[18] and by Penick.[19]

7.3.7 Noise Resistances of Selected Receiving Tubes

Table 7.1 presents information on the noise characteristics of a number of receiving tubes.

7.4 NOISE IN SEMICONDUCTORS[22]

Noise in electron tubes was discussed in order that we could form the basis for proper judgment in the selection of low-noise tubes and circuitry for amplifiers. On the other hand, the study of noise in semiconductors is of interest not only in understanding noise phenomena in transistors and diodes to be used as components in amplifiers but also because semiconductors are more widely used than other materials in infrared radiation detectors. Our study of noise in semiconductors will attempt to uncover and understand the mechanisms of noise sources associated with both bulk and surface properties.

The total mean square noise current in semiconductors has been found experimentally to follow a law of the form

$$\overline{i_N^2} = \left[\frac{K_1 I^\alpha}{f^\beta} + \frac{K_2 I^2}{1 + (f/f_1)^2} + \frac{4kT}{R} \right] \Delta f, \qquad (7.66)$$

where K_1, K_2, f_1, α, and β are constants chosen empirically to fit experimental data. The shape of the curve is shown in Fig. 7.5.

Examination of Eq. 7.66 and Fig. 7.5 shows three distinct regions of the spectrum. At high frequencies the dominant noise is thermal noise,

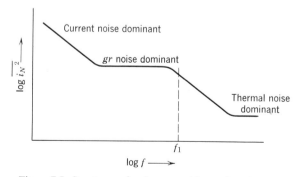

Figure 7.5. Spectrum of noise current in semiconductors.

recognized as the third term in Eq. 7.66. At low frequencies the dominant noise has been found in most semiconductors to be current noise, characterized by a $1/f^\beta$ spectral dependence, where β lies between 1.0 and 1.5. As such it is very similar to flicker noise in electron tubes. At intermediate frequencies a third form of noise appears dominant, of a type described in terms of a characteristic frequency f_1.

Thermal noise, existing at all frequencies, becomes the dominant noise at high frequencies. The mechanism is the same one considered earlier, namely, the random motion of charge carriers. It exists even in the absence of current flow through the semiconductor. For metals, discussed previously, the carriers are electrons. In semiconductors the carriers may be electrons or holes. Although the carriers may be of differing sign or differing effective mass from the free electrons in a metal, the expressions for the thermal noise voltage, current, and power remain the same as those found in a metal, namely, Eqs. 7.8, 7.9, and 7.6, respectively. The arguments leading to the derivations of these equations were based on thermodynamic considerations involving neither the sign nor the magnitude of the charge. The same arguments can be applied to thermal noise in semiconductors, resulting in the same equations.

As was stated, at sufficiently high frequencies thermal noise becomes dominant. In indium antimonide at room temperature, however, it has been found to be the dominant noise at all frequencies,[23] no excess noise having been found even at current densities of 50 amp/cm². This might also be expected in other semiconductors which approach a metallic nature by having energy gaps smaller than that of InSb.

The dominant source of noise at low frequencies in many semiconductors has been called at various times current noise, contact noise, excess noise, $1/f$ ("one over f") noise, and modulation noise. Its power spectrum is characterized by an approximate dependence upon the reciprocal of the frequency and the square of the current. The mechanism of this noise is not clear. We shall discuss several present-day theories.

The third type of noise, which has been called shot noise by some authors and generation-recombination or gr noise by others, is found in some materials at frequencies intermediate between those dominated by current noise and those dominated by thermal noise. Its spectrum can be described in terms of a characteristic frequency, beyond which thermal noise becomes dominant. The mechanism is that of fluctuations in the instantaneous values of the free carrier densities due to the random character of the generation, recombination, and trapping processes. In materials having very large amounts of current noise it may be impossible to detect generation-recombination noise. These materials will be characterized by current noise at low frequencies, passing over directly to thermal noise at high frequencies. A discussion of noise in infrared detectors appears in Section 9.2.

Some authors use the term current noise to encompass all types of noise which depend upon bias current, including gr noise. We use it only to refer to $1/f$ noise.

7.4.1 Current Noise

The power spectrum of current noise is of the same form as that of flicker noise, namely,

$$\overline{i_N^2} = \frac{K_1 I^\alpha \, \Delta f}{f^\beta}, \tag{7.67}$$

where $\alpha \approx 2$, $\beta \approx 1$, and K_1 is a proportionality factor. Several sources of current noise have been suggested. The experiments of Christenson and Pearson[24] demonstrated that intergranular contacts in carbon resistors gave rise to a noise power proportional to the square of the current and inversely proportional to the frequency. Current noise has also been associated with nonohmic contacts to crystals which have wide energy gaps. It has been shown[25] that cadmium sulfide crystals with rectifying contacts are limited by current noise, whereas those with ohmic contacts are not. However, even in single crystals of certain semiconductors, for example, germanium, having ohmic contacts, current noise is found.

The experiments of Montgomery[26] have shown that current noise in Ge is associated with surface conditions. He found that surfaces having high recombination velocities, such as produced by sandblasting, gave the lowest noise, whereas etched surfaces were up to ten times as noisy. By changing the carrier concentrations at the surface through magnetic field effects, he could change the magnitude of the noise. On the other hand, Bess[27] and Morrison[28] have proposed that current noise should be expected to be associated with dislocations. Miller[29] and Torrey and Whitmer[30] have shown that point contact diodes are limited by current noise. Anderson and van der Ziel[31] have shown the same dependence in p-n junctions.

All the above current noise sources are associated with potential barriers, whether they exist at intergranular contacts, at rectifying electrodes, at the semiconductor surface, at dislocations, at point contacts, or at p-n junctions. The absence of current noise in InSb at room temperature may be interpreted as being due to the difficulty of forming potential barriers in the material.

The dependence of the noise voltage upon bias current suggests that current noise is the result of fluctuations in conductivity of the material which modulate the bias current. Petritz[32] has introduced the term "modulation noise," indicating that current noise is due to some effect which modulates the carrier densities and thus the conductivity of the material.

Current noise, having a $1/f$ power dependence, cannot be adequately represented by a model based on a process with a single time constant. It can be formally characterized by the summation of a number of processes having a distribution of time constants, each of which is of the form $f(\tau) \equiv a\tau/(1 + \omega^2\tau^2)$. If each of these processes is multiplied by a weighting factor inversely proportional to τ, the integrated functions has a $1/f$ dependence. In order to interpret this, proposals have been made[33,34,35] that the noise form is due to ions having a distribution of activation energies for diffusion, migrating to a barrier and modulating the barrier height. Rose[36] has proposed that the barrier modulation is due to the presence of electronic charges coming from trapping states distributed in energy over the forbidden band, the lifetime of the charges in the traps being an approximately exponential function of the trap depth. However, van Vliet[37] argues that the free time of the charges will determine the noise spectrum, and long trapping times cannot account for the noise currents at low frequencies. Rollin and Templeton[38] have observed $1/f$ noise down to 2×10^{-4} cps, necessitating lifetimes of the order of hours. In order that the noise power be finite the $1/f$ dependence must have both lower and upper frequency limits. It is difficult to resolve these conflicting viewpoints.

Efforts have been made to attribute the noise to either the minority or majority carrier. The experiments of Montgomery[26] indicated the minority carrier was responsible in Ge. His studies, made mostly on n-type material, related the noise behavior to the fluctuations in hole concentration. On the other hand, Brophy and Rostoker[38] concluded that noise in n-type Ge was due to the majority carriers.

The temperature dependence of current noise has been found to be slight. Montgomery[26] has shown that the noise power in Ge varies over about one order of magnitude between $-200°C$ and $20°C$ but shows no distinct trend.

In contrast to the situation for thermal noise, the magnitude of the current noise voltage depends upon the dimensions of the resistive or semiconducting material. Assume that α in Eq. 7.67 is equal to 2. If the length l of the material in the direction of current flow is increased by a factor n whereas the current I is held constant, the resistance increases by n and the magnitude of $\overline{v_N^2}$ increases by n also. This is because the noise is uncorrelated, that is, the noise in one part of the material does not act in unison with that in another part. Thus the value of $\overline{i_N^2} = \overline{v_N^2}/R^2$ varies as $1/l$. If the cross-sectional area A increases by n and we assume for the moment that I increases also by n, then the current density remains constant and $\overline{i_N{}^2}$ increases by n. On the other hand, if A increases by n but I is held constant, then the current density decreases by n and $\overline{i_N{}^2}$ decreases by n. Thus if the total current I is held constant, the value of $\overline{i_N{}^2}$ varies as $[lA]^{-1}$, that is, as the reciprocal of the volume of the material. The expression for $\overline{v_N{}^2}$ is therefore

$$\overline{v_N{}^2} = \overline{i_N{}^2}R^2 = \frac{K_1 I^2}{f^\beta} R^2 \, \Delta f = \frac{c_1}{lA}\frac{I^2 R^2}{f^\beta} \, \Delta f = \frac{c_1 \rho^2 I^2 l \, \Delta f}{f^\beta A^3} , \qquad (7.68)$$

where c_1 is a factor dependent upon material but independent of dimensions.

From Eq. 7.68 it can be seen that for material of a given resistivity the mean square current noise voltage varies inversely with the cube of the cross-sectional area and therefore with the cube of the thickness. Thus thin films of resistive or semiconducting material will be more likely to be limited by current noise than will bulk material. The derivation above assumed that $\alpha = 2$. If α is greater than 2, the dependence of current noise upon dimensions will be even more pronounced.

A brief mention will be made of correlation. Montgomery[26] states that fluctuations in the minority carrier densities which modulate the conductivity of the material are the sources of current noise. His hypothesis states that the sources of minority carriers are discrete and are distributed over the material in very small regions. Their activity is modulated by some local influence, which therefore modulates the minority carrier density. Experiments[40] have tended to verify this by showing it is possible to observe correlation in the noise at two different points on a filament separated by a distance sufficiently small that the minority carriers generated at the center have not had an opportunity to recombine before reaching the points. Investigations of this correlation may lead ultimately to methods for reducing current noise in photon detectors.

7.4.2 Generation-Recombination Noise

The major source of noise in semiconductors at intermediate frequencies is generation-recombination (gr) noise, characterized by a power spectrum which is constant at low frequencies but decreases rapidly beyond a characteristic frequency related to the inverse of the carrier lifetime. Early experiments with the semiconductors Ge and Si failed to show evidence of gr noise, the very large values of current noise at intermediate frequencies masking the gr noise. The experiments of Herzog and van der Ziel[41] and Mattson and van der Ziel[42] carried out on Ge were among the earliest to detect gr noise. Since then Klaasen and Blok[43] have detected gr noise in thin films of PbS. Generation-recombination noise is the statistical fluctuation in the concentration of carriers in a semiconductor. In this it is analogous to shot noise in electron tubes and has been termed shot noise by some. In photon detectors it is intimately associated with photon noise, the random generation and recombination of carriers due to the random character of the incident photon stream. Van Vliet[37] and van Vliet and Blok[44] have shown that the ultimate performance of photoconductors is limited by approximately equal contributions from photon noise and *gr* noise. This is discussed in more detail in Section 9.4.3.

The analysis of *gr* noise was first performed by Bernamont,[45] although based upon incorrect considerations. More complete determinations have been made by van Vliet,[37] Burgess,[46] and van der Ziel,[47] among others. The complete analysis of *gr* noise involves several discrete areas in which are considered either extrinsic, near intrinsic, or intrinsic semiconductors, and in which the carrier transit time due to the applied field is either less than, equal to, or greater than the lifetime. Also, the recombination mechanism enters in, since an adequate description depends on whether it is monomolecular or bimolecular, that is, whether the recombination rate in, for example, *n*-type material, is proportional to the free electron concentration only or to the product of free electron and bound hole concentrations.

Consider now *gr* noise in an extrinsic semiconductor. The interested reader is referred to the paper by van Vliet[37] for a more complete analysis. Assume an *n*-type semiconductor in which substantially all the current is carried by electrons. The spectrum of the current fluctuations, $P(f)$, is defined from

$$\overline{I^2(t)} \equiv \int_0^\infty P(f)\, df. \tag{7.69}$$

In an *n*-type semiconductor the fluctuations in current can be shown to be related to the fluctuations in number, $Q(f)$, through

$$P(f) = \left(\frac{I_0}{\bar N}\right)^2 Q(f), \tag{7.70}$$

where I_0 is the average current and $\bar N$ is the average total number of free electrons. Assuming the electrons can readily find empty donor states at which they can recombine, the average value $\bar N$ is related to the mean square value $\overline{N^2}$ of the total number of free electrons by

$$\overline{N^2} \equiv \overline{(N - \bar N)^2} = \bar N. \tag{7.71}$$

Van Vliet[37] has shown that in a simple generation-recombination mechanism involving one carrier type the following relation holds:

$$\overline{[N(t+s) - \bar N(t+s)][N(t) - \bar N(t)]} = \overline{[N(t) - \bar N(t)]^2} \exp(-s/\tau), \tag{7.72}$$

where s is an arbitrary time following excitation. This derivation assumes that the lifetime is much shorter than the transit time, that is, the carriers are generated and recombine in the bulk without reaching the electrodes. The factor $\exp(-s/\tau)$ is termed the autocorrelation factor and describes the fact that gr noise is correlated over small distances determined by the carrier lifetime τ and mobility.

The power spectrum can now be determined by the Wiener-Khintchine theorem.[48] This states that the spectrum $Q(f)$ of a quantity $F(t)$ is determined by

$$Q(f) = 4 \int_0^\infty \overline{F(t)F(t+s)} \cos{(\omega s)}\, ds, \qquad (7.73)$$

where $\omega \equiv 2\pi f$. In our case we have

$$\overline{F(t)F(t+s)} = \overline{[N(t) - \bar{N}(t)][N(t+s) - \bar{N}(t+s)]}. \qquad (7.74)$$

Substituting Eqs. 7.72 and 7.74 into 7.73 we have

$$Q(f) = 4 \int_0^\infty \overline{[N(t) - \bar{N}(t)]^2} \exp{(-s/\tau)} \cos{(\omega s)}\, ds. \qquad (7.75)$$

When integrated, Eq. 7.75 takes the form

$$Q(f) = \frac{4\overline{N^2}\tau}{1 + \omega^2\tau}. \qquad (7.76)$$

Introducing Eqs. 7.70 and 7.71 into 7.76 results in

$$P(f) = \frac{4I_0{}^2\tau}{\bar{N}(1 + \omega^2\tau^2)}. \qquad (7.77)$$

Examination of Eq. 7.77 shows the gr noise power spectrum to be proportional to the square of the current through the semiconductor and inversely proportional to the total number of carriers in the semiconductor. Thus we expect a heavily doped semiconductor to exhibit less gr noise than a more pure sample with the same lifetime and equal bias current. In addition, we see that the spectrum is flat to a characteristic frequency related to the reciprocal of the carrier lifetime.

More complete analysis by van Vliet[37] has shown that in near intrinsic semiconductors the power spectrum is

$$P(f) = 4I^2 \frac{(b+1)^2 \bar{N}\bar{P}}{(b\bar{N} + \bar{P})^2(\bar{N} + \bar{P})} \frac{\tau}{1 + \omega^2\tau^2}, \qquad (7.78)$$

where $b \equiv \mu_e/\mu_h$, the mobility ratio, and $\bar{N}$ and $\bar{P}$ are the total numbers of free electrons and free holes, respectively. In intrinsic material this reduces to

$$P(f) = \frac{2I^2\tau}{\bar{N}(1 + \omega^2\tau^2)}. \tag{7.79}$$

7.5 NOISE IN TRANSISTORS AND DIODES

The preceding discussion has been concerned with the sources of noise in semiconductor materials. We have seen that three types exist: current or $1/f$ noise, generation-recombination noise, and thermal noise. In transistors and diodes only current noise and thermal noise have been observed. A typical noise characteristic shows current noise dominating at low frequencies, with thermal noise dominating above some critical frequency. (See Fig. 7.6.) Plotted here is the noise factor as a function of frequency. This parameter, used to describe the noise contribution of a transistor to a circuit, is defined as the ratio of the total noise power in the output from a transistor (neglecting the contribution from any load resistance) to that part of the output noise power due to thermal noise in the source resistance. If only a single value is given, with no frequency or bandwidth mentioned, it usually refers to a 1 cps bandwidth at 1 Kc.

The transition frequency, denoted by f_c in the figure, has been reduced from the region of 1 Mc in the earliest point contact diodes and transistors to the region below 1 Kc in present-day junction diodes and transistors. In addition, the noise factor at 1 Kc has dropped from values in the vicinity of 80 db for point contact devices to 10 db or less for junction devices. From the standpoint of noise characteristics, junction transistors and diodes are superior to point contact ones.

The thermal noise e_{nb} in the transistors can be characterized by the

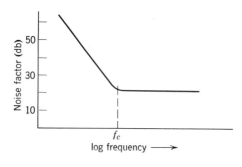

Figure 7.6. Noise characteristic of a transistor or diode.

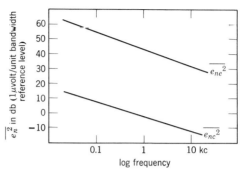

Figure 7.7. Open circuit values of collector noise and emitter noise for a point contact transistor. (Adapted from Ryder and Kircher.[49])

thermal noise formula evaluated for the extrinsic base resistance r_b. Thus we have

$$\overline{e_{nb}^2} = 4kTr_b\,\Delta f. \tag{7.80}$$

The current noise arises predominantly from two sources, the surface and the contacts. In evaluating the noise characteristic of a diode or transistor, it is difficult, if not impossible, to separate the surface contribution from the contact contribution. However, it is usually possible to separate the contribution of one contact from that of another, although some noise correlation has been found to exist between them. Experiments[49,50] have shown the collector to be a much greater source of noise than the emitter. Figure 7.7 shows the open circuit mean square emitter noise $\overline{e_{ne}^2}$ and collector noise $\overline{e_{nc}^2}$ as functions of frequency for a point contact transistor. The same type of relationship between emitter noise and collector noise holds for a junction transistor.

By reducing the collector voltage measured with respect to the base to a value near zero, the collector noise, which is greater than thermal noise at low frequencies, can be reduced greatly. The application of this is termed "hushed transistor" circuitry.[51]

The noise factor of an electron tube is relatively independent of the source impedance and whether the tube is operated grounded plate, grounded grid, or grounded cathode. In a transistor the noise factor is much more dependent upon the source impedance and is slightly dependent upon the operational configuration, whether it be grounded collector, grounded emitter, or grounded base, The noise factors, NF, of the three configurations are given by Shea[52] as:

For grounded emitter

$$NF = 1 + \frac{1}{4kT\,\Delta f r_g}\left[e_{ne}{}^2\left(\frac{r_g + ar_c + r_b}{ar_c - r_e}\right)^2 + e_{nc}{}^2\left(\frac{r_g + r_b + r_e}{ar_c - r_e}\right)^2\right]. \tag{7.81}$$

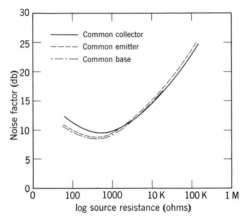

Figure 7.8. Dependence of noise factor on source resistance in three configurations. (Adapted from Hunter.[53])

For grounded base

$$NF = 1 + \frac{1}{4kT\,\Delta f r_g}\left[e_{ne}{}^2 + e_{nc}{}^2\left(\frac{r_g + r_e + r_b}{ar_c - r_b}\right)^2\right].\qquad(7.82)$$

For grounded collector

$$NF = 1 + \frac{1}{4kT\,\Delta f r_g}\left[e_{ne}{}^2\left(\frac{r_g + r_c + r_b}{r_c}\right)^2 + e_{nc}{}^2\left(\frac{r_g + r_b}{r_c}\right)^2\right].\qquad(7.83)$$

In these equations r_g is the source resistance, a is the current amplification factor, r_e is the emitter resistance, and r_c is the collector resistance. Experimental measurements of noise factor as a function of source resistance for the three configurations is shown in Fig. 7.8.

From Fig. 7.8 we see that the optimum source resistance is in the vicinity of 1 KΩ. This is true whether the source be resistive or reactive. The configuration can be seen to have little influence upon the noise factor. However, Dewitt and Rossoff[54] show that the common emitter configuration results in the best possible compromise between noise factor and gain.

7.6 NOISE IN TRANSFORMERS

A noise source which can prove troublesome in low frequency circuits involving transformers is magnetic fluctuation noise, which arises from the Barkhausen effect. If a magnetic field is applied suddenly to a ferromagnetic material, magnetization of the material proceeds not as a smooth function but rather in a series of discrete steps. These steps represent the alignment of individual magnetic domains in the material with the magnetic field. Saturation occurs when all the domains are aligned with the field. In a transformer the alternating current in the primary windings magnetizes the core alternately in one direction, then in the opposite. This gives rise to Barkhausen noise in the core. Since the core is coupled to the secondary windings, the noise appears in the secondary. If the transformer is not well shielded, the noise may be picked up in other parts of the circuit as well. Goldman[55] shows the noise voltage picked up in other parts to be of the form

$$\overline{v_N^2} = c_1 f I_0 \cos (2\pi f t) \, \Delta f, \qquad (7.84)$$

where c_1 is a constant and $I_0 \cos (2\pi f t)$ is the alternating current in the primary windings.

7.7 NOISE IN GALVANOMETERS

One of the means of displaying signals from low impedance infrared detectors consists of coupling the detector directly into a galvanometer. Although this method is severely limited in speed of response, it is useful in systems where observation time may be very great.

The limiting noise in a galvanometer is similar to that of Brownian motion,[56] the mirror and suspension being bombarded by air molecules which cause random movement of the mirror about its equilibrium position. The analysis of galvanometer noise was first made by Ising[57] and is summarized by van der Ziel.[58] Consider the differential equation of a galvanometer.

$$M \frac{d^2\theta}{dt^2} + \frac{A^2 B^2}{R} \frac{d\theta}{dt} + L\theta = ABi(t), \qquad (7.85)$$

where M is the moment of inertia of the coil and mirror, θ is the deflection, $-L\theta$ is the restoring torque, A is the coil cross-sectional area, B is the magnetic induction, R is the sum of the resistance of the galvanometer and the damping resistance, and $i(t)$ is the current to be measured. The steady state deflection of the coil due to a direct current I_0 is given by

$$\theta = \frac{ABI_0}{L}. \qquad (7.86)$$

We refer once again to the equipartition law, which states that each degree of freedom has kT of energy associated with it, divided equally between kinetic and potential energy. Since the potential energy of the system is $\frac{1}{2}\overline{L\theta^2}$, we have

$$\tfrac{1}{2}\overline{L\theta^2} = \tfrac{1}{2}kT, \quad \text{or} \quad \overline{\theta^2} = \frac{kT}{L}. \qquad (7.87)$$

The minimum current $I_{\min}$ required to deflect the galvanometer by an amount $\left(\overline{\theta^2}\right)^{1/2}$ is

$$I_{\min} = \frac{L\left(\overline{\theta^2}\right)^{1/2}}{AB} = \left(\frac{LkT}{A^2 B^2}\right)^{1/2}. \qquad (7.88)$$

If the galvanometer is undamped and no input is present, the first order term in Eq. 7.85 is zero, as is the term on the right. The equation is then of the form

$$\frac{d^2\theta}{dt^2} = -\omega_0^2\theta, \tag{7.89}$$

where ω_0 is defined to be

$$\omega_0^2 \equiv \frac{L}{M}. \tag{7.90}$$

But ω_0 is related also by definition to the reciprocal of the time constant τ_0 of the undamped galvanometer by

$$\omega_0 \equiv \frac{2\pi}{\tau_0}. \tag{7.91}$$

In addition it can be shown that the damping term gives rise to a relationship of the form

$$\frac{A^2B^2}{R} = 2n\omega_0 M, \tag{7.92}$$

where n is the damping constant, equal to unity for critical damping. Introducing these relationships into Eq. 7.88 gives

$$I_{\min} = \left(\frac{\pi kT}{nR\tau_0}\right)^{\frac{1}{2}}. \tag{7.93}$$

In order to reduce $I_{\min}$ the galvanometer must have a large time constant. This is equivalent to stating that a narrow bandwidth is required, which limits the noise.

REFERENCES

1. J. B. Johnson, *Phys. Rev.* **32**, 97 (1928).
2. H. Nyquist, *Phys. Rev.* **32,** 110 (1928).
3. See, for example, A. van der Ziel, *Noise*, Prentice-Hall, Englewood Cliffs, New Jersey (1954), p. 8; R. C. Jones, *Advances in Electronics* **5**, Academic Press, New York (1953), p. 1; J. J. Freeman, *Principles of Noise*, John Wiley and Sons, New York (1958), p. 107; D. A. Bell, *Electrical Noise*, D. Van Nostrand Co., Ltd., London (1960), p. 47.
4. See, for example, R. C. Tolman, *The Principles of Statistical Mechanics*, Oxford University Press, London (1953), p. 93.
5. H. B. Callen and T. A. Welton, *Phys. Rev.* **83,** 34 (1951).
6. W. Schottky, *Ann. Physik* **57**, 541 (1918).
7. See, for example, H. Margenau and G. M. Murphy, *The Mathematics of Physics and Chemistry*, D. Van Nostrand Co., New York (1943), p. 423.
8. See, for example, S. O. Rice, *Bell Syst. Tech. J.* **23**, 282 (1944).
9. See, for example, R. V. Churchill, *Fourier Series and Boundary Value Problems*, McGraw-Hill Book Co., New York (1941), p. 43.

10. C. D. Child, *Phys. Rev.* **32**, 492 (1911).
11. D. O. North, *R.C.A. Rev.* **4**, 441 (1940).
12. A. J. Rack, *Bell Syst. Tech. J.* **17**, 592 (1938).
13. A. H. W. Beck, *Thermionic Valves*, Cambridge University Press, Cambridge (1953), p. 231.
14. A. van der Ziel, *op. cit.*, p. 97; *Adv. in Electronics* **4**, Academic Press, New York (1952), p. 122.
15. A. van der Ziel, *Noise*, Prentice-Hall, Englewood Cliffs, New Jersey (1954), p. 102.
16. W. A. Harris, *R.C.A. Rev.* **5**, 505 (1941); **6**, 114 (1941).
17. W. Schottky, *Phys. Rev.* **28**, 74 (1926).
18. A. C. Rockwood and W. R. Ferris, *Proc. Inst. Radio Engrs.* **17**, 1621 (1929).
19. D. B. Penick, *Bell Syst. Tech. J.* **13**, 614 (1934).
20. G. E. Valley, Jr. and H. Wallman, *Vacuum Tube Amplifiers*, McGraw-Hill Book Co., New York (1948), p. 636.
21. F. Langford-Smith, ed., *Radiotron Designer's Handbook*, Wireless Press, Sydney (1953), p. 938.
22. For a general discussion of noise in semiconductors, see A. van der Ziel, *Fluctuation Phenomena in Semiconductors*, Academic Press, New York (1959).
23. Suits, Schmitz, and Terhune, *J. Appl. Phys.* **27**, 1385 (1956).
24. C. J. Christenson and G. L. Pearson, *Bell Syst. Tech. J.* **15**, 197 (1936).
25. R. W. Smith and A. Rose, *Phys. Rev.* **92**, 857 (1953); R. W. Smith, *Phys. Rev.* **97**, 1525 (1955); Shulman, Smith, and Rose, *Phys. Rev.* **98**, 384 (1955).
26. H. C. Montgomery, *Bell Syst. Tech. J.* **31**, 950 (1952).
27. L. Bess, *Phys. Rev.* **91**, 1569 (1953); *Phys. Rev.* **103**, 72 (1953).
28. S. R. Morrison, *Phys. Rev.* **104**, 619 (1956).
29. P. H. Miller, Jr., *Proc. Inst. Radio Engrs.* **35**, 252 (1947).
30. H. C. Torrey and C. H. Whitmer, *Crystal Rectifiers*, McGraw-Hill Book Co., New York (1948), Chap. VI.
31. R. L. Anderson and A. van der Ziel, *Trans. Inst. Radio Engrs.*, PGED-1, 20 (1952).
32. R. L. Petritz, *Proc Inst. Radio Engrs.* **40**, 1440 (1952).
33. W. Schottky, *Phys. Rev.* **28**, 74 (1926).
34. F. K. DuPré, *Phys. Rev.* **78**, 615 (1950).
35. A. van der Ziel, *Physica* **16**, 359 (1950).
36. A. Rose, *Photoconductivity Conference*, John Wiley and Sons, New York (1956), p. 3.
37. K. M. van Vliet, *Proc. Inst. Radio Engrs.* **46**, 1004 (1958).
38. R. V. Rollin and I. M. Templeton, *Proc. Phys. Soc.* **B66**, 259 (1953).
39. J. J. Brophy and N. Rostoker, *Phys. Rev.* **100**, 754 (1955).
40. Page, Terhune, and Hickmott, *J. Appl. Phys.* **27**, 307 (1956).
41. G. Herzog and A. van der Ziel, *Phys. Rev.* **84**, 1249 (1951).
42. R. H. Mattson and A. van der Ziel, *J. Appl. Phys.* **24**, 222 (1953).
43. F. M. Klaasen and J. Blok, *Physica* **24**, 975 (1958).
44. K. M. van Vliet and J. Blok, *Physica* **22**, 525 (1956).
45. J. Bernamont, *Ann. de Phys.* **7**, 77 (1937).
46. R. E. Burgess, *Brit. J. Appl. Phys.* **6**, 185 (1955).
47. A. van der Ziel, *Noise*, Prentice-Hall, Englewood Cliffs, New Jersey (1954), p. 326.
48. N. Wiener, *Acta Math.* (*Stockholm*) **55**, 117 (1930); A. Khintchine, *Math. Ann.* **109**, 604 (1934).
49. R. M. Ryder and R. J. Kircher, *Bell Syst. Tech. J.* **28**, 367 (1949).
50. J. A. Becker and J. N. Shive, *Elec. Engr.* **68**, 215 (1949).

51. W. K. Volkers and N. E. Pedersen, *Tele-tech* **14**, 82, 156 (Dec. 1955); **15**, 70 (January 1956).
52. R. F. Shea, *Principles of Transistor Circuits*, John Wiley and Sons, New York (1953), p. 440.
53. L. P. Hunter, *Handbook of Semiconductor Electronics*, McGraw-Hill Book Co., New York (1956), p. 11–62.
54. D. Dewitt and A. L. Rossoff, *Transistor Electronics*, McGraw-Hill Book Co., New York (1957), p. 349.
55. S. Goldman, *Frequency Analysis, Modulation and Noise*, McGraw-Hill Book Co., New York (1948), p. 380.
56. R. Brown, *Phil. Mag.* **4**, 161 (1928).
57. G. Ising, *Phil. Mag.* **1**, 827 (1926).
58. A. van der Ziel, *Noise*, Prentice-Hall, Englewood Cliffs, New Jersey (1954), p. 416.

8
Phenomenological description of infrared detection mechanisms

8.1 THE INFRARED DETECTOR

The central element in any infrared detection system is the detector, the device which transduces the energy of the electromagnetic radiation falling upon it into some other form, in most cases, electrical. In this chapter we shall describe the various phenomena used for infrared detection. Mathematical analyses of the four most widely used methods of detection and of the ultimate performance of detectors are found in Chapter 9. The measured performance of selected detectors is left to Chapter 10. Since the operation of many detectors is based upon photoeffects in semiconductors, the discussion shall begin by considering the manner in which photons interact with semiconductors. The reader is referred to Chapter 6 for an introductory discussion of semiconductor physics.

It has long been known that bodies can exchange thermal energy in three ways: by conduction, convection, and radiation. Infrared detectors are concerned with the third method, detecting thermal radiation. Consider a block of a semiconductor isolated from its surroundings in such a way that it loses very little or no energy by conduction or convection. This could be accomplished by supporting it by a very poor thermal conductor, a plastic for example, and enclosing it in an evacuated container. Thus the dominant means of thermal energy exchange is by means of radiation, which we shall assume is transmitted unattenuated by the container.

Three basic interactions of radiation with a semiconductor give rise to photoeffects. First, if the energy of the incident photon is sufficiently great, the annihilation of it by the material can free an electron completely from the material, that is, it will escape from the surface of the semi-conductor. This is known as the external photoeffect, or the photoelectric effect. Second, if the energy of the photon is not sufficient to do this, but is greater than some other value required by the material, the annihilation of it by the material will produce a free electron, a free hole, or both. This is known as the internal photoeffect. Three forms of the internal photoeffect are of primary importance. If the signal detection mechanism is based upon measuring the change in conductivity due to the added carriers, the effect is known as the photoconductive effect. If the carriers are generated at some point at which a potential barrier exists tending to separate the charges and produce a voltage, the effect is known as the photovoltaic effect. If the charges are separated by their action of diffusing in opposite directions in a magnetic field, producing a voltage, the effect is known as the photoelectromagnetic effect. Detectors based upon either external or internal photoeffects are known as photon detectors.

The third method by which radiation interacts with a semiconductor is by means of the heating of the material produced by the radiation. The energy of the radiation, absorbed by the lattice, increases the lattice vibrational energy, that is, the temperature of the material rises. Since the electrical conductivity is dependent upon temperature, measurement of the conductivity change is a measure of the absorbed energy. Detectors based upon this effect are known as bolometers. Bolometers are generally made from semiconductors, which have a negative temperature coefficient of resistivity, that is, the resistivity drops as the temperature rises. However, metal bolometers having a positive temperature coefficient of resistivity are also used.

Bolometers are one example of a class of detectors, known as thermal detectors, whose operation depends upon the heating produced by the radiation. A second example is the radiation thermocouple, formed by joining two metals or semiconductors having differing thermoelectric powers. There will be further discussion of these and other examples of thermal detectors in a later part of the chapter.

A basic distinction between photon and thermal detectors exists in the manner in which they respond to radiation. Photon detectors essentially measure the rate at which quanta are absorbed, whereas thermal detectors measure the rate at which energy is absorbed. Photon detectors, sometimes called photon counters, require incident photons to have more than a certain minimum energy before they can be detected. Thus they are selective detectors of infrared, responding only to those photons of sufficiently short wavelengths. Their response at any wavelength is proportional to the rate at which photons of that wavelength are absorbed.

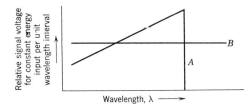

Figure 8.1. Comparison of idealized spectral responses of photon and thermal detectors.

Because the number of photons per second per watt is directly proportional to wavelength, the response of a photon detector for equal amounts of radiant power per unit wavelength interval decreases as the wavelength decreases below that corresponding to the minimum energy. (See curve A of Fig. 8.1.)

On the other hand, thermal detectors respond only to the intensity of absorbed radiant power, disregarding the spectral content of it. Thus they respond equally well to radiant energy of all wavelengths. This behavior is depicted by curve B of Fig. 8.1.

The preceding statements describe ideal behavior, not the actual behavior of every detector in each class. For example, most photon detectors have a quantum efficiency, that is, number of signal events produced per incident photon, which depends on wavelength. Thus their response is not simply a monotonically increasing function of wavelength for equal power input per unit wavelength interval, but exhibits deviations from this. Neither do all thermal detectors possess responses which are completely independent of wavelength. The surface of the sensitive element of a thermal detector may have a low absorptivity in a certain spectral interval so that the energy in that interval cannot be utilized effectively.

An alternate method exists for classifying radiation detectors, into imaging and nonimaging types. The nonimaging or elemental type, when placed at the image plane of an optical system, detects the instantaneous value of the average irradiance on the detector surface. If the optics or the detector are moved so as to scan the image plane, a time sequential "picture" of the radiation distribution in the image plane is produced which corresponds to the distribution in the object plane. On the other hand, the imaging type of detector, located at the image plane, has the entire image focused upon it. The nonuniform distribution of radiation in the image plane produces an infrared "picture" of the entire scene. Readout of the signal may be accomplished in several ways, for example, by electronic means such as are used in various television camera tubes, or by optical means utilizing the reflection or transmission characteristics of the medium upon which the image is focused.

The division into photon and thermal detectors is the more fundamental of the two distinctions, since an imaging type of detector may be simply an extended version of an elemental one. For example, a "mosaic" can be constructed from a large number of elemental detectors arranged in a linear or area pattern. When the mosaic is located in the image plane, the output from each detector represents the intensity in an elemental area of the scene being viewed. The output from the entire mosaic represents the entire scene. Rather than classifying detectors primarily into imaging and nonimaging types, the more fundamental distinction between photon and thermal detectors will be used.

8.2   FIGURES OF MERIT

The infrared system is often designed around the characteristics of the infrared detector. They are defined by certain figures of merit describing the performance of the detector under specified operating conditions. Although a large number of figures of merit exist, the most important ones can be grouped into four categories, providing the answers to the questions:

1. What is the minimum intensity of radiant power falling on the detector which will give rise to a signal voltage equal to the noise voltage from the detector?
2. What signal will be obtained per unit radiant power falling on the detector?
3. How does the signal vary with the wavelength of the radiation?
4. What is the modulation frequency response of the detector?

To answer these questions that relate to a specific detector, it is necessary that the conditions under which the figures of merit are measured be defined as precisely as possible. Many infrared detectors exhibit a signal which is ,a function of the wavelength of the radiation, being sensitive only to radiation in a given spectral interval and exhibiting a response which is dependent on wavelength within that interval. This wavelength dependence, termed, in general, the "spectral response" of the detector, determines how effectively the detector makes use of the radiation from a heated body. We have seen in Section 2.3.2 how the wavelength distribution of the radiation from a heated body depends upon the temperature of the body according to Planck's law. As the body is raised to a higher temperature, the fraction of the radiation lying below (shorter than) any given wavelength increases. This new distribution will coincide to a greater or lesser extent with the region of the spectrum to which the detector is sensitive. Before the first question can be answered, it is necessary to specify the spectral distribution of the radiating source. Otherwise the answer obtained will be ambiguous.

Although specifying the spectral distribution of the source removes some ambiguity, it is not sufficient to make the answer unique. Like all other types of electromagnetic radiation detectors, the ultimate performance of infrared detectors is limited by noise. Depending upon the mode of operation of the detector, the noise mechanism may be any of several discussed in Chapter 7. Because most detectors are based upon photoeffects in semiconductors, the mechanisms encountered most frequently are those which pertain to semiconductors: current noise, thermal noise, generation-recombination noise, and photon noise, discussed in Chapter 9. The measured noise voltage for all noise mechanisms is proportional to the square root of the electrical bandwidth, and for some mechanisms is a function of frequency. If the signal-to-noise ratio for a known source power and spectral distribution is to be given without ambiguity, the electrical frequency interval must be specified.

Some ambiguity still remains. Detectors, especially those made from semiconductors, exhibit signal, noise, and resistance which are strongly temperature dependent. In addition, signal and noise depend upon the area of the detector responding to the radiation. Both operating temperature and sensitive area must be specified. For certain detectors, referred to as being photon noise limited, the optical field of view and background temperature must be specified. This is discussed in detail in Chapter 9. Since some detectors do not respond in a linear manner to the radiant intensity, that is, the signal does not double if the intensity is doubled, the intensity at which the measurements are made should also be specified.

With these considerations in mind, we shall define certain figures of merit which answer the first question. One of the most commonly used figures of merit relating the radiation power capable of producing a signal voltage equal to the noise voltage, that is, a signal-to-noise ratio of unity, is the noise equivalent power, symbolized NEP or P_N. The NEP is defined as the rms value of the sinusoidally modulated radiant power falling upon a detector which will give rise to an rms signal voltage equal to the rms noise voltage from the detector. The temperature of the black body radiation source is usually 500° K. The reference bandwidth is commonly 1 cps or 5 cps. The center frequency is normally 90 cps, 400 cps, 800 cps, or 900 cps. The reference temperature of the detector is either the laboratory ambient, about 22° C (295° K) or the operating temperature if the detector is cooled. The reference area is usually 1 cm². The intensity of radiant power and the field of view are not standardized, but should be specified. The noise equivalent power for a 500° K source, 900 cps chopping frequency and 1 cps bandwidth, is written as NEP (500° K, 900, 1). Note that the detecting capability of the detector improves as the

NEP *decreases.* The NEP is determined experimentally[1] by measuring the signal-to-noise ratio in a specified narrow electrical bandwidth, called the measurement bandwidth. The value obtained for a known amount of radiant power is extrapolated linearly to the power required to give a signal-to-noise ratio of unity. Thus the noise equivalent power P_N is given by

$$P_N \equiv \mathscr{H} A_D \left(\frac{V_N}{V_S}\right) \frac{1}{(\Delta f)^{\frac{1}{2}}}, \tag{8.1}$$

where $\mathscr{H}$ is the rms value of the irradiance falling on the detector of area A_D, and V_N/V_S is the ratio of the rms noise voltage in the bandwidth Δf to the rms signal voltage. Here it is assumed that Δf is sufficiently small so that the noise voltage per cycle within Δf is independent of frequency. The units of NEP are watts/(cps)$^{\frac{1}{2}}$. Common practice is to omit the (cps)$^{\frac{1}{2}}$, referring to NEP in watts.

A second figure of merit describing the performance is the detectivity, symbol D, which is simply the reciprocal of the NEP, the units being (cps)$^{\frac{1}{2}}$/watt. Thus

$$D \equiv \frac{1}{P_N}. \tag{8.2}$$

The same measuring conditions must be specified. The detectivity expresses the rms signal-to-noise voltage ratio obtained per watt of radiant power. Because many types of infrared detectors exhibit a non-linear dependence of signal voltage upon radiant power, the large signal-to-noise ratio predicted for 1 watt of radiant power, a value extrapolated from that measured for, say, a microwatt of radiant power, may not be realized in practice. Since a higher value of detectivity indicates a better detector, the detectivity is perhaps a better choice psychologically than NEP for comparing detectors.

Another common figure of merit is the noise equivalent input, NEI, sometimes referred to as the equivalent noise input, ENI. This is the radiant power per unit area of detector, that is, the irradiance, required to give rise to a signal-to-noise ratio of unity. The conditions of measurement are the same as those for the NEP. Thus the NEI is simply the NEP divided by the detector area.

$$\text{NEI} = \frac{P_N}{A_D} = \frac{1}{A_D D}. \tag{8.3}$$

The units of NEI are watts/cm^2 (cps)$^{\frac{1}{2}}$.

Most detectors exhibit a noise equivalent power which is directly proportional to the square root of the area of the detector. This is illustrated in the theories of several detection mechanisms developed in

Chapter 9. An area independent figure of merit can be obtained by dividing the NEP by the square root of the area, or multiplying the NEI by the square root of the area. The reciprocal of this quantity is known as D^*, pronounced "dee-star."

Thus

$$D^* \equiv \frac{1}{(\text{NEI})A_D^{1/2}} = \frac{A_D^{1/2}}{P_N} = DA_D^{1/2}. \tag{8.4}$$

The measured value of D^* for a detector is written in a manner analogous to that of the NEP, such as D^* (500°K, 900, 1). The units of D^* are cm(cps)$^{1/2}$/watt. The reference bandwidth is always 1 cps.

Confusion has occurred in the use of the term detectivity. The detectivity D, the reciprocal of the noise equivalent power, is no longer a widely used figure of merit, having been replaced by D^*. It has become common usage to refer to D^* as the detectivity. Wherever the term detectivity occurs herein, except in the paragraph where D was defined, it refers to D^*.

A figure of merit which is convenient to use to characterize the performance of a detector whose NEP is proportional to the square root of the area and which is current noise limited is the Jones "S," or simply "S." We have seen that the use of D^* removed the need for specifying the area of the detector. The factor S removes the need for specifying the area and also the measuring frequency for those detectors limited by noise having an rms power which varies inversely with frequency. In Chapter 7 it was shown that the current noise voltage is inversely proportional to the square root of the measuring frequency. The figure of merit S removes this frequency dependence by multiplying the NEP/$A_D^{1/2}$ by the square root of the measuring frequency. Thus

$$S = \frac{P_N}{A_D^{1/2}} f^{1/2} = \frac{f^{1/2}}{D^*}. \tag{8.5}$$

The units of S are watts/cm. This figure of merit can be used only for detectors which are limited by noise having a power spectrum inversely dependent upon frequency, and whose NEP is proportional to the square root of the area. It is no longer widely used.

The figures of merit with which we have been concerned refer to the response of the detector to radiation from a black body source of a specified temperature. Since the responses of many types of detectors are dependent upon the wavelength of the incident radiation, the values of NEP, NEI, D, D^*, and S can be expressed in terms of their responses to monochromatic radiation. In this case the wavelength λ is specified and the figures of merit are written as NEP_λ, NEI_λ, D_λ, D_λ^*, and S_λ. If the

ELEMENTS OF INFRARED TECHNOLOGY

performance at only a single wavelength is given, it is usually that at which the detector performs best.

Although other figures of merit exist, those given above are the ones most widely used to characterize the signal-to-noise performance of an infrared detector. Of these, D^* and D_λ^* are the most popular. The figure of merit D^{**} (dee-double-star), applicable to photon noise limited detectors, is defined in Section 9.4.3.3. The reader is referred to the review articles by Jones,[2] who has made a major contribution to specifying the performance of infrared detectors, for discussion of his quantities M_1, M_2, "the detectivity in the reference condition A," "responsive quantum efficiency" and "detective quantum efficiency."

Having considered the problem of characterizing the signal-to-noise ratio to be expected from a detector for a given intensity of radiant power, we continue to the second question, considering the more simple problem of characterizing the signal voltage per unit radiant power. Since our interest is only in the magnitude of the signal, we can neglect the noise. Because the bandwidth serves only to limit the noise voltage, it is no longer necessary to specify it. In addition, as long as the signal frequency is kept sufficiently low compared to the reciprocal of the response time, discussed in the answer to the fourth question, there is no reason to specify the frequency. A knowledge is still required, however, of the black body temperature, the temperature of the detector, and the area of the detector. Knowing these, we define the responsivity $\mathscr{R}$ to be the rms signal voltage V_S per unit rms radiant power P incident upon the detector.

$$\mathscr{R} \equiv \frac{V_S}{P} = \frac{V_S}{\mathscr{H} A_D}. \tag{8.6}$$

The units of responsivity are volts/watt. Again, the temperature of the black body is usually 500°K. We see that the responsivity is related to the NEP and D^* by

$$\mathscr{R} = \frac{V_N}{P_N(\Delta f)^{1/2}} = \frac{D^* V_N}{(A_D \Delta f)^{1/2}}. \tag{8.7}$$

We also specify a spectral responsivity $\mathscr{R}_\lambda$, the responsivity to monochromatic radiation of wavelength λ. If the value at only a single wavelength is given, it usually refers to that at which the detector exhibits greatest responsivity.

The third question deals with the spectral response of the detector, the manner in which some figure of merit referring to excitation by monochromatic radiation varies with wavelength. The spectral response is displayed as a graph, such as that of Fig. 8.1, having wavelength on the abscissa and a figure of merit of the detector on the ordinate. Frequently the ordinate is given as relative response, defined as the relative signal voltage per unit monochromatic radiant power. Since the absolute value of the radiant power within any given spectral interval is difficult to measure, the ordinate is expressed in terms of relative values of signal voltage per unit monochromatic radiant power referred to that at the wavelength of maximum response, the values extending from zero to one hundred. A much more informative display is obtained by plotting NEP_λ, NEI_λ, or D_λ^* as a function of wavelength.

Although a spectral plot gives much more information concerning the response of a photon detector, a numerical value of interest is the long wave limit or cutoff wavelength of the detector, the longest wavelength to which it is sensitive. This is an ambiguous characteristic, since the actual response of a detector does not suddenly go to zero at some wavelength, but rather decreases over an interval of perhaps one or more microns. To specify the long wave limit, the wavelength is stated at which the response has dropped to a given fraction of the value at the peak. This is taken sometimes as the point at which it has dropped to 50% of the maximum, or to 1% of the maximum, there being no general agreement upon this value. Another limit, termed the "equivalent cutoff wavelength," is determined in the following manner. The detector is exposed first to radiation from a black body at one temperature, then to radiation from a second black body at a different temperature, but having the same area and located at the same position as the first black body. The ratio of the signal obtained from these two sources is measured and set equal to the ratio which would have been obtained by using in the same experiment a hypothetical radiation detector whose response is independent of wavelength up to a certain cutoff wavelength λ_0 and zero beyond. By definition, λ_0 is then the equivalent cutoff wavelength. See Appendix II of Chapter 10 for additional discussion.

Finally we come to the fourth question concerning the frequency response of the detector. One of the factors of paramount importance in the selection of an infrared detector for a given task is the dependence of the responsivity and detectivity upon the rate at which the incident radiation is modulated or "chopped." In most applications the user wishes both figures of merit to be flat, that is, frequency independent, over the range of frequencies in which he is interested. However, many detectors do not exhibit this desirable characteristic.

Consider the responsivity of a detector, that is, the rms output voltage per unit rms radiant flux from a source at a designated temperature. As we have pointed out, this figure considers only the detector signal, ignoring the noise. We have seen in Chapter 6 that semiconductors have associated with them a characteristic time, the carrier recombination time or lifetime. This recombination time limits the rapidity with which a photon detector

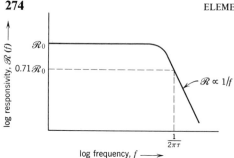

Figure 8.2. Frequency dependence of responsivity.

will respond to chopped radiation. At slow chopping rates, that is, low modulation frequencies, the instantaneous carrier density follows the rise and decay of the radiation upon the detector surface. At sufficiently rapid rates, that is, high frequencies, it can no longer do so. Because the rise and decay times follow in general an exponential law, the frequency response of the signal voltage for unit radiant intensity is similar to that of a low pass filter, that is,

$$\mathscr{R}(f) = \frac{\mathscr{R}_0}{(1 + 4\pi^2 f^2 \tau^2)^{1/2}}, \qquad (8.8)$$

where $\mathscr{R}(f)$ is the responsivity at frequency f, $\mathscr{R}_0$ is the responsivity at zero frequency and τ is the response time, also referred to as the time constant. Equation 8.8 is valid only for those materials in which the recombination is governed by a monomolecular process, that is, one in which the recombination rate is proportional to the instantaneous value of the excess carrier concentration only. However, most materials used in infrared detectors obey this, and Eq. 8.8 is generally applicable. Equation 8.8 is plotted in Fig. 8.2.

At low frequencies such that $f \ll 1/2\pi\tau$ the responsivity is frequency independent. However, at higher frequencies it begins to drop, attaining a value $\mathscr{R} = 0.71\mathscr{R}_0$ at $f = 1/2\pi\tau$. At higher frequencies, at which $f \gg 1/2\pi\tau$, the responsivity is inversely proportional to frequency.

In semiconductors characterized by a simple recombination mechanism, the value of τ corresponds closely to the carrier lifetime. We have seen in Section 6.8 that in conductivity phenomena the lifetime of importance is the majority carrier lifetime, whereas in diffusion processes the minority carrier lifetime is the one of importance. Thus, for example, photoconductors, which are discussed in Section 8.3.2, have response times dictated by the majority carrier lifetime, whereas PEM and photovoltaic detectors have response times dictated by the minority carrier lifetime.

Some detectors are characterized by two time constants, one much longer than the other. In this case the responsivity has the form shown in

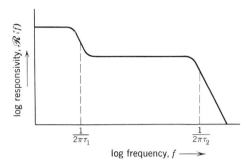

Figure 8.3. Responsivity of a double time constant detector.

Fig. 8.3. Double time constants have been found in some detectors to be associated with different regions of their spectral response. Irradiation with monochromatic radiation of one wavelength may cause one time constant to dominate, whereas another wavelength may cause the other to be predominant. In most applications two time constant behavior is undesirable.

Let us consider now the frequency dependence of the detectivity. Since D^* gives the area independent signal-to-noise ratio per unit radiant flux for unit bandwidth and a defined spectral distribution, it can be obtained by dividing the product of responsivity and square root of area by the noise voltage per unit bandwidth. For white noise limited detectors, the frequency dependence of the detectivity and the responsivity have the same form, since the noise is frequency independent. However, for current noise limited detectors this is not true. In this case, the noise voltage has a $(1/f)^{1/2}$ dependence and D^* has the form

$$D^*(f) = \frac{kf^{1/2}}{(1 + 4\pi^2 f^2 \tau^2)^{1/2}},$$ (8.9)

where $D^*(f)$ is the detectivity at frequency f and k is a proportionality constant. This frequency dependence of current noise limited detectors is shown in Fig. 8.4.

The value of frequency which maximizes the detectivity is obtained by

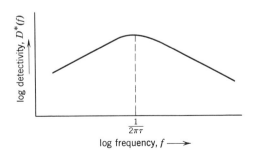

Figure 8.4. Dependence of D^* upon frequency for $1/f$ noise limited detectors.

taking the derivative of $D*$ with respect to f and setting it equal to zero. The resultant frequency f_{max} which maximizes $D*$ is found to be

$$f_{max} = \frac{1}{2\pi\tau}.$$
(8.10)

Thus current noise limited detectors operate with highest detectivity at f_{max}. Any value of detectivity quoted for current noise limited detectors should be accompanied with a notation of the frequency at which it was measured.

There are two basic methods of determining the frequency response. The first is the direct method, in which the response is determined experimentally as a function of the frequency of a mechanical chopper which interrupts the radiation in a periodic manner. A variation of this method involves measuring the response at two frequencies, f_1 on the frequency independent portion of the response curve and f_2 on the frequency dependent portion, in order to determine the response time. The ratio of the two signals V_1 and V_2 is

$$\frac{V_1}{V_2} = \left(\frac{1 + 4\pi^2 f_2^2 \tau^2}{1 + 4\pi^2 f_1^2 \tau^2}\right)^{1/2}.$$
(8.11)

Solving this for τ gives

$$\tau = \frac{1}{2\pi}\left(\frac{V_1^2 - V_2^2}{f_2^2 V_2^2 - f_1^2 V_1^2}\right)^{1/2}.$$
(8.12)

The second method depends upon measuring the carrier lifetime, then identifying the lifetime with the response time. Excited carriers usually decay in an exponential manner. (See Eq. 6.50.) The detector is excited by a very short duration radiation pulse, produced, for example, by techniques involving a high speed rotating mirror[3] or by spark excitation.[4] The decay curve is displayed upon an oscilloscope and the point at which the signal voltage decays to $1/e$ of its maximum value is taken to be the lifetime, this being called the response time of the detector. The spark and rotating mirror techniques, more complicated generally than the mechanical chopper, are reserved for the more difficult problems involving measurement of very short response times.

The importance of specifying all the conditions under which the measured values of detector performance are determined must again be stressed. In some detectors, for example, the response time depends upon the wavelength of the exciting radiation. In many it depends upon the radiation intensity and the detector temperature. As another example, the spectral detectivity of a detector exhibiting multiple time constant behavior depends upon the chopping frequency of the exciting radiation.

It will also depend upon the source intensity and detector temperature in most cases. The field of view and background temperature influence the performance of photon noise limited detectors. Even such a parameter as the ambient humidity is known to influence detector performance. Before making his final choice of detectors, the user should be sufficiently aware of the conditions under which the performance of the detectors has been measured to be able to accurately predict performance in his application.

8.3 PHOTON EFFECTS

Having examined the figures of merit needed to specify the performance of an infrared detector, we can now proceed with a study of the various mechanisms by which detectors operate. We shall consider first the mechanisms used in photon detectors, then those used in thermal detectors. The reader is referred to the books by Smith, Jones, and Chasmar,[5] and by Moss[6] for comprehensive treatments of detection methods and detectors.

8.3.1 The Photoemissive Effect

If radiation of wavelength less than a critical value falls upon the surface of certain materials, it is found that electrons are emitted from the surface. Phototubes using this effect, termed the photoemissive effect, are very widely used. However, the nature of the effect is such that photoemissive detectors operate in the ultraviolet, visible, or very near infrared. The image converter, discussed below, is the only type which has found much application in infrared technology, but its use is limited to the very near infrared. We shall begin our discussion of photon detectors by considering the photoemissive effect. An excellent reference is Hughes and DuBridge.[7] Morton[8] discusses infrared photoemission.

The photoemissive effect, also termed the external photoelectric effect or simply the photoelectric effect, was discovered by Hertz in 1887 and explained by Einstein in 1905. Einstein propounded the existence of quanta, each having a discrete amount of energy equal to hv, where h is Planck's constant and v is the frequency of the radiation. He considered the interaction of quanta with an absorbing solid medium and arrived at his famous equation for the energy E of the emitted electron.

$$E = \tfrac{1}{2}mv^2 = hv - p, \tag{8.13}$$

where $\tfrac{1}{2}mv^2$ is the kinetic energy of the photoelectron and p is a constant determined by the medium. His equation has been verified over a range of spectral frequencies extending from the x-ray region to the near infrared.

The significant factor to be noted in Eq. 8.13 is that the energy of the emitted electron is a function not of the intensity of the incident light but only of the frequency or wavelength. Increasing the intensity increases the number of photoelectrons but not their energies. Note also that excitation begins only when the energy per quantum $h\nu$ exceeds the threshold value p, which can be written in terms of a threshold energy $h\nu_0$ or hc/λ_0. Thus emission occurs only if the photon frequency exceeds ν_0 or the wavelength is less than λ_0. The amount of energy lost by an electron as it leaves the emitter surface is also given by p. We write this energy as $q\phi$, where ϕ, the "work function" of the material under investigation, is determined by the material of the emitter surface, and q is the electronic charge. We can write the threshold in terms of the work function as

$$q\phi = h\nu_0 = \frac{hc_0}{\lambda_0}. \tag{8.14}$$

If we express $q\phi$ in electron volts and λ_0 in microns, the threshold is given by

$$\lambda_0 = \frac{1.2406}{q\phi}. \tag{8.15}$$

Thus in order to have a long wavelength response for the photoemissive effect it is desirable to have a photocathode (electron emitting surface) with a low (small) work function.

The work functions of materials are closely related to the atomic binding forces; the smaller the binding force, the lower the work function. It can be shown that the binding force is reduced as we progress from the top to the bottom of the periodic table, and from the right to the left. Thus we expect that the alkali metals, lithium, sodium, potassium, rubidium, and cesium, falling in the first column, should have low work functions, with cesium, the element at the lower end of the first column, having the smallest work function. (Francium, which is radioactive, has not been considered.) This has been found to be true, the work function of Cs being 1.9 ev. Thus the threshold wavelength for Cs, determined from Eq. 8.15, is $\lambda_0 = 0.65\ \mu$. Since this lies in the visible portion of the spectrum, we see that no elemental photoemitter has an infrared response.

It has been found possible to produce multielement photoemissive surfaces having responses to radiation in the near infrared. The one most widely used in infrared applications is the silver-oxygen-cesium, or Ag-O-Cs surface, also called the S-1 surface. The work function of this surface is slightly less than 1 ev, indicating a response to about $1.2\ \mu$. The method of preparation consists in depositing a thin layer of silver on the inner surface of an evacuated glass envelope. Oxygen is then admitted at a pressure of approximately 500 μ of Hg and a glow discharge is initiated, oxidizing the silver layer so as to change its optical transmission from 10 to 90%. The envelope is then evacuated and more silver evaporated so as to reduce the optical transmission to 50%. After heating the envelope to 180°C, cesium is admitted slowly until the thermionic emission, that is, the electronic emission from the heated surface, passes through a maximum. The flow of Cs is immediately interrupted and the tube baked at 200°C until the thermionic emission passes through a second maximum. The tube is then cooled and can be sensitized further by evaporating a very small quantity of silver onto the cathode, then baking at 100°C. The interested reader is referred to Sommer[9] for additional information. The response of this surface is shown in Fig. 8.5.

A figure of merit for photoemissive surfaces is the quantum efficiency, the number of emitted electrons per incident photon. The quantum efficiency of most photocathodes is very small, lying in the range 10^{-5} to 10^{-1}, and is a function of the wavelength of the exciting radiation. The value at peak sensitivity for the Ag-O-Cs photocathode is about 3×10^{-3}. Most photocathodes exhibit a response peaked near the threshold. At the threshold, where the exciting radiation is absorbed very strongly, incident photons excite electrons near the surface of the photocathode. This is termed the surface or selective photoeffect. At energies just above threshold, that is, at wavelengths just below the cutoff wavelength, the density of electrons per unit energy increases, causing the response to rise. However, at wavelengths shorter than the peak, the efficiency of interaction of the radiation with the electrons decreases and the response drops. At even shorter wavelengths the more penetrating radiation begins to excite electrons at locations lying further from the surface. At this point the response rises again. This is called the volume photoeffect.

It has been found that photocathodes consisting of an alkali metal deposited upon a metallic base exhibit a surface photoeffect dependent upon the polarization of the incident radiation. If the electric vector has a component perpendicular to the surface, the maximum in the response

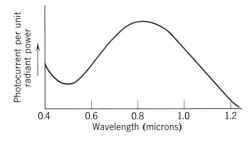

Figure 8.5. Spectral response of the Ag-O-Cs (S-1) photoemissive surface. (Courtesy RCA.)

curve near the photoelectric threshold appears. If there is no perpendicular component, there will be no maximum. For rough surfaces there will always be a perpendicular component, no matter what the orientation of the electric vector. For smooth surfaces this is not so and they exhibit marked polarization effects in the spectral region near the threshold.

8.3.1.1   Photoemissive photocells. The most elemental form of photoemissive cell is the vacuum photocell, consisting of simply a photocathode and an anode in an evacuated enclosure. The photocathode has an extended surface, frequently shaped in the form of a cylindrical section. The anode is a wire extending down the axis of the cylinder. In this geometry very little radiation is lost on account of the shadowing effect of the anode on the cathode.

Vacuum photocells are characterized by a very short response time, less than 10^{-8} sec. This is accompanied by a spectral responsivity dictated completely by the characteristics of the photocathode. If increased responsivity is desired, a phototube containing an inert gas at a low pressure may be used. The electrons emitted by radiation incident on the photocathode are accelerated toward the anode by the applied electric field. At some point in their transit they have acquired sufficient energy to ionize any gas molecules with which they collide. The photoelectrons, plus those electrons produced by ionizing collisions, continue toward the anode. The positive gas ions move toward the photocathode. If they acquire sufficient energy from the field, they may in turn cause impact ionization at the photocathode, freeing more electrons. These in turn move toward the anode, acquiring energy from the field and producing more ions, and so on. In this manner the current may be amplified by a factor of as much as ten. On the other hand, the response time of a gas-filled photocell is much longer than that of a vacuum photocell because of the low mobility of the ions. The response of a gas photocell decreases at frequencies above about 1 kc, the exact value depending upon the tube type.

8.3.1.2   Photomultiplier. High responsivity and excellent frequency response are combined in the multiplier phototube or photomultiplier.[10] The photomultiplier is a vacuum photoemissive detector containing a number of secondary emitting stages termed "dynodes." In the secondary emission process, electrons impinging against certain materials called "secondary emitters" cause additional electrons to be ejected from the surface of the materials. Since these secondary electrons can be more numerous than the primaries, amplification is possible. By arranging a number of surfaces such that secondaries from one are used to produce secondaries from the next, multiplication is obtained. The multiplication factor M of the multiplier is related to the secondary emission ratio δ, that is, the number of secondaries emitted per primary, and the number of stages n by $M = \delta^n$. The factor δ may have values up to ten, whereas the number of stages is frequently ten or twelve. Thus very high multiplication factors are obtainable. The frequency response of a photomultiplier, although far superior to a gas phototube, is poorer than a vacuum photo-tube. The secondary emission process causes a spread in the transit times of the electrons, limiting the frequency response to about 100 Mc.

Those photoemissive surfaces having a low photoelectric work function will also emit electrons by thermal excitation (thermionic emission) which are indistinguishable from those produced by photon excitation. This thermal background sometimes limits the minimum photosignal which can be detected. The background can be greatly reduced by cooling the photo surface to the temperature of dry ice (195°K) or lower.

8.3.1.3   The image converter. Several years prior to World War II the work on photoemissive cathodes by DeBoer, Teves, and others had led to the development of infrared image converter tubes,[11] which "convert" infrared images to visible images. Later work by Kluge,[12] Görlich,[13] and Asao[14] improved the long wavelength response of photoemissive cathodes so that improved tubes were available to both sides for use in World War II.

The mode of operation of the image converter can be described most conveniently by reference to three figures. Figure 8.6 shows (*a*) the spectral radiance of a 2870°K tungsten ribbon source, (*b*) the visibility curve for the eye, and (*c*) the spectral transmission of a Corning type 7-56 filter. Figure 8.7 shows the interior arrangement of the image converter tube. Figure 8.8 shows (*a*) the spectral response of a Ag-O-Cs photoemissive cathode, often designated as type S-1, and (*b*) the spectral luminosity of a zinc cadmium sulfide cathodoluminescent phosphor, often designated as type P2.

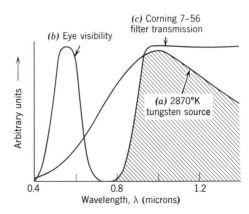

Figure 8.6. Spectral characteristic of source, filter, and eye. (Courtesy Corning Glass Works, Corning, New York.)

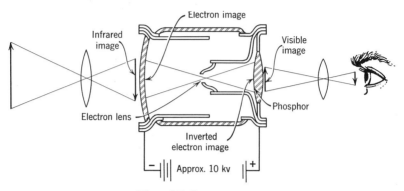

Figure 8.7. Image converter.

An incandescent source operated near 2870°K, equipped with an infrared transmitting filter such as a Corning type 7-56, is used to irradiate the scene to be viewed. The source temperature and filter transmission are chosen so that even though no visible radiation is projected, the scene is flooded with radiation of about 1 μ wavelength. This spectrum of infrared radiation, shown cross-hatched in Fig. 8.6, is reflected by the scene, according to the spectral reflectivity of each object, to an optical system, which focuses the reflected radiation onto the photoemissive cathode of an image converter. (See Fig. 8.7.)

The spectral response of the photocathode (see Fig. 8.8) is such that the incident photons cause an "electron image" of the scene to be formed. The photoemissive current density at each small element of the image corresponds to the irradiance of the infrared image in that portion of the scene. After acceleration down the tube by an applied electric field, the electrons impinge upon a cathodoluminescent phosphor which covers the

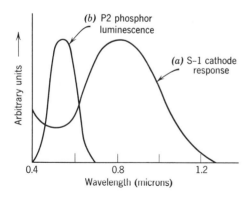

Figure 8.8. Spectral characteristics of photocathode and phosphor. (Courtesy of RCA.)

glass optical surface at the far end of the tube, causing the scene to appear on the phosphor as a visible image having the spectral distribution shown in Fig. 8.8. The image will be inverted because of the action of the electrostatic lens which focuses the electrons on the phosphor. Since the magnification in a typical tube is 0.8, the image will be reduced in size. An eyepiece used to view the image magnifies and erects it.

In order to make image converters most useful, three characteristics need to be optimized: conversion index, resolution, and background brightness.

Conversion index. The conversion index, C.I., provides a measure of the relative intensities of the visible image on the phosphor and the infrared image on the photocathode. It is given by

$$\text{C.I.} \equiv \frac{L_0}{L_i \tau} \, ,$$

where L_0 is the number of lumens of visible radiation emitted by the phosphor, L_i is the number of lumens of $2870°K$ tungsten light used to illuminate the cathode without the Corning type 7-56 filter in place, and τ is the integrated transmittance of the filter. When L_i lumens are emitted by the test source, $L_i \tau$ units of radiation are incident on the cathode. The value of τ has been found to be approximately one-tenth by numerical integration. Increasing the quantum efficiency of the cathode or extending its response toward longer wavelengths tends to increase the conversion index. Increasing the phosphor efficiency, expressed in light output per electrical power input, also increases the conversion index.

Modern image converter tubes[15] have conversion indices as high as 40. This has been achieved by the development of:

1. Very uniform cathodes of high quantum efficiency. Extending the long wavelength limit of the cathode increases the conversion index but not the visual contrast since the thermionic dark current increases as the work function of the cathode decreases. Thermionic emission and field emission (cold emission) increase the background brightness and reduce visual contrast.

2. Techniques which minimize leakage currents within the tube. Leakage limits the maximum voltage which can be applied to increase the energy of the electrons and hence the brightness of the phosphor.

3. Methods for preparing very thin, highly reflecting aluminum films to coat the phosphor. A very thin, continuous film, easily penetrated by the electrons, is effective in reflecting all of the luminescence outward toward the viewing system.

Resolution. The resolution in the image available at the phosphor measures the smallest detail which can be observed in the image. It is specified as the number of white and black line pairs per millimeter of phosphor which can just be resolved or distinguished as separate entities. Some image converters can resolve as high as 40 line pairs per millimeter.

Resolution depends upon image brightness and therefore upon the conversion index. It depends also upon contrast. Phosphor coatings which are too thick reduce resolution. The quality of the electrostatic lens, that is, the electrode tolerances, affects the resolution especially at the periphery of the viewing screen.

Background brightness. The background brightness is a measure of the brightness at the phosphor when the cathode is not being illuminated. There has been some difference of opinion about the importance of attempting to achieve a low background brightness in these tubes. The measurements of Blackwell[16] support the belief that the distance at which objects can be observed and the resolution which can be achieved is unaffected by variations in the average background brightness in the range of average background brightnesses normally encountered. Small bright spots resulting from cold cathode emission are obviously detrimental.

Image converters have been produced in a wide variety of configurations. An excellent review of these devices has been written by Klein.[17] The reports of two symposia[18,19] discuss image converters and other electronic imaging devices.

8.3.1.4 The Phothermionic image converter. The Phothermionic image converter[20] (see Fig. 8.9), is a thermal imaging device based upon the temperature dependence of photoelectric emission.

An infrared image is formed by an optical system upon the photo-emissive surface of the tube, termed the "retina," producing a temperature distribution on the surface related to that in the scene being viewed. A light beam from a flying spot scanner is swept across the photoemissive surface. Since the photoelectric emission of the surface is temperature dependent, the temperature distribution corresponding to the infrared

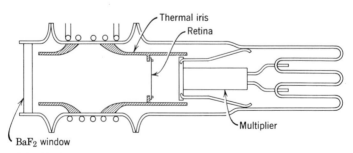

Figure 8.9. Phothermionic image converter. (After Garbuny, Vogl, and Hansen.[20])

image on the photocathode causes the photoemission to vary from point to point on the photocathode. The photoelectrons are collected by a secondary emission multiplier. The output signal modulates the intensity of the beam of a kinescope which is synchronized with the scanning beam. Since the performance depends upon the heating effects of the radiation, the spectral response is relatively independent of wavelength.

Limitations to the performance of the several tubes constructed arose primarily from shot noise in the emitted photoelectrons, nonuniformities in the retina, and lateral heat conduction through the retina. Of these, the first two were found to offer the greatest difficulties. In one of the last tubes produced, the limiting factor was due to retina nonuniformity. Using this tube, with $f/1.5$ optics and 4 Mc bandwidth, a temperature difference in the object field of 15°C caused a variation in the emitted current equal to the "noise" due to the nonuniformities of the retina.

8.3.1.5 The image dissector. The image dissector is a television camera tube which uses an S-1 photoemissive cathode having its spectral peak at 0.8 μ. It is capable of producing a television picture, using only near infrared radiation, with a better "gray scale" than other tubes. The term gray scale is used to specify the number and magnitude of the gradations of brightness appearing in the image. If the scene can be irradiated at high brightness, an image dissector will also provide very high resolution. However, the dissector is a "nonstorage" type of pick-up tube, that is, it does not integrate the signal from photons which strike each picture element over a period equal to the time required to scan the entire scene once, that is, the frame time. Instead, the photons which arise in each little element of a scene are utilized only during the brief instant that the element is being scanned.

Figure 8.10 shows the principal parts of the tube. A lens is used to focus the scene on the cathode. Just as in the image converter tube, electrons are emitted from the vacuum side of the cathode with a distribution in numbers corresponding to the energy distribution in the optical image. They are accelerated down the tube by a series of rings which produce an electric field whose gradient is parallel to the axis of the tube. Three sets of coils surround the tube. Two are deflection coils excited by sawtooth waves which scan the electron image past the aperture of a secondary emission multiplier. The outermost coil is a long solenoid used to focus the image. As the electron distribution is scanned past the multiplier aperture, electrons corresponding to each small element of the scene enter in succession to produce a current variation at the multiplier output which is related to the scene brightness.

In a typical application the image is divided into about 130,000 elements and 30 images are transmitted each second. Therefore each element is

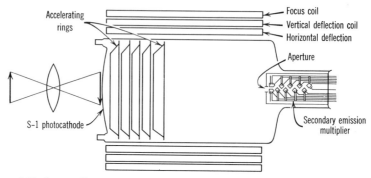

Figure 8.10. Image dissector. (Courtesy of ITT Federal Laboratories, Fort Wayne, Indiana.)

supplying current to the aperture for about 1/4,000,000 of a second each frame time. The total cathode current, even for brilliantly illuminated scenes, is of the order of 10^{-4} amp. Thus each picture element (at a highlight) is emitting about 10^{-9} amp or about 6×10^9 electrons per second. During the 2.5×10^{-7} sec that it is being scanned past the aperture, a bright element in the image will then supply 1500 electrons to the first stage of the multiplier. The shot noise fluctuations in this number will amount to about $(1500)^{1/2}$ or approximately 40, imposing a stringent upper limit on the attainable signal-to-noise ratio. This ratio is further decreased by shot noise added by the first multiplier stage. For moderate scene illumination and especially for the darker elements in the image, shot noise causes a very marked deterioration of the reproduced scene. Because of this very fundamental limitation, the image dissector has not been used very much in infrared applications.

8.3.1.6   The image orthicon. The image orthicon should be included in any discussion of infrared image tubes even though its spectral response does not extend very far into the infrared. Its extraordinary ability to provide useful images at very low levels of illumination permits its use as an aid to "seeing in the dark." This type of tube has an extremely complicated structure which requires complex procedures for the fabrication and mounting of its component parts and precise control of the vacuum exhaust schedule in order to achieve good performance. The image orthicon contains a thermionic cathode, a photoemissive cathode, a high resistivity target, and a secondary emission multiplier, all of which must be processed in the same envelope.

Figure 8.11 shows the operation of an image orthicon. A scene is imaged onto the photoemissive cathode, causing electrons to be emitted according to the energy distribution in the image. Acquiring 300 ev from an electric field, they bombard a glass target. Since the secondary emission ratio is greater than unity, electrons emitted by the bombardment of the target are more numerous than those doing the bombarding and the target becomes charged positively. The positive charge distribution corresponds to the electron distribution and therefore to the distribution of energy in the image on the photocathode.

A beam of low velocity electrons scans the target, depositing charge to neutralize the positive charge built up on it. The target is only about 3 μ thick and has a resistivity at room temperature of approximately 10^{11} ohm cm. In typical applications the transverse resistance from picture element to picture element is much larger than the resistance through the target so that the side of the target toward the scanning beam quickly assumes the potential of the side toward the photocathode. The amount of charge deposited modulates the amount of beam current which is reflected by the target back to a five-stage secondary emission multiplier having a gain of 500 which surrounds the thermionic cathode. The return beam is smaller at those instants when it is passing over target elements which have a large positive charge than when it is scanning across ones having a smaller charge.

Some types of television camera tubes, including the image orthicon, suffer from a defect called "rainback" of the secondary electrons released from the target by the electrons of the scanning beam. If these secondary electrons fall back or "rain back" upon adjacent areas of the target, they modify the charge pattern that was stored there by the illumination from the scene being viewed. In order to eliminate or, at least, minimize rainback, a target screen is placed very close to the target. The screen is maintained at a bias positive with respect to the target in order to collect

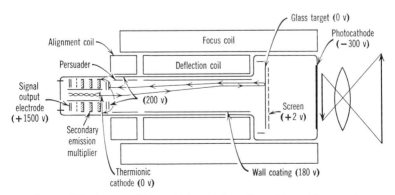

Figure 8.11. Image orthicon. (Adapted from Zworykin and Morton.[21])

the secondary electrons. Some formidable technical problems had to be solved in order to make and mount it since it is of very fine mesh (500 to 1000 openings per linear inch), is almost 75% open, yet it is stretched and mounted so that the spacing between target and screen is only a few thousandths of an inch and very uniform.

If the image orthicon is to be used at very low levels of illumination, the spacing between target and screen is made greater than in tubes intended for use in television studio work. Less charge is then stored on the target for a given potential difference between target and screen. This smaller charge can be scanned off in one frame time by the use of a smaller current in the scanning beam. At low levels of illumination the noise is proportional to the square root of the current in the scanning beam so that a modified tube to provide a better signal-to-noise ratio with a larger target-to-screen spacing is used for these applications.[21]

According to Zworykin, Ramberg and Flory,[22] "Recognizable pictures can be obtained for a scene brightness of the order of 0.02 foot lamberts, which corresponds to light objects in bright moonlight." Their statement is probably quite conservative since Rotow[23] reports that a standard image orthicon was able to reproduce scenes whose brightness was only 10^{-3} foot lamberts. R. K. H. Gebel[24] has reported upon the performance of a modified image orthicon, called an intensifier image orthicon, used for detection at very low light levels and has shown that extremely weak sources can be detected if the tube is used with a properly chosen optical system. The focal length of the optical system should be such that the image of the source which is to be detected should just cover the smallest area resolvable at the photocathode. He shows that under these conditions the intensifier image orthicon is theoretically capable of detecting a point source of white light which provides approximately 5.8×10^7 photons per second on the smallest resolvable area on the cathode. Since one lumen of white light provides 1.4×10^{16} photons per second, this camera tube will, in theory, detect a source having an illuminance of 4.1×10^{-9} lumens.

Gebel has used an intensifier image orthicon, video amplifier, and cathode ray tube display to detect celestial bodies, including two of the satellites of Jupiter, during daylight. These satellites have stellar magnitudes of five and six, which represent the limit of the detection capability of the average naked eye when dark adapted.

8.3.2 Photoconductivity

The photon effect most widely used in infrared detection is photoconductivity. From the early thallous sulfide or thalofide cell of Case, the lead sulfide cells of World War II, the other lead salt detectors and compound semiconductor detectors, to the modern low temperature impurity photoconductors, this has been the detection mechanism most extensively applied. The reason for this is that photon effects in general are faster and more efficient than thermal effects, and photoconductivity is the most simple of the photon effects. That is to say, photoconductive detectors as a class are easier to produce and easier to use than detectors based upon the other photon effects. A mathematical analysis of photo-conductivity is found in Section 9.3.1.

8.3.2.1 Elemental detector. We have seen in Chapter 6 that a semi-conductor in thermal equilibrium at any given temperature contains free electrons and holes. The concentrations of these are determined by the temperature, effective masses, and location of the Fermi level, and are subject to the condition that the product of the free electron and hole concentrations equals the square of the intrinsic concentration. It is possible to change these concentrations by allowing photons of sufficient energy to be absorbed by the semiconductor. In intrinsic excitation their energy goes into freeing hole-electron pairs. In the extrinsic case, free holes and bound electrons, or bound holes and free electrons, are produced.

The requirement imposed upon the photon energy is that it be sufficient to cause excitation. If $\Delta\mathscr{E}$ is the minimum energy required, whether this represents the width of the forbidden band or the activation energy required for extrinsic excitation, then we have, analogous to Eq. 8.14,

$$\Delta\mathscr{E} = h\nu_0 = \frac{hc_0}{\lambda_0},$$ (8.16)

where ν_0 is the lowest detectable photon frequency, λ_0 is the longest detectable wavelength, and c_0 is the speed of light. Frequencies greater or wavelengths shorter than these will, of course, also produce excitation, the excess energy lifting the carriers into higher lying free states corresponding to kinetic energy imparted to the carriers. For $\Delta\mathscr{E}$ expressed in electron volts and λ_0 in microns, the long wavelength threshold is

$$\lambda_0 = \frac{1.2406}{\Delta\mathscr{E}}.$$ (8.17)

Perhaps the most fundamental relationship describing photoconductivity states that at equilibrium the number of carriers excited by photons equals the product of the generation rate and the lifetime,[25]

$$N = F\tau,$$ (8.18)

where N is the total number of carriers excited by photons, F is the number excited (generated) per unit time, and τ is the lifetime, the length of time an excited carrier remains free. This relationship is one of extreme generality. It applies also, for example, to the population of a country. If the birth rate and death rate are equal, the number of persons, that is, the population, equals the birth rate multiplied by the average lifetime.

We can rewrite Eq. 8.18 in a more appropriate manner. The carriers excited by photons are termed excess carriers, since they exceed the number of free carriers normally present due to thermal generation. Symbolizing the excess electron concentration by Δn, the excess hole concentration by Δp, and the generation rate per unit volume by g, we have for intrinsic excitation

$$\Delta n = \Delta p = g\tau. \tag{8.19}$$

These added carriers increase the conductivity of the semiconductor. By referring to Section 6.7, the conductivity σ_p upon being irradiated is

$$\sigma_p = (n + \Delta n)q\mu_e + (p + \Delta p)q\mu_h$$
$$= q\mu_h[bn + p + \Delta n(b + 1)], \tag{8.20}$$

where n and p are the free electron and hole concentrations due to thermal excitation, μ_e and μ_h are the electron and hole mobilities, b is the mobility ratio, and q is the electronic charge.

The relative change in conductivity $\Delta\sigma/\sigma$ is given by

$$\frac{\Delta\sigma}{\sigma} \equiv \frac{\sigma_p - \sigma}{\sigma} = \frac{\Delta n(b + 1)}{bn + p}. \tag{8.21}$$

To obtain a large relative change upon irradiation of a given intensity, it is necessary to reduce the concentration of thermally excited carriers, termed the "thermal background." This may be accomplished by reducing the temperature of the semiconductor, a technique frequently used in infrared detection. For extrinsic photoconductors it is also necessary to cool the semiconductor to a temperature sufficiently low so that the impurity levels of interest are not thermally ionized. If the levels are all thermally ionized, there will be no bound carriers available for excitation by photons. Operation at lower temperatures serves also to reduce thermal and generation-recombination noise. Developments in photoconductive detectors have been directed toward lower temperature operation with its associated higher detectivity and longer wavelength response.

Consider now the operation of a photoconductive detector in its electrical circuit. We do not propose at this point to study the amplifier, but only to discuss the input circuit. In order to detect the change in conductivity of a photoconductor upon exposure to radiation, it is necessary to supply a bias battery and load resistor. (See Fig. 8.12.)

The photoconductor of dark resistance R is placed in series with a load resistor R_L and a battery of emf V_0. Infrared radiation, which is modulated

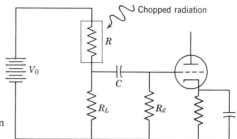

Figure 8.12. Photoconductor in measuring circuit.

or "chopped" by some means, for example, by passing a toothed wheel over a circular aperture,[26] is absorbed by the photoconductor, causing its resistance to decrease. This causes an increase in the current flowing in the circuit and a greater fraction of V_0 to appear across R_L. The voltage appearing across R_L is impressed upon the grid of the first stage of the amplifier. The resistor R_g self-biases the grid of the tube, whereas the capacitor C serves to block dc from the grid.

In the absence of radiation the voltage V_{R_L} appearing across R_L is given by

$$V_{R_L} = V_0 \frac{R_L}{R + R_L},$$ (8.22)

where $R_L/(R + R_L)$ is termed the "bridge factor." When radiation is absorbed by the photoconductor, its resistance decreases by an amount ΔR. The change in voltage ΔV_{R_L} appearing across R_L is given by

$$\Delta V_{R_L} = \frac{-V_0 R_L \Delta R}{(R + R_L)^2}.$$ (8.23)

The maximum value of ΔV_{R_L} is obtained under the condition $R_L = R$, the load resistor matching the photoconductor. Under this matching condition and assuming that ΔR is much less than $(R_L + R)$, that is, we are interested in small signals, Eq. 8.23 becomes

$$\Delta V_{R_L} = -\frac{V_0 \Delta R}{4R}.$$ (8.24)

Thus for maximum responsivity, that is, maximum signal voltage per unit radiant flux, we wish to match the load to the detector. However, this will not affect the detectivity, which is a measure of the signal-to-noise ratio at the detector, for the noise divides just as the signal does, the ratio remaining independent of the value of the load.

8.3.2.2 The infrared vidicon. We have seen in Section 8.3.1.5 that nonstorage types of image tubes operate at a disadvantage by comparison with storage type tubes, particularly if the resolution requirements are

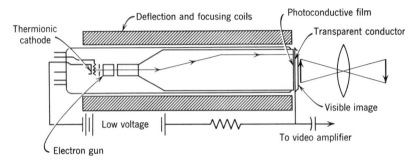

Figure 8.13. Vidicon camera tube. (Adapted from Zworykin and Morton.[21])

high. By utilizing the radiation on each picture element during the entire viewing time instead of just during the time that the element is being scanned, storage type tubes resolving m by n elements integrate each element mn times as long as nonstorage type tubes.

The vidicon is a storage type electronically scanned photoconductive television camera tube. Figure 8.13 is a schematic representation of its component parts. An electron gun incorporating a thermionic cathode is mounted at one end of the tube. The coils which surround the structure are used to focus and deflect the beam from the gun so that it scans across the optically polished faceplate at the other end of the tube in a series of closely spaced, parallel lines, termed a "raster." The faceplate is coated on the side toward the evacuated interior of the tube with a layer of tin oxide, a transparent conductor, so that the entire inner surface of the faceplate forms a good electrical contact to the ring which is sealed to the periphery of the plate. A thin film or target of photoconductive material is deposited on top of the transparent conductor. A lens is used to image the scene onto the photoconductive film.

The principle of operation of the vidicon is easiest to understand if we examine the behavior of a single element of the target. Consider the element to be an electrical capacitor, one plate of which is formed by the surface exposed to the vacuum and the other the conducting film deposited on the faceplate.

That side of the capacitor which is in intimate contact with the transparent conductor is maintained at a small positive potential relative to the cathode of the electron gun. If the capacitor under consideration happens to be an element of the scene which is dark, that is, not irradiated, then the conductivity of that small portion of the film will be very low and electrons will accumulate there from the scanning beam. Soon the vacuum side of this capacitor will come approximately to cathode potential and it then repels the charges in the low velocity beam. On subsequent scans the beam passes over this picture element without any change in the situation. If, however, that element begins to receive radiation as the scene changes, then, because of photoconductivity, that element takes on the characteristics of a capacitor with a leaky dielectric. The vacuum side of the film tends to assume the potential of the transparent conductor, slightly positive relative to cathode. On the next passage of the scanning beam, charge is deposited, returning it to cathode potential. The brief pulse of displacement current which flows in the film is coupled to a video amplifier, as shown in the figure, and ultimately to a kinescope, or picture tube. The pulse height depends upon the extent of the recharging of the capacitor, which depends, in turn, upon the integrated leakage during the time between scans. Since this depends upon the average irradiance of that element, the beam deposits charge in accordance with the brightness of the scene at each element and therefore causes a fluctuating voltage to appear at the input to the video amplifier.

A typical image measures about 14 by 19 mm with a picture element corresponding to a rectangle 37 by 50 μ by one-half μ thick. The RC product for this capacitor should be approximately 1/30 sec. If less, the charge will leak away so rapidly that the film will not integrate over a frame time. If A is the capacitor plate area, d is the separation between the plates, and introducing the permittivity of free space $\varepsilon_0 = 8.85 \times 10^{-12}$ farad/m, we have

$$RC = \frac{K_e \varepsilon_0 A}{d} \frac{\rho d}{A} = K_e \varepsilon_0 \rho = \frac{1}{30}. \qquad (8.25)$$

For a typical film the relative capacitivity (dielectric constant) K_e is about 3. Thus the resistivity ρ of the material in the film must be of the order of 10^{11} ohm cm.

In addition to the requirement for high resistivity targets, infrared vidicons[27] also require photoconductive layers possessing an energy band structure such that only a small excitation energy is needed to produce free carriers. Infrared vidicons must therefore have their target layers maintained at a temperature T such that the energy available for thermal excitation of the carriers is small by comparison with hc/λ_0, where λ_0 is the long wavelength cutoff. Redington and van Heerden[28] point out, however, that the requirements for high resistivity demand operating temperatures considerably lower than those which would be needed to assure that extrinsic photoconductivity be observable. Doped silicon and germanium photoconductors have been studied by them for use as infrared sensitive layers in vidicons. These materials suffer the disadvantage of having low optical absorption for extrinsic excitation, that is, for wavelengths longer than the absorption edge. Use of thick layers degrades the

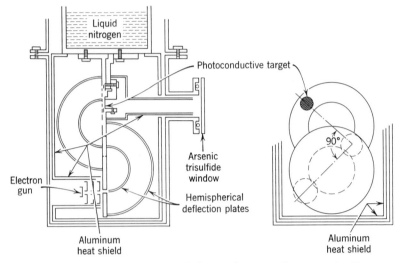

Figure 8.14. Infrared vidicon. (After Redington and van Heerden.[28])

resolution of the images reproduced by the tube since the resistance through the film then becomes comparable with the resistance between adjacent elements so that the charge patterns produced on the film during operation diffuse laterally, smearing the image. The use of optical systems of low f/number (fast systems) is desirable from the standpoint of high irradiance for a given scene brightness. However, such systems have a small depth of focus, which also militates against the use of thick layers.

In spite of these difficulties, Redington and van Heerden[28] have achieved very good performance using gold-doped silicon. For wavelengths near 1 μ, approximately 3% of the photons were able to free electrons to produce photoconductivity. Even with f/16 optics, objects heated to 140° C produced an image. The use of f/2 optics would perhaps permit the imaging of the human body by its own radiation.

A completely different tube geometry is needed for vidicons sensitive in the infrared since, with the arrangement shown in Fig. 8.13, the film would constantly "see" the thermal radiation from the electron gun. Figure 8.14 shows the arrangement devised by them which permits an electron beam to be scanned across the film, yet prevents the heated electron gun from radiating directly toward the photoconductor. Application of saw-tooth shaped deflection voltages between two pairs of hemispherical plates causes the electron beam to be scanned across the film at normal incidence. Cooling the target to 77° K permits operation with materials sensitive to 5 μ. Liquid hydrogen cooling, 20° K, moves the threshold wavelength to 20 μ. Cooling to liquid helium temperature, 4.2° K, would reduce thermal excitation sufficiently to permit the use of extrinsic photoconductors sensitive out to approximately 100 μ.

Electronic scanning possesses a number of fundamental advantages over mechanical scanning as a means of producing thermal images. All these advantages are related to the fact that scanning by an electron beam is nearly inertialess. Tubes employing this feature will undoubtedly be the subject of even more intensive development.

8.3.2.3 The electron mirror tube. The electron mirror tube was first conceived in Germany prior to World War II. The principle of operation is that of imaging a scene on a photoconductive layer to produce a surface of nonuniform charge density which reproduces the radiant image, as in the vidicon. A major difference between the electron mirror tube and the vidicon is the fact that it provides a visual image directly and not a fluctuating current which is used to modulate the current density of the scanning beam in a television picture tube or kinescope. Figure 8.15 shows the arrangement of the component parts of the tube.

In contrast to the scanning beam used in the vidicon, the electron mirror tube uses a "flood" beam which passes through holes at the center of both the plane mirror and the fluorescent screen and is spread out uniformly by deflection plates over a large area near the photoconductive film. The potential on the vacuum side of the film varies from point to point in accordance with the energy distribution in the radiant image. The electrons arrive at the film with approximately zero energy. Those arriving opposite areas where the irradiance is high are repelled by the charge on the film, whereas those arriving where the irradiance is low reach the film. The deflection plates cause the reflected electrons to be

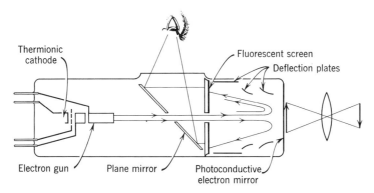

Figure 8.15. Electron mirror tube. (After Canada.[29])

imaged upon the fluorescent screen, the light output of which is viewed directly. The visible image on the fluorescent screen thus corresponds to the radiant image focused on the photoconductive film.

The relatively broad distribution of energy of the electrons in the flood beam has two deleterious effects. First, it sets a lower limit on the potential difference at the mirror film which can be detected as a brightness difference at the fluorescent screen. Second, the resolution capabilities of the tube are degraded because the focusing ability of the electrostatic reflecting system is dependent upon both the velocity of the electrons as they leave the cathode and the distance of the imaged point from the tube axis. These considerations and others related to difficulties in fabrication are responsible for the lack of success that has been associated with the electron mirror tube.

8.3.3 The Photovoltaic Effect

The second internal photoeffect of interest is known as the photovoltaic effect. As the name implies, the action of photons in this case produces a voltage, which can be detected directly without need for bias supply or load resistor. In addition to the photovoltaic detector, we shall also mention the photodiode and phototransistor, the operations of which are based upon this effect. A mathematical analysis of the photovoltaic effect is given in Section 9.3.3.

We have seen in Section 6.10 that rectification occurs at p-n junctions within the body of a semiconductor. Mention was also made of rectification at metal-semiconductor contacts. Since metal-semiconductor photovoltaic detectors are not widely used in infrared detection, whereas p-n junction detectors are, our study of the photovoltaic effect shall be confined to that occurring at p-n junctions. Consider again the energy level diagram characteristic of a p-n junction. (Fig. 8.16.)

In the absence of radiation falling upon the junction, the Fermi levels in the p and n regions are aligned. An internal electric field E, the direction

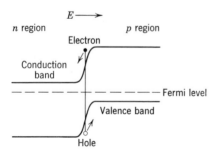

Figure 8.16. Carrier excitation at a p-n junction.

of which is shown by the arrow, exists at the potential barrier. A photon of wavelength sufficiently short to cause intrinsic excitation, when absorbed at the barrier, produces a free hole-electron pair. The action of the electric field at the barrier separates the pair, moving the electron into the n-type material and the hole into the p-type material. Since excess electrons have moved into the n-type region, and excess holes into the p-type region, the n-type region becomes charged negatively and the p-type positively. As long as radiation falls upon the junction, electron-hole pairs will be formed and separated by the internal field at the junction. If the semiconductor ends are short circuited by an external conductor, a current will flow in the circuit as long as radiation falls upon the junction. On the other hand, if the ends are open circuited by a high impedance voltage measuring device, a voltage will exist as long as radiation falls upon the sample.

Our discussion has concerned the action of radiation which is absorbed at the barrier. What of radiation which is absorbed by the semiconductor in the n or the p regions? If radiation falls on a point remote from the barrier, there will be no photovoltaic effect. If, however, excess carriers are generated sufficiently close to the barrier to be able to diffuse there and be separated, then a photovoltaic effect will be observed. It might be argued that the action of the electric field at the barrier will cause drift in the field to be superimposed upon the diffusion process. However, in general, the field is very localized and does not contribute to the velocity of carriers moving toward the barrier. The magnitude of the photovoltaic effect for a constant intensity of the exciting radiation will be a function of the distance from the point of excitation to the barrier. Those carriers which are excited a diffusion length away from the barrier and happen to diffuse toward the barrier will suffer recombination along the way with only $1/e$ of the original number arriving at the barrier and contributing to the photovoltage. Those excited closer to the barrier will arrive in greater numbers. By moving a small spot of radiation toward the barrier from either side and observing the photovoltage as a function of displacement, it is possible to generate a sensitivity profile. (See Fig. 8.17.)

The diffusion process is an ambipolar one, governed in strongly extrinsic material by diffusion of the minority carriers. As seen in Eq. 6.76, the excess carrier concentration decreases in an exponential manner with distance. We can determine the minority carrier diffusion lengths by observing the points in the sensitivity profile on either side of the barrier at which the relative response has decreased to $1/e$ of the value at the barrier. Since the diffusion length is related to the mobility and lifetime through Eq. 6.77, knowledge of the mobility enables us to determine the minority carrier lifetimes. It should be noted that the approximation has

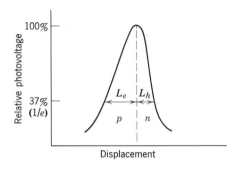

Figure 8.17. Sensitivity profile of a *p-n* junction.

been made here and throughout the discussion that the junction width is very much less than the carrier diffusion lengths.

We have seen in Section 6.10 that junctions prepared by overcompensating a melt during pulling can be revealed by cutting a face perpendicular to the plane of the junction. (See Fig. 6.20.) In photovoltaic detectors using grown junctions, the radiation is parallel to the plane of the junction. The shape of the sensitive area is therefore that of a rectangle of width equal to the sum of the minority carrier diffusion lengths and of whatever length is desirable, usually many times the width. Such long and narrow sensitive areas are of limited use in military infrared systems, but are of interest in spectroscopic applications where the aspect ratio may be chosen to match that of the beam emerging from the exit slits. On the other hand, we have seen that diffusion techniques produce broad area junctions. Photovoltaic detectors with diffused junctions have the radiation incident perpendicular to the plane of the junction. These broad area detectors find much wider application.

The photovoltaic effect about which we have been speaking generates a voltage in the absence of any external electric field, such as, for example, may be provided by a bias battery. It is also possible, however, to detect

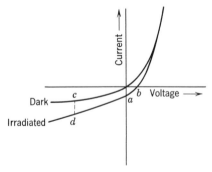

Figure 8.18. Rectification characteristic of a photodiode.

the radiation by applying a bias voltage to the junction in the reverse direction and detecting the change in current between dark and irradiated conditions. Such a device is termed a photodiode. Let us consider the rectification characteristic of a photodiode. (See Fig. 8.18.)

In the absence of radiation, the current-voltage characteristic is that labeled "Dark." Actually, the curve may not pass through the origin but be displaced slightly below and to the right by thermal effects. The effect of radiation falling on the junction is shown by the curve labeled "Irradiated." The point a corresponds to the short circuit current,whereas b corresponds to the open circuit voltage.

Let us bias the detector in the reverse direction in the dark by applying positive polarity to the n side and negative to the p. The current and voltage are those at point c. Radiation of the proper wavelength falling on the junction causes an increase in leakage current over the top of the barrier, the reverse current changing to point d. This increase in current is detected, as in the photoconductive effect, by placing a load resistor in series with the detector and measuring the increase in voltage appearing across it in the presence of radiation on the detector.

A third device is the n-p-n or p-n-p phototransistor. The explanation of this device involves the concept of transistor action which is not within the scope of this textbook. The interested reader is referred to Shive[30] for a description and explanation. The n-p-n or p-n-p phototransistors have not found wide application in infrared detection.

8.3.4 The Photoelectromagnetic Effect

The third internal photoeffect, which has been investigated extensively but used to only a limited extent in infrared detection, is the photoelectromagnetic, or PEM, effect.[31] The nature of the effect is such that in most semiconductors operating either at ambient temperature or cooled, the response is inferior to that of the photoconductive effect. However, in materials characterized by high mobilities and short lifetimes, the PEM response may be greater than the photoconductive one. As the name implies, the PEM effect takes place in a semiconductor immersed in a magnetic field. (See Fig. 8.19.)

Photons of wavelength shorter than the absorption edge are intensely absorbed at the irradiated surface, termed the front surface. These give rise to excess hole-electron pairs which diffuse from the surface into the interior. Since they are moving in a transverse magnetic field, the carriers experience a force in a direction normal to the plane of the magnetic field and the Poynting vector and are separated, the holes deflecting toward one end of the sample and the electrons toward the other. If the sample ends are short circuited externally, a current will flow as long as radiation is

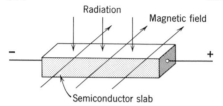

Figure 8.19. The PEM effect.

absorbed at the front surface. If the sample ends are open circuited, a voltage will appear at them as long as radiation is absorbed. The photo-signal due to the PEM effect is given in terms of either the short circuit current or the open circuit voltage. The theory of the effect, developed in Section 9.3.2, shows that the short circuit current and open circuit voltage are dependent upon the mobilities, lifetimes, magnetic induction, and front and back surface recombination velocities.

The PEM effect is basically a diffusion effect in a magnetic field. As such, the lifetime of interest is the ambipolar lifetime of the electron-hole pair, which is largely dictated by the shorter of the two carrier lifetimes. We shall see in the derivation of the theory of the PEM effect in Chapter 9 that the diffusion length of importance is the ambipolar diffusion length in the presence of the magnetic field. It is important that surface recom-bination be small at the front surface so that carriers generated there may diffuse into the bulk before recombining. The sample thickness for optimum response is of the order of the ambipolar diffusion length. Some of the carriers generated at the front surface will reach the back. It is important that they recombine at the back surface. If they do not, the charge gradient from the front to the back surface will be reduced, thus reducing the diffusion current.

The spectral response of the PEM effect resembles that of the other internal photoeffects, the long wavelength threshold again given by Eq. 8.17. The background of thermally excited carriers does not affect the PEM response as much as it does the photoconductive response. This is because only those carriers which are generated at the front surface can diffuse into the bulk, producing the photosignal. The major effect of the thermally excited carriers is their influence upon the bulk resistivity and thus upon the open circuit voltage.

8.3.5   The Dember Effect

We shall consider briefly another internal photoeffect known as the Dember effect. Although it has not found application in infrared detectors, it is of interest as another detection mechanism. The reader is referred to Moss[6] for a more detailed treatment.

The Dember effect is another photodiffusion phenomenon like the PEM effect, but does not require a magnetic field. As in the PEM effect, radiation of wavelength shorter than the absorption edge falling upon the surface of a semiconductor slab is intensely absorbed, producing excess electron-hole pairs in a thin layer at the surface. The large electron-hole concentration gradient causes these to diffuse into the bulk, moving normal to the surface. The electrons, which in general are more mobile than holes, will advance farther in a given interval than will the holes. This charge separation gives rise to an electric field directed from the front surface to the back, dark surface. The direction of the field is such as to oppose the flow of electrons and aid the flow of holes. Thus in the open circuit condition an equilibrium voltage will appear with the front surface positive and the back negative. If the front and back surfaces are connected in an external circuit, a short circuit current will flow as long as radiation is absorbed at the front surface.

It is apparent that materials having large mobility ratios should exhibit a greater Dember effect than those with small ratios. It can be shown that a low value of recombination velocity on the front surface and a high value of electron diffusion length are needed to achieve large values of open circuit voltage.

8.3.6 The Filterscan Tube

A thermal imaging device named Filterscan, based upon free carrier absorption in silicon, has been described by Lasser, Cholet, and Emmons.[32] The tube incorporates a silicon target scanned in raster fashion by an electron beam. (See Fig. 8.20.)

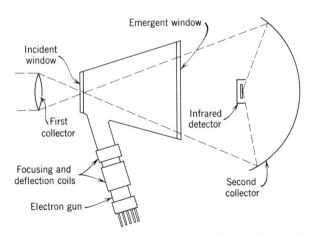

Figure 8.20. Free carrier absorption imaging tube. (After Lasser, Cholet, and Emmons.[32])

As seen in Section 4.2.1, a major optical absorption mechanism in semiconductors at wavelengths longer than the absorption edge is that due to free carriers. The optical absorption coefficient was shown to be proportional to the conductivity and thus to the density of free carriers present in the semiconductor. In the device considered here, an electron beam is used to inject electrons into silicon. Because of the high beam energy, approximately 30 kev, each electron impinging upon the silicon yields 2800 electron-hole pairs. This localized increase in carrier concentration causes the optical absorption coefficient to increase at the point upon which the beam impinges. For wavelengths greater than that of the absorption edge, approximately 1 μ in silicon, the action of the beam thus causes a decrease in transmission. By proper choice of material resistivity and thickness, injection by the beam causes at least 99 % absorption. Thus an opaque spot can be moved across the transparent semiconductor by scanning the beam.

In operation, a scene is imaged upon the silicon target by an optical system. A second system images the silicon target upon an elemental infrared detector. Scanning the electron beam across the target causes changes in the instantaneous value of the radiant power falling upon the detector, corresponding to the radiation distribution in the image focused upon the target.

The advantages of the device are the simplicity of its construction and the use of electronic scanning. The disadvantage is that it is a nonstorage type of system, and therefore is much less sensitive than a storage imaging system such as the vidicon. Lasser, Cholet, and Emmons[32] found that for a signal-to-noise ratio of two in a 30 kc bandwidth, a temperature difference of 125° C between the emitting object and its background was required. A better elemental detector would have improved the system.

8.3.7 Phosphors

Phosphors[33] are dielectrics capable of converting energy into visible radiation. The type of phosphor is characterized by the method of excitation. Most familiar is the cathodoluminescent phosphor, found on the faceplates of cathode ray kinescopes or television picture tubes. This type of phosphor converts the energy of incident electrons into photons having energy mainly in the visible region. Photoluminescent phosphors are excited by incident photons, usually of wavelength in the blue or ultraviolet regions and luminesce, that is, emit visible radiation, of longer wavelength than the exciting radiation. Both types are sensitive to electron and photon excitation, but each operates most efficiently under one type of excitation. Other types of photon excitation, such as by means of x- and gamma radiation, will also cause phosphors to luminesce. A third type of phosphor, excited by the action of an electric field, a phenomenon known as electroluminescence,[34] is of recent interest. Electroluminescent phosphors find application in electro-optical display devices and light amplifiers. We shall limit our discussion in this section to photoluminescent phosphors, since these have found some application in infrared detection. A photoluminescent phosphor consists of a host crystal, frequently zinc sulfide, having certain activator elements added in minute amounts. These elements introduce impurity levels in the forbidden energy band of the ZnS lattice. Radiation incident upon the lattice ionizes certain of the levels, raising electrons to higher energies. In many cases they will not be made free, but rather remain bound in the same center or in other centers at higher energies. The electrons then decay to lower energies, emitting photons of wavelength characterized by the energy loss.

The action of infrared radiation upon certain phosphors may be either to stimulate radiation or to quench it. The infrared stimulation process consists of initially irradiating the phosphor with photons capable of exciting it. This light is turned off and the luminous output drops to a very low value. Infrared radiation is then allowed to fall upon the phosphor, at which time the luminous output rises, then decays. Figure 8.21 illustrates the process.

Table 8.1 lists the spectral properties of several infrared stimulable phosphors.

The other action of infrared upon phosphors is that of quenching luminescence. In this case, a primary source of radiation is used to excite the phosphor. The radiation is then turned off and the luminescence

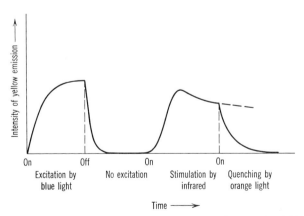

Figure 8.21. Typical intensity-time relationships for an infrared stimulable phosphor. (After Leverenz.[33])

TABLE 8.1. Spectral characteristics of some useful IR-stimulable phosphors
(After Leverenz.[33])

Peak Wavelengths of

Phosphor	Common Designation	Excitation, Å	Emission, Å	Stimula-tion, Å
1. Cub.-Sr(S:Se):(flux)* Sm:Eu	B1	4,600	5,700	9,300
2. Cub.-SrS:(flux):Sm:Eu	Std. VI	4,800	6,300	10,200
3. Cub.-Ca(S:Se):(flux): Sm:Eu		>4,800	>6,300	11,000
4. Cub.-CaS:(flux): Sm:Eu		>4,800	6,600	11,700
5. Cub.-SrS:(flux):Sm:Ce	Std. VII	2,900 3,500	4,800 5,400	10,200
6. Hex.-ZnS:Cu:Pb(SO₄): [NaCl(2)]	101	3,700	4,880	7,500 13,200

* Where (flux) is marked, the flux usually contains both O and F, for example, as in SrSO₄ or LiF.

decays exponentially with a long time constant. Infrared radiation is now allowed to fall upon the phosphor. The action of the infrared is to increase the decay rate, "quenching" the luminescence. Figure 8.22 illustrates this process. Note that the use of infrared stimulation in an imaging application provides a bright image on a dark background, whereas quenching provides a dark image on a bright background.

The phenomenon of infrared stimulation has been used in a near infrared viewing device termed a metascope, shown schematically in Fig. 8.23.

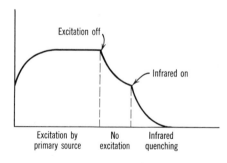

Figure 8.22. Infrared quenching. (After Leverenz.[33])

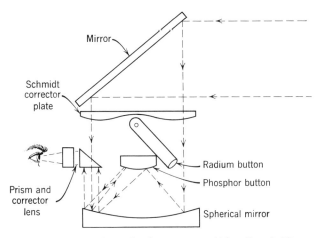

Figure 8.23. Schematic of metascope. (After Canada.[29])

A phosphor button is located at the focal point of an optical system incorporating a spherical mirror with a Schmidt corrector plate. A radium source moved into position in front of the phosphor excites the phosphor, "charges it," so to speak. The source is moved away and an infrared image is focused upon the button. Infrared stimulation occurs, providing a bright image which decays. The process is repeated upon moving the radium source again into place.

O'Brien[35] has reviewed some of the properties and applications of infrared sensitive phosphors. They provide a simple way of viewing a scene by infrared radiation without need for electrical power and with a minimum of mechanical complexity. On the other hand, they lack sensitivity and speed of response, can be used only in the near infrared to about 1.3 μ, and require the use of an infrared source to irradiate the target. Although they find little military use in imaging applications, being greatly inferior to the image converter, they can be used in signaling systems and as indicators of infrared irradiation from hostile sources.

8.3.8 Microwave Effects

Whereas most methods of infrared detection depend upon making electrical contact directly to the detector element and amplifying an audio frequency voltage appearing at the contacts, other methods for detecting the radiation and amplifying the signal have been explored. One of these is the microwave radiometer described by Dicke.[36] Although the radiation from a heated body is mostly confined to the infrared and visible portions of the spectrum, the long wavelength "tail" of the Planck distribution function extends, of course, to all possible wavelengths. Thus a small fraction of the total radiation from a heated body lies in the microwave region of the spectrum. The Dicke radiometer consists of a microwave receiver operating in the 1 cm region. By the use of a very narrow bandwidth obtained by a homodyne (phase sensitive) technique, Dicke[36] detected a temperature change of 0.5° C and a power of 10^{-16} watts. The limiting noise appeared to originate in the receiver.

Another mechanism for the detection of infrared which utilizes microwave effects has been developed.[37] Infrared radiation falling upon a semiconductor changes not only its electrical conductivity but also its relative capacitivity (dielectric constant), since the capacitivity is a function of the free carrier concentration (see Section 3.2.2). The detector utilizes a small piece of germanium suspended in one arm of a "magic tee" microwave cavity. A second arm contains a second piece of germanium identical to the first. Microwave energy incident into a third arm of the tee is reflected from the arms containing the germanium elements into a signal output arm. The energy reflected from an arm containing one element is 180 degrees out of phase with that reflected from the arm containing the other, so that with the elements subjected to identical conditions, no power is coupled into the output arm. However, if infrared radiation of sufficiently short wavelength is absorbed by one of the elements, the excess carriers produced change the electrical conductivity and dielectric constant of the element sufficiently to unbalance the arms, resulting in power being coupled into the output arm. In order to obtain a relatively greater imbalance for a given radiation power and to reduce noise, the detector is cooled to the temperature of liquid nitrogen, 77° K.

8.3.9 Narrow Band Quantum Counters

A great deal of attention has been centered upon certain narrow band sources and detectors referred to as infrared masers, irasers, and lasers. Section 2.9 clarified the terminology and discussed the principles of operation of irasers used as oscillators, that is, sources. In this section we shall consider two types of detectors falling in the category of narrow band quantum counters. Gelinas[38] describes and discusses the ultimate performance of each. The first type is the iraser, that is, an infrared detector whose operation is based upon the stimulated emission of radiation.

The principle of operation of an iraser source, discussed in Section 2.9, is based upon ideas conceived by Schawlow and Townes[39] and by Dicke.[40] The operation of an iraser detector is similar to that of an iraser source, depending upon population inversion of discrete energy levels. A three-level system is shown in Fig. 8.24.

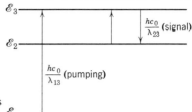

Figure 8.24. Transitions between energy levels in an iraser.

Pumping radiation of wavelength λ_{13} causes the population of level 3 to increase at the expense of that of level 1. This causes a population inversion between levels 3 and 2. Signal radiation of wavelength λ_{23} induces transitions from level 2 to level 3 and from level 3 to level 2. Because of the population inversion there will be more photons emitted than absorbed. The emitted radiation of wavelength λ_{23} is detected by a photon detector operating in a spectral region overlapping λ_{23}. Note that the iraser operated in this manner acts as an amplifier of radiation of wavelength λ_{23}.

The second type of narrow band quantum counter has been referred to by Bloembergen as an IQ (infrared quantum) detector.[41] The operation of this type makes use of discrete energy levels, but does not depend upon population inversion. Consider the energy level scheme depicted in Fig. 8.25.

Bloembergen suggests a four-level scheme such that hc_0/λ_{12}, the energy involved in transitions between levels 1 and 2, is much greater than kT. If λ_{12} is of the order of 100 μ, then T is required to be about $2°$ K or less. In the absence of irradiation, only level 1 is occupied, that is, the temperature is so low that the probabilities of occupation of levels 2 and 3 are extremely small. Thus pumping radiation of wavelength λ_{23} incident on the crystal is not absorbed and provides no excitation as long as signal radiation of wavelength λ_{12} is absent. The pump is then said to be "idling." When signal radiation is present, electrons are excited from level 1 to

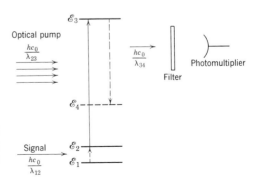

Figure 8.25. Energy levels and transitions in an infrared quantum counter. (Adapted from Bloembergen.[41])

level 2, where they are immediately excited to level 3 by the pump. The pump also stimulates transitions from level 3 to level 2, resulting in an emission of photons of wavelength λ_{23}. It may be difficult to detect these in the presence of the intense pump frequency. Therefore a fourth energy level is introduced between levels 2 and 3 to which transitions are made from level 3. Photons of wavelength λ_{34}, in the visible or ultraviolet, emitted in the decay from level 3 to level 4, are transmitted by an optical filter tuned to λ_{34} and are detected by a photomultiplier. The IQ counter is thus a detector of monochromatic long wavelength infrared radiation, having an output of visible or ultraviolet radiation which is detected by a photomultiplier.

A third approach to a narrow band quantum counter has been discussed by Lubin.[42] The principle of operation resembles somewhat that of the IQ detector, but the pumping radiation is replaced by an electron beam. The beam energy, which must be controlled precisely, causes excitation from a low lying state to an upper one. Signal radiation excites electrons to a third, higher state, where they decay to the lowest state, emitting ultraviolet photons which are detected by a scintillation crystal and photomultiplier.

The reader interested in additional information on narrow band quantum counters is referred to the papers by Prokhorov,[43] Sanders,[44] Maiman,[45] Shimoda et al,[46] and Tager and Gladun.[47] The ultimate performance of narrow band quantum counters is derived in Section 9.4.4.

8.3.10 Photographic Film

Certain dyes have been found to activate silver halide emulsions, extending their spectral response to about 1.2 μ. Such infrared sensitive photographic film finds military application in aerial photography, where it is used to detect installations camouflaged against detection by visible radiation, and in photography under haze limited conditions and those of very low ambient light intensity. It has found commercial application in such diverse fields as medicine and astronomy. The reader interested in the mechanism of the photographic effect in alkali halides is referred to Mott and Gurney.[48] Jones[49] has made an extensive study of the factors limiting the performance of photographic film. Infrared photography is discussed in detail by Clark.[50] Larmore[51] has reviewed some properties and applications.

8.4 THERMAL EFFECTS

The detection mechanisms discussed thus far have fallen into the category of photon effects, in which incident photons of sufficiently short wavelength are absorbed at lattice sites, giving rise to excess charge carriers. Because photon detectors respond to the rate at which photons are absorbed, their spectral responses were seen to be dependent upon the energy of the photons, that is, the wavelength of the radiation.

We shall consider in this section the other broad category of detection mechanisms, termed thermal effects. As stated previously, thermal detectors make use of the heating effect of radiation. As such they are simply energy detectors, their responses being dependent upon the radiant power which they absorb, but independent of the spectral content of the radiation except insofar as the spectral content influences the absorption of the radiation. Associated with all thermal detectors is some form of thermal mass, the temperature of which changes when it absorbs radiation. It is apparent that large temperature changes per unit radiant power absorbed are associated with small thermal masses; thus the sensitive elements of most thermal detectors are small.

The most well-known forms of thermal detectors are the bolometer and the thermocouple. Our discussion of thermal detectors will include these, together with the Golay cell, Evaporograph, and absorption edge image converter. Only the bolometer has seen widespread military use, although the thermocouple has found extensive commercial application in infrared spectroscopy.

8.4.1 Metal and Thermistor Bolometers

A bolometer is a radiant power detecting device, the operation of which is based upon measuring the temperature change in resistance of a material due to the heating effect of absorbed radiation. The simplest form of bolometer is a short length of fine wire. At the temperature at which it is operated, it has a given resistance. Radiation allowed to fall upon it is partially absorbed, causing the temperature of it to rise. The resistance change due to the change in temperature is a measure of the radiant power absorbed.

Since bolometers are power detectors, they are capable of detecting radiation of all wavelengths. They are widely used as detectors of microwave power. In this application they are placed within a waveguide down which microwave energy is propagated. Because of their relatively slow speed of response, to be discussed later, they are not used as microwave demodulating devices but simply for determining average power levels. They are useful also in power measuring equipment in the audio and radio frequency regions. Here the electrical power dissipated in a resistive element causes its temperature to rise and infrared radiation to be emitted, which is detected by a bolometer. Since the emitted radiation is related to the temperature rise, the signal from the bolometer is related to the power dissipated in the resistive element.

Bolometers may be of three types: metal, semiconductor, and super-conductor. Metal and semiconductor bolometers are operated at ambient temperatures, whereas the superconducting bolometer must be cooled to temperatures near absolute zero. We shall consider metal and semiconductor bolometers in this section, and the superconducting bolometer in the next.

The resistance of a metal increases linearly with small changes in temperature, following a law of the form

$$R = R_0[1 + \gamma(T - T_0)], \tag{8.26}$$

where R is the resistance at temperature T, R_0 is the resistance at temperature T_0, and γ is approximately independent of temperature. The value of γ is of the order of 0.5% per degree Centigrade for many metals. Metal bolometers may be in the form of small diameter wires or thin films. A form of bolometer used in microwave work consisting of an encapsulated platinum wire is known as a "barretter."

Semiconductors exhibit a much more pronounced dependence of resistance upon temperature than do metals. We have seen from Eq. 6.42 that in the intrinsic region their resistivity is an exponential function of the reciprocal of the absolute temperature. Thus the resistance of a semiconductor drops as its temperature rises. This exponential dependence, much more pronounced than the linear behavior of metals, has caused semiconductors to be much more widely used than metals for bolometers.

Semiconductor bolometers are known as "thermistor" bolometers, thermistor denoting thermally sensitive resistor. Thermistor materials were developed originally at Bell Telephone Laboratories.[52] The two materials in common use, known as No. 1 and No. 2, are sintered oxides of manganese, cobalt, and nickel, used in bolometers in the form of flakes about 10 μ thick. These are mounted upon heat dissipating substrates known as "thermal sinks." Radiation falling on a flake warms it. If the radiation is removed, the flake returns to its original temperature with a decay time dependent upon the thermal conductance between the flake and the sink, and radiation interchange with its surroundings. Fast response is obtained by mounting the thermistor directly on a solid sink, the bolometer being termed "solid backed." The flake may also be gas or vacuum backed, with poorer transfer of heat from the flake and slower response. Although this is accompanied by a greater detectivity, most applications require the faster response of the solid backed flake.

Since the resistance of a thermistor flake is an exponential function of the temperature, the thermistor is a nonlinear circuit element, the resistance being a function of the applied voltage. Consider the current-voltage characteristic of a thermistor shown in Fig. 8.26.

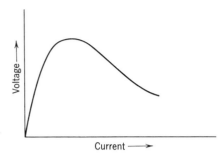

Figure 8.26. Current-voltage characteristic of a thermistor.

At low applied voltages and low currents the power dissipation in the thermistor is small and heating effects are negligible. Thus the dynamic resistance (rate of change of voltage with respect to current) is constant and the thermistor acts as a linear (ohmic) device. However, as the current is increased, the power dissipation becomes appreciable and Joulean heating occurs. As the thermistor heats up, its resistance decreases. Finally, with increased current the dynamic resistance passes through zero and acquires negative values. In this region the voltage drop across the thermistor decreases as the current increases. Unless a ballast resistor is used, the thermistor will burn itself out when operated in the negative resistance region.

Although these nonlinear properties have many interesting applications, such as in negative resistance amplifiers, we shall consider only application to infrared detection. The reader is referred to Shive[53] for a thorough discussion of thermistors. Thermistor bolometer infrared detectors[54] are operated on the left (low current) side of the maximum in Fig. 8.26, usually at a bias voltage of about 0.6 of that at the peak of the curve. To compensate for changes in the ambient temperature, usually two identical flakes are mounted in the same housing, with a shield over one to prevent the signal radiation from falling upon it. They are used in a bridge circuit as shown in Fig. 8.27.

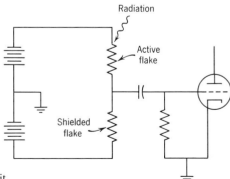

Figure 8.27. Bolometer circuit.

In the absence of a signal, the bridge is balanced, since the flakes are identical. Radiation falling on the "active" flake reduces the resistance of it, unbalancing the bridge. As with other types of infrared detectors, the radiation is chopped to improve amplifier performance and aid in target discrimination. A mathematical analysis of bolometer operation is found in Section 9.3.4.

8.4.2 The Superconducting Bolometer

A particularly interesting type of bolometer is based upon the phenomenon of superconductivity. Certain materials, termed superconductors, including lead, tin, tantalum, niobium nitride, and niobium stannide, exhibit a sharp change in resistivity at temperatures near absolute zero. Figure 8.28 depicts the resistivity of a superconductor as a function of temperature.

As the temperature of the superconductor is reduced from room temperature, the resistance decreases in the normal fashion for metals. However, as the temperature approaches a critical value T_0, the resistance sharply drops to zero. Below the critical temperature the material is said to be superconducting. That the resistance in the superconducting state is truly zero has been dramatically demonstrated in experiments with ring-shaped superconductors, in which induced electrical currents have circulated for years without any decrease in magnitude.

The rapid change in resistance from the normal to the superconducting state is referred to as the superconducting transition. This occurs over a temperature range of tenths or hundredths of a degree. Thus the slope of the resistance-temperature curve is extremely steep. Transition temperatures range from near 1°K for some materials to about 15°K for niobium nitride.

Bolometers based upon superconductors have been described by Andrews et al.,[55] Milton,[56] and Fuson.[57] The bolometer consists of a thin

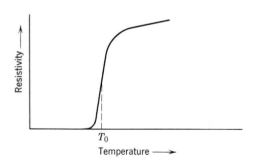

Figure 8.28. Superconducting transition.

flake, wire, or film of a superconductor, mounted upon a heat sink within a dewar vessel. The temperature of the superconductor is maintained by a control system at T_0 near the center of the transition region. Infrared radiation falling upon it warms it slightly, increasing the resistance of it. This change in resistance is measured by an external bridge circuit.

The problem of stable operation at the transition point is severe. Since the transition takes place over a very small temperature interval, the operating temperature must be controlled with extreme precision, maintaining the superconductor within about $0.0001°$ K. Fluctuations in the temperature control show up as noise in the bolometer. This precise control can be accomplished for bolometers operating above $4.2°$ K by equipping the bolometer with a resistance heater, evacuating the enclosure, and suspending the bolometer above a liquid helium bath. The helium vapors tend to reduce the bolometer temperature, whereas passing current through the heater will raise it. A servo system is used to maintain the proper temperature with the required precision.

A superconducting bolometer has the advantage over thermistor and metal bolometers of reduced thermal noise because of the low operating temperature, reduced heat capacity of the bolometer due to the low temperature, and a steep resistance-temperature curve. On the other hand, the problems associated with low temperature operation and precise temperature control are severe. Although the superconducting bolometer holds promise of being a high performance detector, the realization of this is extremely difficult. It remains a laboratory device.

8.4.3 The Thermocouple and Thermopile

The radiation thermocouple was one of the earliest infrared detectors, yet it is still used as the detector in modern infrared spectrometers. It consists of a low mass junction between two metals having different thermoelectric powers, which is mounted on a blackened receiver. Radiation absorbed by the receiver causes the junction temperature to rise. Since the thermoelectric emf is proportional to the temperature rise, measurement of the emf gives a measure of the intensity of the radiation.

A widely used form of the thermocouple is the radiation thermopile, consisting simply of a series connection of a number of junctions. These are arranged in a housing as shown in Fig. 8.29. The active and reference junctions are maintained within the same enclosure at the same temperature, the reference junctions being shielded from the radiation to be measured. Radiation falling on the active junctions raises their temperature slightly with respect to that of the reference junctions. Variations of the ambient temperature do not greatly affect performance.

The thermopile has a higher responsivity than the thermocouple

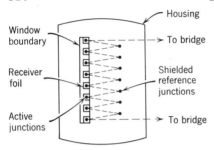

Figure 8.29. Radiation thermopile. (Courtesy of The Eppley Laboratory, Inc., Newport, Rhode Island.)

because of the use of multiple junctions. However, its response time is long and it is not suitable for ac amplification techniques. On the other hand, the small mass of the thermocouple results in a shorter response time, roughly several hundredths of a second. Evacuated enclosures having infrared transmitting windows are frequently employed to improve detectivity by reducing convection losses. The design of radiation thermocouples is described very well by Hornig and O'Keefe.[58] The ultimate performance of radiation thermocouples has been discussed by Fellgett.[59] Smith, Jones, and Chasmar[5] give an extremely thorough description and performance analysis.

8.4.4   The Golay Cell

The Golay pneumatic detector,[60] named after its inventor, is a thermal detector of high detectivity, slow response, and somewhat fragile construction. It consists of a gas filled chamber connected by a passage to a flexible membrane having a reflecting film upon the side opposite to the passage. (See Fig. 8.30.) Within the chamber is an absorbing film, a

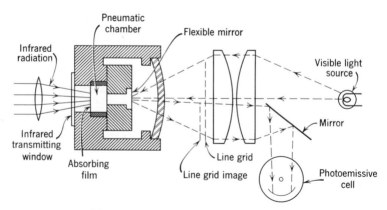

Figure 8.30. Golay detector. (After Golay.[60])

membrane of low thermal mass. Radiation incident upon the cell through an infrared transmitting window impinges upon the absorbing film. The energy absorbed is transmitted to the gas, resulting in an increase in the gas temperature and pressure. This causes the flexible mirror to bulge slightly. In order to detect the mirror movement, a line grid is projected upon the mirror and reflected back along the incident path to a photocell. Displacement of the mirror, due to the increased gas pressure within the cell, causes a shift in the reflected image of the line grid, causing a change in the amount of light falling upon the photocell.

The pneumatic detector approaches the ideal detector limited by photon noise, discussed in Chapter 9. However, it deviates from the ideal because of thermal coupling of the pneumatic chamber to the walls, incomplete absorption of the incident radiation by the absorbing film, and window reflection losses. In addition, if a chopped radiation input is used, more of the signal is lost. These factors combined lead to a detectivity an order of magnitude lower than the photon noise limited ideal detector. Its slow response, the time constant of the order of 10^{-2} sec, combined with fragile construction, render it of little use in military applications. It has found some utility in infrared spectroscopy.

8.4.5 Evaporograph

The last two thermal detectors which we shall discuss, the Evaporograph and the absorption edge image converter, are thermal imaging devices. The Evaporograph includes a vacuum pump, optical system, cell, and camera. Its operation is based upon the selective evaporation of an oil film from a membrane upon which an infrared image is projected. The cell, the detecting element, is shown in Fig. 8.31. A nitrocellulose membrane separates the two chambers of the cell, both of which are

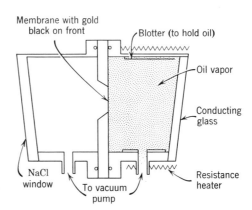

Figure 8.31. Evaporograph cell. (Courtesy Baird-Atomic.)

evacuated by a vacuum pump. The chamber to the right in the figure contains an oil reservoir which is electrically heated to vaporize the oil. An optical system focuses an infrared image through a rock salt window in the other chamber upon the front side of the membrane, selectively heating the membrane in the pattern of the image. The oil vapor in the back chamber condenses selectively on the membrane, the warmer areas receiving less oil. The thickness of the condensed film is such as to show interference colors when illuminated by white light. This colored image is viewed through a window on the back of the cell.

In operation, a clearing light evaporates all the oil condensed on the membrane. This light is turned off, the shutter remaining closed, and the oil film is allowed to build up by condensation on the membrane. When the film appears dark blue, the shutter is opened and radiation focused upon the membrane begins to cause the oil film to alter in thickness. If the temperature of the viewed object is above 80°C, the radiation causes the film to evaporate from the membrane. If the temperature is below 80°C, the radiation is not sufficient to evaporate all the oil but merely to slow the rate of condensation. In either case, the distribution in thickness of the condensed film, corresponding to the distribution in intensity of the infrared image, is determined by viewing the interference colors.

The speed of response of the Evaporograph varies from a fraction of a second for large temperature differences in the scene to several seconds if the scene contains only small temperature differences of the order of a few tenths of a degree. The manufacturer states that the formation of the image of a man at room temperature takes about 15 sec, whereas a soldering iron at 600°F requires a fraction of a second.

8.4.6 Absorption Edge Image Converter

The second thermal imaging device is the absorption edge image converter, described by Harding, Hilsum, and Northrop.[61] Operation of the device is based upon utilizing the temperature dependence of the location of the absorption edge of a semiconductor. Figure 8.32 depicts the optical absorption of a semiconductor measured at two temperatures.

At a temperature T_1 the absorption coefficient has a value a at a wavelength λ_0. As the semiconductor is warmed to T_2, the absorption coefficient at λ_0 increases to the value b. If the absorption edge is steep and the temperature dependence of the edge is great, a slight warming of the semiconductor substantially decreases the transmission of radiation having wavelength λ_0. (It should be noted that in some semiconductors the absorption edge shifts to shorter wavelengths upon heating.)

The absorption edge image converter built by Harding, Hilsum and Northrop[61] used amorphous selenium as the semiconductor. This has the

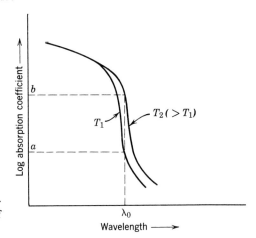

Figure 8.32. Temperature dependence of the absorption edge of a semiconductor.

advantage of a relatively steep absorption edge, located in the visible spectrum at the sodium D line, 5890 Å, having a pronounced temperature dependence. The tube was constructed as shown in Fig. 8.33.

Infrared radiation emitted by the object to be viewed is focused by a parabolic mirror through a rock salt infrared transmitting window onto a film of amorphous selenium 1 μ thick. In order to increase the absorption of radiation, a layer of chromium is added to the film on the side opposite to the parabolic mirror. The film is in an evacuated enclosure to eliminate convection losses. The radiant image on the film heats it locally, reducing the transmission in the heated region. This is observed by passing light from a sodium lamp of 5890 Å wavelength, corresponding to point λ_0 in Fig. 8.32, and viewing the transmitted radiation. The authors obtained recognizable photographs of a teapot 60°C above ambient and of a man's

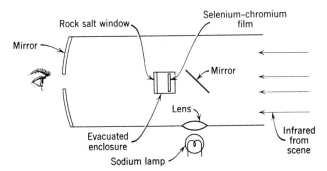

Figure 8.33. Absorption edge image converter. (After Harding, Hilsum, and Northrop.[61])

head. The time constant is stated to be 0.5 sec. The theory and applications of the absorption edge image converter are discussed by Hilsum and Harding.[62] They show that bodies 10°C above ambient can be imaged.

REFERENCES

1. For experimental methods used to determine the figures of merit see: A. J. Cussen, *The Procedures Used in the Study of Properties of Photoconductive Detectors*, U.S. Dept. of Commerce, National Bureau of Standards, ATI 194147; Potter, Pernett, and Naugle, *Proc. Inst. Radio Engrs.* **47**, 1503 (1959); Jones, Goodwin, and Pullan, *Standard Procedure for Testing Infrared Detectors and for Describing Their Performance*, Office of the Director of Defense Research and Engineering, Washington, D.C. (Sept. 12, 1960).
2. R. C. Jones, *Advances in Electronics* V, Academic Press, New York (1953), p. 1, *Advances in Electronics* XI, Academic Press, New York (1959), p. 87.
3. Garbuny, Vogl, and Hansen, *Rev. Sci. Instr.* **28**, 826 (1957).
4. J. S. Blakemore, *Phys. Rev.* **110**, 1301 (1958).
5. Smith, Jones, and Chasmar, *The Detection and Measurement of Infrared Radiation*, Oxford University Press, London (1957).
6. T. S. Moss, *Optical Properties of Semiconductors*, Butterworths, London (1959); *Photoconductivity in the Elements*, Butterworths, London (1952); see also *Advances in Spectroscopy* **1**, 175, Interscience Publishers, New York (1959), H. W. Thompson, ed.
7. A. L. Hughes and L. A. DuBridge, *Photoelectric Phenomena*, McGraw-Hill Book Company, New York (1932).
8. G. A. Morton, *Proc. Inst. Radio Engrs.* **47**, 1467 (1959).
9. A. Sommer, *Photoelectric Cells*, Chemical Publishing Company, Brooklyn (1947).
10. For a discussion of photomultipliers, see *DuMont Multiplier Photo-Tubes*, Allen B. DuMont Laboratories, Clifton, New Jersey (1956).
11. Holst et al., *Physica* **1**, 297 (1934).
12. W. Kluge, *Z. Physik* **95**, 734 (1935).
13. P. Görlich, *Z. Tech. Physik* **18**, 460 (1937).
14. S. Asao, *Proc. Phys.-Math. Soc. Japan* **22**, 448 (1940).
15. R. S. Wiseman and M. W. Klein, *Proc. Inst. Radio Engrs.* **47**, 1604 (1959).
16. H. R. Blackwell, *J. Opt. Soc. Amer.* **36**, 624 (1948).
17. M. W. Klein, *Proc. Inst. Radio Engrs.* **47**, 904 (1959).
18. J. D. McGee and W. L. Wilcock, ed., *Photo-Electronic Image Devices, Advances in Electronics* XII, Academic Press, New York (1960).
19. *Image Intensifier Symposium*, ERDL, Ft. Belvoir, Virginia, Oct. 6–7, 1958, AD 220160 (Unclassified).
20. Garbuny, Vogl, and Hansen, *Westinghouse Research Laboratories Final Report, Thermal Imaging Program*, Contract AF 33(616)-2282 (1956).
21. V. K. Zworykin and G. A. Morton, *Television*, John Wiley and Sons, New York (1954).
22. Zworykin, Ramberg, and Flory, *Television in Science and Industry*, John Wiley and Sons, New York (1958).
23. A. A. Rotow, *R.C.A. Review* **17**, 425 (1956).
24. R. K. H. Gebel, *Aero. Research Lab. Report*, Wright Air Development Center, WCLJH-19 (June, 1957); *Wright Air Development Center Technical Note 58–324*

(October, 1958); *Image Intensifier Symposium*, ERDL, Ft. Belvoir, Virginia, Oct. 6–7, 1958, AD 220160 (Unclassified), p. 215

25. A. Rose, *Proc. Inst. Radio Engrs.* **43**, 1850 (1955); *Progress in Semiconductors*, 2, Heywood, London (1957).

26. R. B. McQuistan, *J. Opt. Soc. Amer.* **48**, 63 (1958); **49**, 70 (1959).

27. G. A. Morton and S. V. Forgue, *Proc. Inst. Radio Engrs.* **47**, 1607 (1959).

28. R. W. Redington and P. J. van Heerden, *J. Opt. Soc. Amer.* **49**, 997 (1959); see also Redington, Saum, and van Heerden, *Trans. Inst. Radio Engrs. Electron Devices* ED-6, No. 3 (July 1, 1959).

29. A. H. Canada, *Infrared, Its Military and Peacetime Uses*, Data Folder No. 87516, General Electric Company, Utica, N.Y. (1951).

30. J. N. Shive, *J. Opt. Soc. Amer.* **43**, 239 (1953).

31. P. W. Kruse, *J. Appl. Phys.* **30**, 770 (1959).

32. Lasser, Cholet, and Emmons, *Proc. Inst. Radio Engrs.* **47**, 2069 (1959).

33. See, for example, H. W. Leverenz, *Luminescence of Solids*, John Wiley and Sons, New York (1950).

34. G. Destriau and H. F. Ivey, *Proc. Inst. Radio Engrs.* **43**, 1911 (1955).

35. B. O'Brien, *J. Opt. Soc. Amer.* **36**, 369 (1946).

36. R. H. Dicke, *Rev. Sci. Instr.* **17**, 268 (1946).

37. G. Doundoulakis, *Proc. IRIS* **5**, No. 1, 167 (1960)(Unclassified); *Space/Aeronautics*, April, 1959, p. 147.

38. R. W. Gelinas, *Infrared Detection by Ideal Irasers and Narrow Band Counters*, RAND Report P-1844 (Oct. 26, 1959); *Masers and Irasers*, RAND Report P-1585, (Dec. 30, 1958); *Proc. IRIS* **5**, No. 1, 151 (1960) (Unclassified).

39. A. L. Schawlow and C. H. Townes, *Phys. Rev.* **112**, 1940 (1958); see also A. L. Schawlow, *Quantum Electronics*, Columbia Univ. Press (1960), p. 553, C. H. Townes, ed.

40. R. H. Dicke, U.S. Patent 2,851,652 (Sept. 9, 1958).

41. N. Bloembergen, *Phys. Rev. Lett.* **2**, 84 (1959).

42. A. Lubin, *Quantum Electronics*, Columbia Univ. Press (1960), p. 589, C. H. Townes, ed.

43. A. M. Prokhorov, *J. Exp. Theo. Phys.* **34**, 1658 (1958).

44. J. H. Sanders, *Phys. Rev. Lett.* **3**, 86 (1959).

45. T. H. Maiman, *Phys. Rev. Lett.* **4**, 564 (1960); *Nature* **187**, 493 (1960).

46. Shimoda, Takahasi, and Townes, *J. Phys. Soc. Japan* **12**, 686 (1957); see also Shimoda, *J. Phys. Soc. Japan* **14**, 966 (1959).

47. A. S. Tager and A. D. Gladun, *J. Exp. Theo. Phys.* **8**, 560 (1959).

48. N. F. Mott and R. W. Gurney, *Electronic Processes in Ionic Crystals*, Oxford University Press, London (1953).

49. R. C. Jones, *J. Opt. Soc. Amer.* **45**, 799 (1955); **48**, 874 (1958).

50. W. Clark, *Photography by Infrared*, John Wiley and Sons, New York, 2nd Edition (1946). Out of print.

51. L. Larmore, *Proc. Inst. Radio Engrs.* **47**, 1487 (1959).

52. W. H. Brattain and J. A. Becker, *J. Opt. Soc. Amer.* **36**, 354 (1946).

53. J. N. Shive, *The Properties, Physics, and Design of Semiconductor Devices*, Van Nostrand Company, Inc., Princeton (1959).

54. *See:* R. DeWaard and E. M. Wormser, *Thermistor Infrared Detectors; Properties and Developments*, NAVORD 5495, Part 1 (April 30, 1958), AD 160106 (Unclassified); *Proc. Inst. Radio Engrs.* **47**, 1508 (1959).

55. D. H. Andrews, *Phys. Soc. (London)*, Rept. 1–2, 56 (1946); Andrews, Milton, and deSorbo, *J. Opt. Soc. Amer.* **36**, 518 (1946).

56. R. M. Milton, *Chem. Rev.* **39,** 419 (1946).
57. N. Fuson, *J. Opt. Soc. Amer.* **38,** 845 (1948).
58. D. F. Hornig and B. J. O'Keefe, *Rev. Sci. Instr.* **18,** 7 (1947); **18,** 474 (1947).
59. P. B. Fellgett, *Proc. Phys. Soc.* **B62,** 351 (1949).
60. M. J. E. Golay, *Rev. Sci. Instr.* **18,** 346 (1947); **18,** 357 (1947); **20,** 816 (1949).
61. Harding, Hilsum, and Northrop, *Nature* **181,** 691 (1958).
62. C. Hilsum and W. R. Harding, *S.E.R.L. Tech. J.* **10,** No. 1, 26 (1960).

9
Mathematical analyses of selected detection mechanisms and of the photon noise limit

9.1 INTRODUCTION

In the previous chapter the various photon and thermal effects used in infrared detection were described. Four of these, the photoconductive, photoelectromagnetic, photovoltaic, and bolometer effects, are much more widely used than the others. In the first part of this chapter we shall consider in detail the theories of these effects. The primary aim is to establish criteria needed to enable the designer of infrared detectors to optimize detector performance through adjustment of material parameters. It will be possible to use the expressions developed to compare different materials operating in the same mode, or to study one material in several modes, to determine the optimum material or mode for a given application.

The latter part of the chapter is devoted to determining the performance of ideal detectors. The ideal photon detector, capable of counting every photon of wavelength shorter than the cutoff wavelength which reaches the detector, has no internally generated noise. The term "Blip" (background limited infrared photoconductor) has been coined by Burstein and Picus[1] to describe an ideal photoconductor. The ideal thermal detector, also noiseless, is capable of absorbing all radiation falling upon it, and is coupled to its surroundings by radiation exchange only. The noise which ultimately limits the performance of both types is that arising from the background radiation falling upon the detector. Since the detectors are noiseless, the electrical signals from them show random fluctuations due to the random arrival of the photons from the thermal background surrounding them, and, for thermal detectors only, the random emission of photons from the detector to the background. Even if a background at a temperature of absolute zero were to surround thermal detectors, the electrical signal from them would exhibit photon noise due to the random emission of photons from them to the background.

Practically speaking, those detectors which are photon noise limited are operated at low temperatures, of the order of 77°K or below. They frequently are surrounded by an enclosure, also cooled to the operating temperature. Through an aperture in the enclosure they see the detector window and the radiation penetrating the window, originating from the target being viewed and from the background which surrounds the target.

Since the enclosure is at a low temperature, it does not emit radiation of great intensity, and thereupon does not contribute greatly to the photon noise. Radiation from the window and from the target and background passing through the window, however, is much more intense, and is therefore the predominant source of the photon noise. The photon noise level, being a function of the total radiant power, is therefore a function of the aperture size. The noise level can be reduced by reducing the aperture. This, however, reduces the viewing angle, and therefore the utility of the detector. We shall derive in Section 9.4 the relationship between viewing angle, background temperature, and spectral response.

Under certain circumstances relating to background emission and to the spectral response and time constant of the detector, it is possible for an ideal detector to be limited, not by photon noise from the background but by signal fluctuations. Since the detector in its electrical circuit has a certain "counting time" related to the electrical bandwidth, it is necessary in order to be reasonably sure of detecting a source that at least an average of one photon per counting time interval reach the detector. The probability of detection increases as the average number of photons per counting time interval increases. This problem of signal fluctuations is treated in Section 9.4.5.

9.2 NOISE IN DETECTORS

If the detector is not limited by photon noise or signal fluctuations, then it will be limited by some noise arising in the detector. We shall restrict our discussion here to the elemental detectors, choosing only those based upon the widely used semiconductor effects. Considered first are photon detectors.

We have seen in Chapter 7 that the fundamental noise arising in conducting materials, whether metals or semiconductors, is thermal or

Johnson noise, due to the random motion of electrical charges in the materials. This noise was seen to depend upon the temperature and measurement bandwidth, and to be frequency independent, the rms value of the thermal noise voltage v_N being given by

$$v_N = (4kTR \, \Delta f)^{1/2}, \tag{9.1}$$

the symbols having their usual significance. We recall that any noise over and above thermal was referred to as excess noise.

Thermal noise limits some of the detectors which operate without benefit of bias voltage, such as the PEM type of detector. Many photoconductive detectors, however, are limited by current noise, which as recalled from Eq. 7.68 has an inverse frequency spectrum described by

$$v_N = \left(\frac{c_1}{lA} I^2 R^2 \frac{\Delta f}{f^\beta} \right)^{1/2}, \tag{9.2}$$

where c_1 is a constant, lA is the volume of material, and β is a factor usually near unity. However, β may have values much less or much greater than unity in certain photoconductive materials. Although photovoltaic detectors are operated without bias, in certain instances their noise spectra, instead of being white as anticipated, show a $1/f$ dependence. Detectors which are current noise limited at low frequencies may become thermal or photon noise limited at high frequencies. It is desirable, of course, to have this transition take place at as low a frequency as possible.

Some detectors are also limited by generation-recombination noise. Our discussion in Section 7.4.2 shows the rms noise current to be inversely proportional to the total number of current carriers within the semiconductor. (See Eq. 7.77). Spicer[2] has considered the performance of gr noise limited detectors. His analysis shows the peak NEP decreases exponentially, that is, improves, as the forbidden energy gap width increases. That is to say, for detectors limited by gr noise, those with the longest cutoff wavelengths exhibit the poorest peak NEP. We shall see that this is also the case with photon noise limited detectors.

As with photon detectors, thermal detectors are also background limited in their ultimate performance. If this ultimate performance is not realized, they may be temperature noise limited. Since thermal detectors depend upon a change in temperature to indicate a radiation signal, any random fluctuations in temperature will show up as spurious signals, or noise. These random fluctuations occur through coupling of the detector to its surrounding by thermal energy exchange. Generally, this takes place by conduction to some heat sink. If, however, conduction and convection interchanges are negligible, then the principal interchange is by means of radiation. We shall show in Section 9.4.2 that under this condition in which radiation exchange is dominant, temperature noise becomes identified with photon noise.

Most thermal detectors, however, are not temperature noise limited. Thermistor bolometers are limited by current noise at very low frequencies, going over to thermal noise at higher frequencies. If not in evacuated housings they may be limited by "swish" noise caused by motion of air within the housings. The superconducting bolometer appears limited by some form of excess noise of undetermined origin. Thermocouples, operated without bias, are thermal noise limited. Since much less effort has been expended in the development of thermal detectors, they do not approach the photon noise limit as closely as do photon detectors.

9.3 MATHEMATICAL THEORY OF SELECTED DETECTION MECHANISMS

We have seen that a large number of methods exist for transducing the energy of infrared radiation into some other form, such as electrical. We shall consider in this section the mathematical theory pertaining to the design and performance of elemental detectors based upon the four most widely applied mechanisms: the photoconductive, photovoltaic, photoelectromagnetic, and bolometer effects. The reader interested in theories of other effects, such as the Dember and thermoelectric effects, is referred to the books by Moss[3] and by Smith, Jones, and Chasmar.[4] We shall develop expressions for $D*$ in terms of material parameters, such as carrier concentrations, mobilities, surface recombination velocities, and lifetime, in terms of physical parameters including the sample dimensions, and in terms of parameters appropriate to the mode of operation, including the electric field in the photoconductive mode and magnetic field in the PEM. The expressions for $D*$ developed in each of the cases will be broken into surface, bulk, and mode terms in order that the effects of each may be clearly seen. The resulting expressions, when applied to a specific material, can be used to determine which of the three photon detection modes will be superior to the others for any given material and temperature. Similar expressions developed by Hilsum and Simpson[5] have been applied to indium antimonide, indium arsenide, and lead sulfide. Since thermal detection mechanisms are fundamentally different from photon mechanisms, the development of the theory of bolometers will be independent of the other three developments, but in this case the detectivity expression will also be determined.

The following analyses of the photon effects are based upon a number of assumptions pertaining to each mode of operation.

1. The sample is uniform throughout.
2. The sample is of infinite extent in the plane perpendicular to the incident radiation.
3. The incident radiation is distributed uniformly over the front surface.
4. The incident radiation is totally absorbed with unit quantum efficiency in a depth short compared to the carrier diffusion lengths.
5. The magnetic induction **B** in the PEM mode is uniform throughout the sample.
6. The joule heating in the photoconductive mode is not excessive.
7. The barrier width in the photovoltaic mode is small compared to the minority carrier diffusion lengths.
8. The effect of surface states may be ignored.
9. The incident radiation intensity is sufficiently low so that the concentration of optically excited carriers remains much smaller than that of thermally excited carriers.
10. The excess carriers recombine directly.

We shall initially consider operation in the photoconductive and the PEM modes. The sample orientation applicable to either of these modes is shown in Fig. 9.1. The sample has a finite thickness in the y-direction, but is infinite in the x- and z-directions. Radiation falling on the front surface, $y = 0$, is absorbed there, giving rise to free electrons and holes which diffuse into the bulk, where they are acted upon by electric and magnetic fields. They may recombine at either the front surface, $y = 0$, the back surface, $y = d$, or in the bulk.

Our analysis begins by reiterating the expressions developed in Chapter 6 pertaining to diffusion and drift of electrons and holes in electric and magnetic fields. We have

$$\mathbf{J}_e = nq\mu_e \mathbf{E} + q D_e \, \nabla n - \mu_e \mathbf{J}_e \times \mathbf{B}, \qquad (6.81)$$

$$\mathbf{J}_h = pq\mu_h \mathbf{E} - q D_h \, \nabla p + \mu_h \mathbf{J}_h \times \mathbf{B}, \qquad (6.82)$$

$$\frac{\partial p}{\partial t} = (g - r) - \frac{1}{q} \, \nabla \cdot \mathbf{J}_h, \qquad (6.73)$$

and
$$\frac{\partial n}{\partial t} = (g - r) + \frac{1}{q} \, \nabla \cdot \mathbf{J}_e. \qquad (6.74)$$

In these equations $\mathbf{J}_e$ and $\mathbf{J}_h$ are the electron and hole current density vectors, n and p are the total electron and hole concentrations, μ_e and μ_h are the electron and hole mobilities, D_e and D_h are the electron and hole diffusion constants, g and r are the generation and recombination rates, q is the electronic charge, $\mathbf{E}$ is the electric field vector, and $\mathbf{B}$ is the magnetic induction vector.

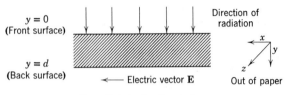

Figure 9.1. Sample orientation for photoconductivity and the PEM effect.

Applying these equations to the sample orientation under consideration results in

$$J_{ex} = nq\mu_e E_x - \mu_e B J_{ey},$$ (9.3)

$$J_{ey} = nq\mu_e E_y + q D_e \frac{dn}{dy} + \mu_e B J_{ex},$$ (9.4)

$$bJ_{hx} = pq\mu_e E_x + \mu_e J_{hy} B,$$ (9.5)

$$bJ_{hy} = pq\mu_e E_y - qb D_h \frac{dp}{dy} - \mu_e J_{hx} B,$$ (9.6)

where the x and y subscripts indicate the scalar components of the vector quantities. Combining the equation of continuity, Eq. 6.74, and the direct recombination expression, Eq. 6.52, results in

$$\frac{dJ_{ey}}{dy} = \frac{q}{\tau} \frac{(np - n_0 p_0)}{n_0 + p_0},$$ (9.7)

where n_0 and p_0 are the electron and hole concentrations at thermal equilibrium and τ is the lifetime associated with direct recombination. We require also that the total current in the y-direction be zero, since there is no closed circuit in that direction. Thus

$$J_{ey} = -J_{hy}.$$ (9.8)

Finally, we must introduce the boundary conditions at the front surface, $y = 0$, and at the back surface, $y = d$. These are determined by considering the net current flowing away from the front surface and to the back surface as a result of excitation at the front surface and recombination at both surfaces. Recalling the definition of surface recombination velocity as the recombination rate per unit area per unit excess carrier concentration just below the surface, we write for the front surface

$$J_{ey} = -qN + es_1 \frac{(np - n_0 p_0)}{(n_0 + p_0)}, \qquad (y = 0)$$ (9.9)

and for the back

$$J_{ey} = -qs_2 \frac{(np - n_0 p_0)}{(n_0 + p_0)}, \qquad (y = d)$$ (9.10)

where s_1 and s_2 are the surface recombination velocities at front and back, and N is the number of photons per unit front surface area per second absorbed by the detector. We require for our analysis one more basic equation, Eq. 3.1, which describes the divergence of the electric field due to the net charge. In our example this becomes

$$\frac{dE_y}{dy} = \frac{q}{\varepsilon} [(p - p_0) - (n - n_0)], \tag{9.11}$$

where ε is the absolute capacitivity.

9.3.1 Photoconductivity

Equations 9.3 through 9.11 are the basic equations describing excitation, motion, and recombination within a semiconductor in the presence of electric and magnetic fields. We shall now use these to derive the expression for the detectivity of a photoconductor. Our procedure will be to determine first the expression for J_{ey} in terms of the other parameters. This will involve solving a second order differential equation in J_{ey}. Having done this, we will be able to evaluate the expression for the photoconductive short circuit current and, in the next section, the PEM short circuit current. These will lead to expressions for the corresponding open circuit voltages. Assuming a noise mechanism, we will be able to derive expressions for D_λ^*.

The differential equation for J_{ey} is obtained by taking the derivative of Eq. 9.7 and substituting it into Eq. 9.4, then solving Eqs. 9.3 through 9.8. By so doing we obtain

$$J_{ey}'' - \frac{nba}{L_e^2 n_0(a+1)} \left[1 + \left(\frac{\mu_e}{b}\right)^2 B^2 + \frac{p}{nb}(1 + \mu_e^2 B^2) \right] J_{ey} = 0, \tag{9.12}$$

where L_e, the electron diffusion length, is given by (see Eq. 6.78)

$$L_e = (D_e \tau)^{1/2} = \left(\frac{\mu_e kT}{q} \tau\right)^{1/2}, \tag{9.13}$$

and we have defined $a \equiv n_0/p_0$ and $b \equiv \mu_e/\mu_h$. Equation 9.12 is nonlinear because n and p are functions of y. However, for small signals, assumption (9), n and p are approximately given by n_0 and p_0. Introducing this assumption into Eq. 9.12 results in

$$J_{ey}'' - \frac{J_{ey}}{(L_D^*)^2} = 0, \tag{9.14}$$

where we define L_D^* as

$$L_D^* \equiv L_e \left[\frac{1 + a}{(1 + \mu_e^2 B^2) + ba(1 + \mu_e^2 B^2 b^{-2})} \right]^{1/2}. \tag{9.15}$$

Thus we see that $L_D{}^*$ is a multiple of the electron diffusion length L_e. Because $L_D{}^*$ depends on both electrons and holes moving in a magnetic field, it is termed the "ambipolar diffusion length in a magnetic field." For the photoconductive effect, in which $B = 0$, the value of $L_D{}^*$ is

$$L_D{}^* = L_e\left(\frac{1+a}{1+ba}\right)^{1/2}, \qquad (B = 0). \tag{9.16}$$

The solution of Eq. 9.14 is

$$J_{ey} = \alpha \exp(-y/L_D{}^*) + \beta \exp(y/L_D{}^*). \tag{9.17}$$

The values of α and β are determined by substituting the boundary conditions, Eqs. 9.9 and 9.10, into Eq. 9.17. They are found to be

$$\alpha = -\exp(2d/L_D{}^*)\left(1 + \frac{s_2\tau}{L_D{}^*}\right)\frac{qN}{r}, \tag{9.18}$$

and

$$\beta = \frac{qN}{r}\left(1 - \frac{s_2\tau}{L_D{}^*}\right), \tag{9.19}$$

where r is defined as

$$r \equiv \exp(2d/L_D{}^*)\left(1 + \frac{s_1\tau}{L_D{}^*}\right)\left(1 + \frac{s_2\tau}{L_D{}^*}\right) - \left(1 - \frac{s_1\tau}{L_D{}^*}\right)\left(1 - \frac{s_2\tau}{L_D{}^*}\right),$$
$$\equiv 2\exp(d/L_D{}^*)\left[\left(1 + \frac{s_1 s_2\tau^2}{L_D{}^{*2}}\right)\sinh\frac{d}{L_D{}^*} + \left(\frac{s_1\tau}{L_D{}^*} + \frac{s_2\tau}{L_D{}^*}\right)\cosh\frac{d}{L_D{}^*}\right]. \tag{9.20}$$

We can now proceed to determine $i_{S,PC}$, the photoconductive short circuit current per unit width. This is defined as the increment in current per unit width in the presence of irradiation over that in the dark, determined with an electric field applied to the sample and the ends short circuited. Thus $i_{S,PC}$ is given by

$$i_{S,PC} = E_x\,\Delta\sigma = E_x q\mu_e\int_0^d\left(n - n_0 + \frac{p - p_0}{b}\right)dy, \tag{9.21}$$

where $\Delta\sigma$ is the change in electrical conductivity upon irradiation. To determine the value of the integral we consider Eqs. 9.7 and 9.11. For small signals Eq. 9.7 becomes

$$J_{ey}' \approx \frac{q}{\tau}\frac{[p_0(n - n_0) + n_0(p - p_0)]}{(n_0 + p_0)}. \tag{9.22}$$

Substituting Eq. 9.11 into 9.22 and rearranging terms we arrive at

$$(n - n_0) + \left(\frac{p - p_0}{b}\right) = \frac{\tau}{q}\left(1 + \frac{1}{b}\right)J_{ey}' - \frac{\varepsilon}{q}\frac{n_0}{(n_0 + p_0)}\left(1 - \frac{p_0}{n_0 b}\right)E_y'. \tag{9.23}$$

Equation 9.23 may be substituted into 9.21 to determine $i_{S,PC}$. However, the integral in E_y' goes to zero since the value of E_y at either surface equals zero. Thus the integral for $i_{S,PC}$ is

$$i_{S,PC} = E_x \mu_e \tau \left(1 + \frac{1}{b}\right) \int_0^d J_{ey}' \, dy. \tag{9.24}$$

Since we have already determined the analytic expression for J_{ey}, Eq. 9.17, we may differentiate and substitute into the integral, subject to the condition of no magnetic induction, $B = 0$. We find

$$i_{S,PC} = E_x \mu_e \tau \left(1 + \frac{1}{b}\right) \{\alpha[\exp(-d/L_D^*) - 1] + \beta[\exp(d/L_D^*) - 1]\}. \tag{9.25}$$

Substituting in the values of α and β determined from Eqs. 9.18 and 9.19, and r from 9.20, we find

$$i_{S,PC}$$
$$= \frac{E_x \mu_e \tau \left(1 + \frac{1}{b}\right) q N \{\sinh(d/L_D^*) + s_2 \tau / L_D^* [\cosh(d/L_D^*) - 1]\}}{[1 + s_1 s_2 \tau^2 / L_D^{*2}] \sinh(d/L_D^*) + [s_1 \tau / L_D^* + s_2 \tau / L_D^*] \cosh(d/L_D^*)}. \tag{9.26}$$

From Thevenin's theorem[6] we know the open circuit voltage per unit width v_0 to be equal to the short circuit current per unit width multiplied by the resistance R. Thus we have for small signals

$$v_0 = i_{S,PC} R = \frac{i_{S,PC} l}{w d n_0 q \mu_e (1 + 1/ba)}, \tag{9.27}$$

where l is the sample length in the direction of current flow, w is the width, and d as before is the thickness.

We shall now determine the value of D_λ^*. We must assume a noise mechanism. For our first example we shall assume it to be thermal noise, since an analytic expression exists for thermal noise. The thermal noise voltage v_N is

$$v_N = (4kTR \, \Delta f)^{1/2}. \tag{9.1}$$

In terms of the radiant power P_λ absorbed on the surface of area lw, the flux density N of absorbed photons of wavelength λ is

$$N = \frac{P_\lambda}{lw} \frac{\lambda}{hc_0} = \mathscr{H}_\lambda \frac{\lambda}{hc_0}, \tag{9.28}$$

where $\mathscr{H}_\lambda$ is the irradiance. The value of D_λ^* is obtained by determining the value of P_λ required to give $v_0 w$ equal to v_N for unit bandwidth, then dividing this into the square root of the detector area lw. This results in

$$D^*_{\lambda,\mathrm{PC}} = \frac{E_x \lambda q^{3/4}}{2hc_0(kT)^{3/4}} \frac{\mu_e^{1/4} \tau^{3/4}}{n_i^{1/2}} A_1 B_1, \qquad (9.29)$$

where

$$A_1 \equiv \frac{\sinh m + \alpha_2[\cosh(m) - 1]}{m^{1/2}[(1 + \alpha_1 \alpha_2)\sinh m + (\alpha_1 + \alpha_2)\cosh m]}, \qquad (9.30)$$

$$B_1 \equiv \frac{(b + 1)a^{1/4}}{b^{1/2}(a + 1)^{1/4}(ba + 1)^{1/4}}. \qquad (9.31)$$

Here we have $m \equiv d/L_D^*$, $\alpha_1 \equiv s_1 \tau/L_D^*$, $\alpha_2 \equiv s_2 \tau/L_D^*$, and n_i, the intrinsic concentration, is given by $n_i = n_0/a^{1/2}$. (See Eq. 6.17.)

Equation 9.29 is the analytic expression for the value of D_λ^* of a thermal noise limited detector operated in the photoconductive mode where the detector material has an electron mobility μ_e, lifetime τ, intrinsic concentration n_i, ambipolar diffusion length L_D^*, front and back surface recombination velocities s_1 and s_2, mobility ratio b, ratio of free electron to free hole concentration a, and thickness d, in the presence of an electric field of magnitude E_x and radiation of wavelength λ absorbed with unit quantum efficiency at the front surface with no reflection losses. The value of D_λ^* is expressed in terms of a dimensionless thickness factor A_1, involving the thickness, ambipolar diffusion length, and recombination velocities, and a dimensionless bulk factor B_1, involving the mobility ratio and concentration ratio. By studying Eq. 9.29 we can determine the requirements needed to maximize the value of $D^*_{\lambda,\mathrm{PC}}$. We require first a high electron mobility and a low intrinsic concentration. We need the longest possible carrier lifetime, and the highest possible applied field, up to the point at which overheating occurs (or until current noise predominates). The value of D_λ^* will be greatest for wavelengths just short of the absorption edge, for this will give us the maximum permissible value of λ. Assuming that μ_e, τ, and b are independent of temperature, we want to operate at the lowest possible temperature.

Finally, we must maximize the values of A_1 and B_1. Consider first B_1. The maximum value of B_1 is determined by setting the derivative of B_1 with respect to a equal to zero. We find

$$a = \frac{1}{b^{1/2}} \text{ for } B_{1\max}. \qquad (9.32)$$

Therefore the photoconductor must be doped on the p side sufficiently to make the ratio of hole to electron concentration equal to the square root of the mobility ratio. The maximum value of B_1 is, from Eq. 9.31,

$$B_{1\max} = \frac{(b + 1)}{b^{1/2}(b^{1/2} + 1)^{1/4}}. \qquad (9.33)$$

For $b = 1$, $B_{1\max} = 1.68$. As b approaches infinity $B_{1\max}$ varies as $b^{1/4}$. Since $D_\lambda{}^*$ varies as $\mu_e^{1/4}B_1$ it is important for photoconductive detectors to have high electron mobilities and high mobility ratios. In a practical case μ_e, τ, and b are dependent upon impurity concentration. Therefore the maximum value of B_1 may be attained under conditions slightly different from $a = b^{-1/2}$.

No simple expression exists for the conditions needed to maximize A_1. However, Fig. 9.2 shows the dependence of A_1 upon m for selected values of α_1 and α_2. Examining Fig. 9.2 we see that when $\alpha_1 = \alpha_2 = 0$, which

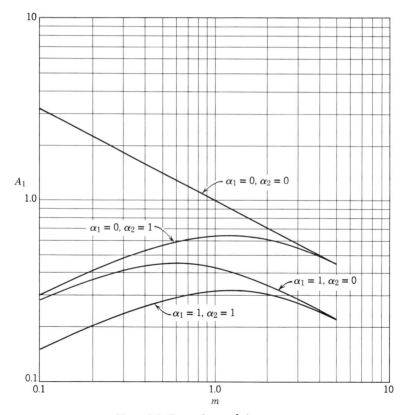

Figure 9.2. Dependence of A_1, upon m.

may be achieved by making both surface recombination velocities equal to zero, the semiconductor element should have the smallest possible thickness, at least to the point where the assumption of infinite absorption is no longer valid. This invalidation takes place at a thickness less than two or three times the reciprocal of the optical absorption constant for the wavelength of interest. On the other hand, for the three other cases considered, an optimum thickness does exist, of value about that of the ambipolar diffusion length. If the sample is five times as thick, the condition of the back surface becomes unimportant, as shown by the meeting of the curves for $\alpha_1 = 0$, $\alpha_2 = 0$, and $\alpha_1 = 0$, $\alpha_2 = 1$ at that point, and the meeting of the curves $\alpha_1 = 1$, $\alpha_2 = 0$, and $\alpha_1 = 1$, $\alpha_2 = 1$ at that point. This simply means that very few carriers reach the back surface to recombine there. For very thin samples, of thickness about one-tenth the ambipolar diffusion length, the condition of both surfaces is important, both contributing approximately equally, witnessed by the close approach of the curves for $\alpha_1 = 0$, $\alpha_2 = 1$ and $\alpha_1 = 1$, $\alpha_2 = 0$ at that point. In this case the sample is so thin that the numbers diffusing to the back surface are almost equal to those leaving the front surface.

Consider now the expression for D_λ^* for a photoconductive detector limited by current noise. In Section 7.4.1 the dependence of current noise voltage upon material dimensions was discussed. Although the magnitude of the current noise voltage could not be determined, the dependence upon dimensions was found to be

$$\overline{v_N^2} = \frac{c_1}{lA} I^2 R^2 \frac{\Delta f}{f^\beta}, \tag{7.68}$$

where c_1 depends upon the material but is independent of dimensions, and A, the cross-sectional area, is equal to wd. Introducing $E_x l = IR$, and following the procedure used in determining D_λ^* under thermal noise limitations, we equate $v_0 w$ with $\left(\overline{v_N^2}\right)^{1/2}$ for unity bandwidth from Eq. 7.68. Solving for P_λ through Eq. 9.28 we divide it into the square root of the area to determine D_λ^*.

$$D_{\lambda,\text{PC}}^* = \frac{\lambda q^{1/4} \tau^{3/4}}{c_1^{1/2} hc_0(kT)^{1/4} \mu_e^{1/4} n_i} \cdot \frac{f^{\beta/2}}{(1 + 4\pi^2 f^2 \tau^2)^{1/2}} \cdot A_1 B_{11}, \tag{9.34}$$

where A_1 is given by Eq. 9.30 and plotted in Fig. 9.2 and

$$B_{11} \equiv \frac{(b + 1)a^{1/2}}{(a + 1)^{1/4}(ba + 1)^{3/4}}. \tag{9.35}$$

In Eq. 9.34 we have introduced the factor $(1 + 4\pi^2 f^2 \tau^2)^{1/2}$ into the denominator to account for the dependence of the signal voltage upon frequency. The value of B_{11} exhibits a maximum at

$$a = \frac{(b^2 + 14b + 1)^{1/2} - (b - 1)}{4b}, \quad \text{for } B_{11} \text{max.} \tag{9.36}$$

Thus the dependence of D_λ^* for current noise limited photoconductors upon thickness, determined by A_1 is the same as for thermal noise limited photoconductors, but the dependence upon purity through B_{11} differs from the dependence for thermal noise limited photoconductors through B_1.

9.3.2 The Photoelectromagnetic Effect

The analysis of the PEM effect follows a pattern similar to that for photoconductivity. However, as we shall see, it is somewhat more complicated because of magnetoresistive effects. The application of a magnetic field to a semiconductor causes the measured resistance to increase, a phenomenon known as magnetoresistivity. We shall solve for the PEM short circuit current in a manner similar to that for the photoconductive short circuit current. However, the open circuit voltage involves the magnetoresistivity, depending upon material parameters, magnetic induction, and dimensions.

We begin by considering the PEM short circuit current per unit width, $i_{S,PEM}$, that current per unit width which would flow in the presence of radiation if the sample ends were short circuited. It is defined as

$$i_{S,PEM} \equiv \int_0^d (J_{ex} + J_{hx}) \, dy. \tag{9.37}$$

The solution to this is found by relating J_{ex} and J_{hx} to J_{ey} through Eqs. 9.3, 9.5, and 9.8, subject to the condition of no electric field, that is, $E_x = 0$. Since J_{ey} has already been determined, Eq. 9.17, the expressions for J_{ex} and J_{hx} in terms of J_{ey} are substituted into Eq. 9.37. In this case we obtain

$i_{S,PEM}$
$$= \frac{-\mu_e B(1 + 1/b)qNL_D^*[1 - \cosh(d/L_D^*) - (s_2\tau/L_D^*)\sinh(d/L_D^*)]}{\left(1 + \frac{s_1 s_2 \tau^2}{L_D^{*2}}\right)\sinh(d/L_D^*) + [s_1\tau/L_D^* + s_2\tau/L_D^*]\cosh(d/L_D^*)}. \tag{9.38}$$

The open circuit voltage per unit width, v_0, is given again by the product of the short circuit current per unit width and the resistance. However, in this case the resistance is that in the magnetic field, that is, the magnetoresistance. Since we do not know a priori the functional representation for the magnetoresistance, we must solve directly for the open circuit voltage. This is determined by setting the current equal to zero.

$$\int_0^d (J_{ex} + J_{hx}) \, dy = 0. \tag{9.39}$$

We again obtain a differential equation in J_{ey}, given by

$$J_{ey}'' + \frac{\mu_e^2}{L_e^2}B\left(1 + \frac{1}{b}\right)E_x\tau J_{ey}' - a\frac{\left[n\left(b + \frac{\mu_e^2 B^2}{b}\right) + p(1 + \mu_e^2 B^2)\right]}{L_e^2 n_0(a+1)}J_{ey}$$

$$= -\frac{B\mu_e^2 q(b+1)}{L_e^2(a+1)b}n_0 E_x. \quad (9.40)$$

We again make the small signal approximation. We may neglect the first order term if the coefficient of it is small compared to the geometric mean of the coefficients of the second order and zero order terms.

$$\mu_e B E_x\left(1 + \frac{1}{b}\right) \ll \frac{L_e}{\mu_e \tau}\left[\frac{a}{(a+1)}\left(b + \frac{\mu_e^2 B^2}{b}\right) + \frac{(1 + \mu_e^2 B^2)}{(a+1)}\right]^{1/2}. \quad (9.41)$$

This will usually be valid even for those semiconductors in which magnetoresistive effects are important. The solution of Eq. 9.40 is

$$J_{ey} = \gamma \exp\left(-y/L_D*\right) + \zeta \exp\left(y/L_D*\right) - ML_D*^2, \quad (9.42)$$

where

$$M \equiv -\frac{B\mu_e^2 q(b+1)}{L_e^2(a+1)b}n_0 E_x, \quad (9.43)$$

and γ and ζ, determined by introducing the boundary conditions, Eqs. 9.9 and 9.10, are found to be

$$\gamma = ML_D*^2 \frac{\exp\left(d/L_D*\right)}{\left(1 - \frac{S_2\tau}{L_D*}\right)} - \exp\left(2d/L_D*\right)\frac{\left(1 + \frac{S_2\tau}{L_D*}\right)eN}{r}$$

$$- \exp\left(2d/L_D*\right)\frac{\left(1 + \frac{S_2\tau}{L_D*}\right)}{\left(1 - \frac{S_2\tau}{L_D*}\right)}\frac{ML_D*^2}{r}$$

$$\times \left[\exp\left(d/L_D*\right)\left(1 + \frac{S_1\tau}{L_D*}\right) - \left(1 - \frac{S_2\tau}{L_D*}\right)\right], \quad (9.44)$$

$$\zeta = L_D*^2$$

$$\times \frac{M\left[\exp\left(d/L_D*\right)\left(1 + \frac{S_1\tau}{L_D*}\right) - \left(1 - \frac{S_2\tau}{L_D*}\right)\right] + qN/L_D*^2\left(1 - \frac{S_2\tau}{L_D*}\right)}{r}.$$

$$(9.45)$$

Introducing Eq. 9.42 into Eqs. 9.3 and 9.5, together with Eq. 9.8, we determine J_{ex} and J_{hx}. Substituting these into Eq. 9.39 and solving, we determine the value of v_0, the open circuit voltage per unit width, from $v_0 = E_x l/w$. This results in

$$v_0 = \frac{i_{S,\text{PEM}} R_0}{\psi} = \frac{i_{S,\text{PEM}} l}{w d n_0 q \mu_e \left(1 + \dfrac{1}{ba}\right)\psi}, \quad (9.46)$$

where

$$\psi \equiv 1 - \frac{a(b+1)^2 \mu_e^2 B^2 \chi L_D^{*2}}{b(ab+1)L_e^2(1+a)}, \quad (9.47)$$

and

$$\chi \equiv 1 - \frac{L_D^*[\exp(d/L_D^*) - 1]}{dr}\left[\exp(d/L_D^*)\left(1 + \frac{s_1 \tau}{L_D^*}\right)\right.$$

$$\left. + \exp(d/L_D^*)\left(1 + \frac{s_2 \tau}{L_D^*}\right) - \left(1 - \frac{s_1 \tau}{L_D^*}\right) - \left(1 - \frac{s_2 \tau}{L_D^*}\right)\right],$$

$$= 1 - \frac{L_D^*}{d}\frac{\left[2[\cosh(d/L_D^*) - 1] + \left(\dfrac{s_1 \tau}{L_D^*} + \dfrac{s_2 \tau}{L_D^*}\right)\sinh(d/L_D^*)\right]}{\left[\left(1 + \dfrac{s_1 s_2 \tau^2}{L_D^{*2}}\right)\sinh(d/L_D^*) + \left(\dfrac{s_1 \tau}{L_D^*} + \dfrac{s_2 \tau}{L_D^*}\right)\cosh(d/L_D^*)\right]}.$$

$$(9.48)$$

In order to determine the value of D_λ^* we assume again that the limiting noise is thermal noise. Since there is no bias current, there is no possibility of the PEM detector being limited by current noise. Proceeding as before, we determine the value of P_λ required to give $v_0 w$ equal to v_N for unit bandwidth, and divide this into the square root of the detector area, resulting in

$$D_{\lambda,\text{PEM}}^* = \frac{\lambda \mu_e^{3/4} B q^{1/4} \tau^{1/4}}{2hc_0(kT)^{1/4} n_i^{1/2}} \frac{A_2}{\psi^{1/2}} B_2, \quad (9.49)$$

where

$$A_2 \equiv \frac{\alpha_2 \sinh(m) + \cosh(m) - 1}{m^{1/2}[(1 + \alpha_1 \alpha_2)\sinh(m) + (\alpha_1 + \alpha_2)\cosh(m)]}, \quad (9.50)$$

and

$$B_2 \equiv \frac{(b+1)(1+a)^{1/4} a^{1/4}}{b^{1/2}(1+ba)^{1/2}\left[1 + \mu_e^2 B^2 + ba\left(1 + \dfrac{\mu_e^2 B^2}{b^2}\right)\right]^{1/4}}. \quad (9.51)$$

Equation 9.49 is the analytic expression for D_λ^* in the PEM mode at the wavelength λ, shorter than the absorption edge, in terms of the electron mobility μ_e, the lifetime τ, intrinsic concentration n_i, and magnetic induction B. Since we would like to make D_λ^* large, we want large values of mobility, lifetime, and magnetic induction, combined with a low value of intrinsic concentration. The value of D_λ^* is also in terms of the dimensionless parameters A_2, B_2, and ψ. We would like to maximize A_2 and B_2, yet minimize ψ. The situation is complicated by the presence of magnetoresistive phenomena. If we assume that the magnetoresistivity is small, that is, $\mu_e B \ll 1$, then $\psi \approx 1$ and we have

$$D_{\lambda,\mathrm{PEM}}^* = \frac{\lambda}{2hc_0} \frac{\mu_e^{3/4} B q^{1/4} \tau^{1/4}}{(kT)^{1/4} n_i^{1/2}} A_2 B_2', \qquad (\mu_e B \ll 1) \qquad (9.52)$$

where

$$B_2' \equiv \frac{(b+1)(1+a)^{1/4} a^{1/4}}{b^{1/2}(1+ba)^{3/4}}. \qquad (9.53)$$

Note that the expression for A_2 remains the same. In this case, where the magnetoresistivity is negligible, the maximum value of B_2' is obtained for

$$a = \frac{1-b}{b} + \left[\left(\frac{1-b}{b}\right)^2 + \frac{1}{b}\right]^{1/2}.$$

The function A_2 includes terms in $m \equiv d/L_D^*$, the ratio of thickness to ambipolar diffusion length. It is plotted in Fig. 9.3 for selected values of the parameters α_1 and α_2.

Examining Fig. 9.3, we note that for the four combinations of α_1 and α_2 chosen, the same four used in Fig. 9.2, an optimum value of m exists which will maximize A_2 and thus D_λ^*. These four optimum values of m indicate the proper thickness is of the order of the ambipolar diffusion length. The highest values of A_2 are obtained for the condition $\alpha_1 = 0$, $\alpha_2 = 1$, in contrast to the photoconductive mode, in which $\alpha_1 = 0$, $\alpha_2 = 0$ gave the highest A_2. The reason for this, of course, is in the differing mode of operation. Whereas the potential difference at the electrodes in the photoconductive mode arises from the electric field applied, in the PEM mode it is due to the motion of the carriers across the magnetic field lines. In order for the carriers generated at the front surface to move across the field, a charge gradient must exist. By allowing some of the carriers to recombine at the back surface, the diffusion potential can be made large and thus give the rest of the carriers a greater rate of diffusion. If the recombination velocity at the back surface is low, some of those carriers reaching there will not recombine, but will turn around and diffuse back toward the front surface. The action of the magnetic field on these carriers will deflect them opposite to the direction of those leaving the front surface. Thus those reflected from the back will tend to cancel the electric field set up by those leaving the front. For proper operation in the PEM mode, some of the photoexcited carriers must be sacrificed at the back surface to aid the diffusion of the others.

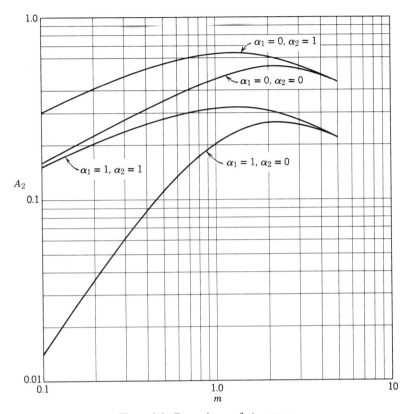

Figure 9.3. Dependence of A_2 upon m.

The above description is based upon the behavior under the condition $\mu_e B \ll 1$. For larger values of $\mu_e B$ the exact expression for $D^*_{\lambda,\mathrm{PEM}}$ is that of Eq. 9.49 instead of Eq. 9.52. Although ψ is dimensionless, it involves terms in m, α_1, α_2, b, and a, among others, and therefore is not easily represented in graphical form.

9.3.3 The Photovoltaic Effect

The third photon effect we shall discuss is the photovoltaic effect. Separate analyses are necessary for the diffused junction detector and the grown junction detector, since in the former the direction of the radiation is perpendicular to the junction, whereas in the latter it is parallel to the junction. We shall consider only the diffused junction type, since that is more widely used than the grown junction. Our analysis concerns only the motion of the minority carriers, since, as we have pointed out in

Section 6.10, it is their behavior which controls the performance of the detector. The model of the p-n junction is shown in Fig. 9.4.

Radiation of wavelength shorter than the absorption edge, falling on the front surface on the p region of the sample, is absorbed in an infinitesimally thick region at the surface, producing both electrons and holes. Some of the minority carriers, that is, the electrons, recombine at the front surface; the others diffuse toward the junction. Their appearance at the junction lowers the potential barrier there. This allows minority carriers to diffuse across the barrier in both directions, electrons toward the n region and holes toward the p region. If electrodes placed on the front and back surfaces are tied together, a short circuit current will flow in the external circuit. If the electrodes are not connected, an open circuit voltage will be present. We shall determine the expression for the open circuit voltage, and assuming a thermal noise mechanism, deduce the expression for $D_\lambda{}^*$.

We begin by considering the expression for the diffusion of minority carriers from the surface into the bulk, the minority carrier form of Eq. 8.7. We assume here that the current across the junction is due only to minority carrier flow, as is the case with ideal junctions. The continuity equation, expressing the fact that the minority carrier generation rate minus the recombination rate equals the divergence of the minority carrier flow, is

$$\frac{dJ_{ey}}{dy} = \frac{q}{\tau_e} \Delta n_p = D_e \frac{d^2(\Delta n_p)}{dy^2}. \tag{9.54}$$

Here we have defined $\Delta n_p \equiv n_p - n_{p0}$, where n_p is the nonequilibrium electron concentration in the p region, n_{p0} is the electron concentration in the p region at thermal equilibrium, and τ_e is the electron lifetime in the p region. At the front surface we have the same boundary condition as before, Eq. 9.9.

$$J_{ey} = q D_e \frac{d(\Delta n_p)}{dy} = -qN + qs_1(\Delta n_p), \quad (y = 0). \tag{9.55}$$

However, in this case we do not consider recombination at the back surface. Instead, the surface of interest is the junction to which the electrons diffuse. Recalling our discussion of p-n junctions in Section 6.10, we saw that the total current density was equal to the minority carrier

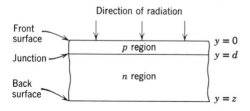

Figure 9.4. Sample orientation for the photovoltaic effect.

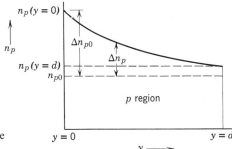

Figure 9.5. Relationship between the parameters in the p-region.

diffusion current across the boundary. Only those minority carriers which have sufficient energy to surmount the potential barrier at the junction can contribute to the current. In our example, since we consider the thickness of the p region to be variable, we express the electron current across the barrier as

$$J_{ey} = q D_e \frac{d(\Delta n_p)}{dy}, \qquad (y = d), \qquad (9.56)$$

subject to the boundary condition at the barrier that

$$n_p = n_{p0} \exp{(qV/kT)}, \qquad (y = d). \qquad (9.57)$$

Equation 9.57 can be obtained from Eqs. 6.83 and 6.84, where we have here identified n_p with Z of Chapter 6. In order to clarify the relations between the various parameters, we show concentration as a function of distance in Fig. 9.5.

The solution of Eq. 9.54 is

$$\Delta n_p = \xi \exp{(-y/L_e)} + \eta \exp{(y/L_e)}, \qquad (9.58)$$

where $L_e = (D_e \tau_e)^{1/2}$, the electron diffusion length in the p material.

The values of ξ and η, found by evaluating Eq. 9.58 under the boundary conditions, Eqs. 9.55 and 9.57, are

$$\xi = \frac{\left\{ (1 - \alpha_1)n_{p0}[\exp{(qV/kT)} - 1] + \dfrac{NL_e}{D_e}\exp{(d/L_e)} \right\}}{[(1 - \alpha_1)\exp{(-d/L_e)} + (1 + \alpha_1)\exp{(d/L_e)}]}, \qquad (9.59)$$

and

$$\eta = -\frac{NL_e}{D_e(1 - \alpha_1)}$$
$$+ \frac{(1 + \alpha_1)n_{p0}[\exp{(qV/kT)} - 1] + \left(\dfrac{1 + \alpha_1}{1 - \alpha_1}\right)\dfrac{NL_e}{D_e}\exp{(d/L_e)}}{[(1 - \alpha_1)\exp{(-d/L_e)} + (1 + \alpha_1)\exp{(d/L_e)}]}, \qquad (9.60)$$

where we have defined as before $\alpha_1 \equiv s_1 \tau_e / L_e$.

The expression for the electron current density across the barrier, found by substituting Eqs. 9.59 and 9.60 into 9.58 and 9.56, is

$$J_e = \frac{qD_e}{L_e} \left\{ \frac{n_{p0}[\exp(qV/kT) - 1][\alpha_1 \cosh(m_1) + \sinh(m_1)] - NL_e/D_e}{\cosh(m_1) + \alpha_1 \sinh(m_1)} \right\},$$

$$(9.61)$$

where $m_1 \equiv d/L_e$. Equation 9.61 is the electron contribution to the current flow in the p-n junction under irradiation. We must now consider the hole flow. Once again only the minority carriers are the important ones, in this case, holes in the n region. Thus we must solve the continuity equation for holes, analogous to Eq. 9.54.

$$D_h \frac{d^2(\Delta p_n)}{dy^2} = \frac{\Delta p_n}{\tau_h}, \qquad (9.62)$$

where $\Delta p_n \equiv p_n - p_{n0}$, p_n being the nonequilibrium hole concentration in the n region, p_{n0} being the value at thermal equilibrium, and τ_h being the hole lifetime in the n region. The boundary conditions in this case are that at the junction

$$p_n = p_{n0} \exp(qV/kT), \qquad (y = d) \qquad (9.63)$$

and at the back surface, located at $y = z$, which is a relatively large distance,

$$p_n = p_{n0}. \qquad (y = z) \qquad (9.64)$$

The solution to Eq. 9.62 follows the same pattern as before.

$$\Delta p_n = \phi \exp(-y/L_h) + \nu \exp(y/L_h). \qquad (9.65)$$

Introducing the boundary conditions into 9.65, we find

$$\phi = \frac{-p_{n0} \exp(2z/L_h)[\exp(qV/kT) - 1]}{\exp(d/L_h)[1 - \exp(2z/L_h) \exp(-2d/L_h)]}, \qquad (9.66)$$

and

$$\nu = \frac{p_{n0}[\exp(qV/kT) - 1]}{\exp(d/L_h)[1 - \exp(2z/L_h) \exp(-2d/L_h)]}. \qquad (9.67)$$

Writing the expression for the hole current density across the junction as

$$J_h = -qD_h \frac{d(\Delta p_n)}{dy}, \qquad (y = d) \qquad (9.68)$$

and evaluating in terms of Eqs. 9.65, 9.66, and 9.67, we arrive at

$$J_h = \frac{qD_h}{L_h} p_{n0}[\exp(qV/kT) - 1] \coth(m_2), \qquad (9.69)$$

where we have written $m_2 \equiv (z - d)/L_h$.

The total current density across the junction is found by adding the electron and hole components, Eqs. 9.61 and 9.69.

$$J_{\text{tot}} = \frac{q D_e}{L_e} \left\{ \frac{n_{p0}[\exp(qV/kT) - 1][\alpha_1 \cosh(m_1) + \sinh(m_1)] - NL_e/D_e}{\cosh(m_1) + \alpha_1 \sinh(m_1)} \right\}$$

$$+ \frac{q D_h}{L_h} p_{n0}[\exp(qV/kT) - 1] \coth(m_2). \quad (9.70)$$

We can obtain the expression for the short circuit current density J_s by setting $V = 0$. This results in

$$J_s = \frac{-qN}{\cosh(m_1) + \alpha_1 \sinh(m_1)}. \quad (9.71)$$

The open circuit voltage V_0 is obtained by setting $J_{\text{tot}} = 0$, resulting in

$$V_0 = \frac{kT}{q} \ln \left\{ \frac{N}{\left[\dfrac{D_e}{L_e} n_{p0}[\alpha_1 \cosh(m_1) + \sinh(m_1)] \right.} + 1 \right\}$$
$$\left. + \frac{D_h}{L_h} p_{n0}[\cosh(m_1) + \alpha_1 \sinh(m_1)] \coth(m_2) \right]$$

$$(9.72)$$

The junction resistance R, defined as the slope of the current-voltage curve at zero voltage, is given by

$$\frac{1}{R} \equiv A_D \left. \frac{dJ_{\text{tot}}}{dV} \right|_{V=0}$$

$$= \frac{A_D q^2}{kT} \left\{ \frac{L_h p_{n0}}{\tau_h} \coth(m_2) + \frac{L_e}{\tau_e} n_{p0} \left(\frac{\alpha_1 \cosh(m_1) + \sinh(m_1)}{\cosh(m_1) + \alpha_1 \sinh(m_1)} \right) \right\}, \quad (9.73)$$

where A_D is the front surface area, which is equal to the junction area.

Substituting Eqs. 9.71 and 9.73 into 9.72, we note that the open circuit voltage is given by

$$V_0 = \frac{kT}{q} \ln \left[1 - RA_D J_s \frac{q}{kT} \right]. \quad (9.74)$$

For small values of $RA_D J_s(q/kT)$ this reduces to

$$V_0 = -RA_D J_s. \quad (9.75)$$

The value of $D^*_{\lambda,\text{PV}}$ is determined by equating the open circuit voltage with the thermal noise voltage, then determining the value of P_λ, the monochromatic radiant power using Eq. 9.28, from the equality, and dividing this into the square root of the area. Since we are interested only in small signals, we may use Eq. 9.75, resulting in

$$D^*_{\lambda,\mathrm{PV}} = \frac{q\lambda R_0^{1/2}}{2(kT)^{1/2}hc_0[\cosh(m_1) + \alpha_1 \sinh(m_1)]}, \qquad (9.76)$$

where $R_0 \equiv A_D R$, or

$$D^*_{\lambda,\mathrm{PV}} = \frac{\lambda A_3}{2hc_0}, \qquad (9.77)$$

where

$$A_3 \equiv \frac{1}{\left[\theta_2 \coth(m_2) + \theta_1\left(\dfrac{\alpha_1 \cosh m_1 + \sinh m_1}{\cosh m_1 + \alpha_1 \sinh m_1}\right)\right]^{1/2}(\cosh m_1 + \alpha_1 \sinh m_1)},$$

and   (9.78)

$$\theta_1 \equiv \frac{L_e n_{p0}}{\tau_e} ; \qquad \theta_2 \equiv \frac{L_h p_{n0}}{\tau_h}. \qquad (9.79)$$

In order to maximize the value of $D^*_{\lambda,\mathrm{PV}}$ we want to utilize radiation of wavelength just short of the absorption edge and we want to maximize the factor A_3, which is proportional to $R_0^{1/2}(\cosh m_1 + \alpha_1 \sinh m_1)^{-1}$. We note from Eq. 9.78 that A_3 is in terms of m_1, a function of the distance from the front surface to the junction, and of m_2, a function of the distance from the junction to the back surface. Since each of these may be adjusted independently, the condition required of m_2 in order that A_3 be a maximum is that $\coth m_2$ approach its smallest permissible value, which is unity. This is achieved for large values of m_2. Thus the distance in the n region from the junction to the back surface should be large compared to the hole diffusion length in the n region. If the distance is twice the diffusion length, the value of $\coth m_2$ is 1.037, very close to unity. Thus two diffusion lengths or more are required. Consider the requirement for m_1 in order to maximize A_3. Examining Eq. 9.78 we note that the highest value of A_3 is obtained when m_1 becomes zero. When m_1 is zero and m_2 is much greater than unity, Eq. 9.78 reduces to

$$A_3 = \frac{1}{[\theta_2 + \alpha_1\theta_1]^{1/2}}. \qquad (9.80)$$

If in addition the front surface recombination velocity can be reduced to zero, then A_3 becomes

$$A_3 = \frac{1}{\theta_2^{1/2}}. \qquad (9.81)$$

Thus to maximize A_3 we must minimize θ_2. For a given material this means minimizing p_{n0}. Since the product of the majority and minority carrier concentrations in any region is a constant, this means we must dope the n region behind the junction so that it will be strongly n-type. If the front surface recombination velocity is greater than zero, that is, α_1 is

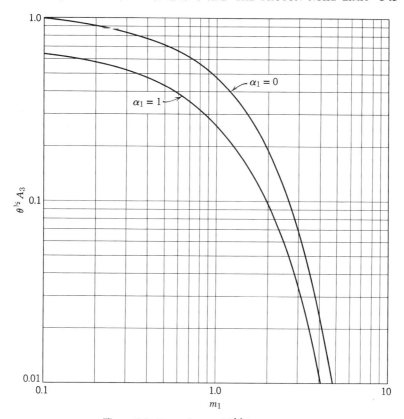

Figure 9.6. Dependence of $\theta^{1/2}A_3$ upon m_1.

greater than zero, then it is important to dope the region in front of the barrier so that it will be strongly p-type, in order that n_{p0}, the minority carrier concentration in this region, be kept low.

To sum it up, for operation in the photovoltaic mode with the radiation perpendicular to the junction, it is necessary that the region between the surface upon which the radiation is incident and the junction surface be as thin as possible, as long as the assumption of infinite absorption at the front surface is a good approximation. It is also desirable that the front surface recombination velocity be as low as possible. The region between the junction surface and the back surface should be at least two minority carrier diffusion lengths thick. The recombination velocity at the back surface is unimportant. Both regions should be heavily doped in order that the minority carrier concentrations be low.

In Fig. 9.6 we have plotted the function $\theta^{1/2}A_3$ as a function of m_1, where θ is defined from $\theta = \theta_1 = \theta_2$ for simplicity. We have assumed coth m_1 is unity and selected two values of α_1. Note that $\theta^{\frac{1}{2}}A_3$ increases as m_1 decreases and as α_1 decreases from unity to zero.

In Table 9.1 we compare the performance of intrinsic detectors in the photoconductive, PEM, and photovoltaic modes based upon thermal noise limited D_λ^* values calculated from Eqs. 9.29, 9.52, and 9.77. Here

TABLE 9.1. Comparison of computed D_λ^* values for intrinsic detectors operating in the photoconductive, PEM, and photovoltaic modes at room temperature

Material	Mode	Assumed Peak Wavelength (microns)	D_λ^* (cm cps$^{\frac{1}{2}}$/watt)
Si	PC	1.0	6.42×10^{15}
	PEM	1.0	4.35×10^{11}
	PV	1.0	2.26×10^{13}
Ge	PC	1.5	2.54×10^{14}
	PEM	1.5	7.36×10^{11}
	PV	1.5	1.65×10^{11}
PbS	PC	2.2	3.99×10^{12}
	PEM	2.2	2.18×10^{10}
	PV	2.2	7.94×10^{10}
PbSe	PC	3.4	4.44×10^{10}
	PEM	3.4	1.37×10^{9}
	PV	3.4	2.47×10^{6}
PbTe	PC	5.9	1.25×10^{13}
	PEM	5.9	6.56×10^{11}
	PV	5.9	8.79×10^{10}
InAs	PC	3.6	6.30×10^{12}
	PEM	3.6	6.97×10^{11}
	PV	3.6	6.56×10^{7}
InSb	PC	6.6	4.43×10^{11}
	PEM	6.6	1.41×10^{10}
	PV	6.6	1.60×10^{7}
Te	PC	3.6	1.84×10^{12}
	PEM	3.6	8.00×10^{10}
	PV	3.6	2.04×10^{10}

we have arbitrarily assumed values of $m = 1$, $\alpha_1 = 0$, and $\alpha_2 = 0$ for the photoconductive mode and $m = 1$, $\alpha_1 = 0$, and $\alpha_2 = 1$ for the PEM mode. The optimum values of B_1 and B_2 have been chosen. These in turn dictate the values of a for the photoconductive and PEM modes in terms of the mobility ratios. The material parameters n_i, μ_e, b, and τ have been taken from Table 6.3. The values of E_x and B have been assumed to be 100 v/cm and 10,000 gauss, respectively. For the photovoltaic mode we have assumed A_3 is given by Eq. 9.81, that is, $\alpha_1 = 0$, $m_1 = 0$, and m_2 is much greater than unity. In order to arrive at a value of p_{n0} we have assumed it is 1 % of the intrinsic concentration for each material.

9.3.4 The Bolometer

Bolometer operation has been treated theoretically by Shive,[7] Jones,[8] Chasmar, Mitchell and Rennie,[9] Smith, Jones, and Chasmar,[4] and Lovell,[10] among others. Our discussion of it is aimed at clarifying the operation of the bolometer for the user so that he may make better use of it. In contrast to detectors based upon photon effects, where new materials are constantly being investigated, most bolometers make use of thermistor material, composed of the oxides of manganese, cobalt, and nickel. Changes in bolometer design are usually made by changing the heat sink material or the adhesion material, rather than the radiation sensing material.

As discussed in Sections 8.4.1 and 8.4.2, bolometers may be not only the thermistor type but may be also of metal or of a superconducting material. Since only the thermistor bolometer finds widespread use, our analysis shall be directed mainly toward it, although we shall consider the metal bolometer in less detail. We shall derive the expressions for $D_\lambda{}^*$ based upon thermal noise limitations, justified by the fact that thermistor bolometers, while showing some current noise, approach very closely to the thermal noise limitation. The following conditions are assumed:

1. All the radiation incident upon the detector is absorbed uniformly throughout the volume.
2. The material is uniform throughout the volume.
3. End effects due to nonuniform Joulean heating can be ignored.

First we shall discuss the circuit arrangement of the bolometer and load resistor, Fig. 9.7.

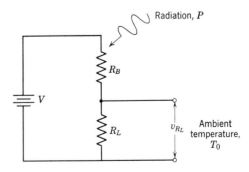

Figure 9.7. Bolometer circuit.

The bolometer is in series with R_L, the load resistor, and the bias supply of emf V. If the circuit is opened so that no current flows, and if no signal radiation is present, the bolometer is at the ambient temperature T_0. Closing the circuit causes current to flow, such that the Joulean heating in the bolometer element R_B increases its temperature to T_1. At the same time its resistance changes to a value characteristic of that at T_1. If radiation now falls upon the bolometer, its temperature changes by ΔT to the new value T. This results in a resistance change in the bolometer, causing a change in the voltage appearing across R_L. Since chopped radiation is used, the value of v_{R_L}, the signal voltage, represents only that due to the radiant signal.

The dynamic analysis of the operation is based upon use of the heat transfer equation for the bolometer. In the absence of radiation, but with current flowing in the circuit, the equation is

$$C\frac{dT}{dt} + K_0(T_1 - T_0) = i^2 R_B. \qquad (9.82)$$

The first term in Eq. 9.82 represents the influence of the heat capacity C upon the rate of change of bolometer temperature, and is zero in the steady state condition. The second term represents conduction from the bolometer at temperature T_1 to its surroundings at T_0 through a medium of average thermal conductance K_0. In most cases this conduction will be to the thermal sink, although in ideal operation it represents only radiation interchange with the surroundings. In the latter case K_0 will be a function of T. The term on the right is the Joulean heating of the bolometer.

In the presence of signal radiation such that P watts are incident on it, the bolometer changes its temperature to T by the incremental amount ΔT. The heat transfer equation becomes

$$C\frac{d\,\Delta T}{dt} + K\,\Delta T = \frac{d(i^2 R_B)}{dT}\Delta T + P. \qquad (9.83)$$

Here K is the thermal conductance when the bolometer is at the temperature T. The first term on the right is

$$\frac{d(i^2 R_B)}{dT}\Delta T = \frac{d}{dT}\left[\frac{V^2 R_B}{(R_B + R_L)^2}\right]\Delta T = \frac{V^2(R_L - R_B)}{(R_L + R_B)^3}\frac{dR_B}{dT}\Delta T. \qquad (9.84)$$

The quantity α describing the dependence of bolometer resistance upon temperature is defined as

$$\alpha \equiv \frac{1}{R_B}\frac{dR_B}{dT}. \qquad (9.85)$$

For a semiconductor R_B is given by (see Eq. 6.42)

$$R_B = R_{B0}\exp\left(\beta/T\right). \qquad (9.86)$$

Thus
$$\alpha = -\beta/T^2. \tag{9.87}$$

For a metal, which has a linear dependence of resistance upon temperature, that is,

$$R_B = R_{B0}[1 + \gamma(T - T_0)], \tag{9.88}$$

we have

$$\alpha = \frac{\gamma}{1 + \gamma(T - T_0)}. \tag{9.89}$$

Returning to Eq. 9.83 and introducing Eqs. 9.84 and 9.85, we have

$$C\frac{d(\Delta T)}{dt} + K\,\Delta T = \frac{V^2 R_B \alpha}{(R_L + R_B)^2}\left(\frac{R_L - R_B}{R_L + R_B}\right)\Delta T + P. \tag{9.90}$$

However, the Joulean heating in the bolometer in the steady state is related to the conduction losses by

$$\frac{V^2 R_B}{(R_L + R_B)^2} = K_0(T_1 - T_0). \tag{9.91}$$

Therefore, we may write Eq. 9.90 as

$$C\frac{d(\Delta T)}{dt} + K_e\,\Delta T = P, \tag{9.92}$$

where we define

$$K_e \equiv K - K_0(T_1 - T_0)\alpha\left(\frac{R_L - R_B}{R_L + R_B}\right). \tag{9.93}$$

The solution to Eq. 9.92 is determined by assuming P to be a periodic function described by

$$P = P_0 \exp(j\omega t). \tag{9.94}$$

The value of ΔT is then

$$\Delta T = \Delta T_0 \exp\left(-\frac{K_e}{C}t\right) + \frac{P_0 \exp(j\omega t)}{K_e + j\omega C}. \tag{9.95}$$

The first term in Eq. 9.95 represents a transient, whereas the second is a periodic function. If K_e is positive, that is, if

$$K > K_0(T_1 - T_0)\alpha\left(\frac{R_L - R_B}{R_L + R_B}\right), \tag{9.96}$$

then the transient term goes to zero with time and only the periodic function remains. However, if

$$K < K_0(T_1 - T_0)\alpha\left(\frac{R_L - R_B}{R_L + R_B}\right), \tag{9.97}$$

then the first term in Eq. 9.95 increases exponentially with time, and the bolometer overheats and burns up. If we assume R_L to be much greater than R_B and the thermal conductance to not greatly change with temperature, that is, $K \approx K_0$, then the unstable burnout condition is attained when

$$\alpha(T_1 - T_0) > 1. \tag{9.98}$$

For metals, in which α decreases with temperature, the inequality described by Eq. 9.98 is not fulfilled and "self-burnout" does not occur. However, for semiconductors, self-burnout can occur at large bias currents. Then the inequality for self-burnout is

$$\frac{-\beta}{T_1^2}(T_1 - T_0) > 1. \tag{9.99}$$

The value of $\alpha = -\beta/T_1^2$ for thermistor materials is about 0.040. Thus for operation of thermistor bolometers at an ambient of 300°K, the critical temperature is about 325°K. If the temperature is raised slightly above this, by Joulean heating or heating due to the radiant signal, self-burnout occurs.

Returning to the discussion of Eq. 9.95, we see that under conditions in which the transient term goes to zero, we may write the steady state solution of ΔT as

$$\Delta T = \frac{P_0}{(K_e^2 + \omega^2 C^2)^{1/2}}. \tag{9.100}$$

The change in resistance ΔR_B corresponding to the temperature change ΔT is, from the definition of α,

$$\Delta R_B = \alpha R_B \Delta T = \frac{\alpha R_B P_0}{(K_e^2 + \omega^2 C^2)^{1/2}}. \tag{9.101}$$

The signal voltage v_{R_L} appearing across the load resistor is

$$v_{R_L} = \frac{d(iR_L)}{dR_B}\Delta R_B = \frac{-V}{(R_B + R_L)^2}R_L \Delta R_B$$
$$= -\frac{K_0^{1/2}(T_1 - T_0)^{1/2}}{(R_B + R_L)R_B^{1/2}}R_L \Delta R_B, \tag{9.102}$$

where we have introduced Eq. 9.91. Substituting Eq. 9.101 into 9.102, we find

$$v_{R_L} = -\frac{K_0^{1/2}(T_1 - T_0)^{1/2}R_L \alpha R_B^{1/2} P_0}{(R_B + R_L)(K_e^2 + \omega^2 C^2)^{1/2}}. \tag{9.103}$$

By assuming a thermal noise mechanism, a thermal conductance independent of temperature in the region of interest, a chopping frequency

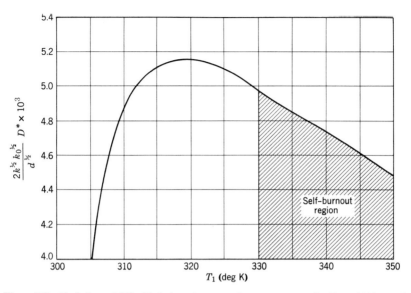

Figure 9.8. Variation of D^* with bolometer operating temperature for $T_0 = 300°$K and $\beta = 3600°$K.

approaching zero, and $R_L \gg R_B$, the open circuit value of D^* for thermistor bolometers is

$$D^* = \frac{(T_1 - T_0)^{1/2}\beta d^{1/2}}{2k^{1/2}T_1^{5/2}k_0^{1/2}[1 + (T_1 - T_0)\beta T_1^{-2}]},$$ (9.104)

where d is the thickness of the insulating layer between the thermistor flake and the heat sink and k_0 is the thermal conductivity of the layer. We note that the value of D^* is independent of wavelength, since we are considering a thermal detector. It is also independent of detector area and resistance. The dependence upon operating temperature T_1 is not obvious. However, Fig. 9.8 shows the relative value of D^* as a function of T_1, assuming an ambient of 300°K and a β value of 3600°K, typical of thermistor material. We note that the optimum value of D^* is obtained for $T_1 = 320°$K. Self-burnout, predicted by Eq. 9.97, occurs at $T_1 \geq 330°$K.

9.4 THEORETICAL PERFORMANCE OF PHOTON NOISE LIMITED DETECTORS

Photon noise is the term describing fluctuations in the instantaneous value of the number of photons emitted by a radiating source. The emission of photons by the body is a quasi-random event in the sense that the Bose-Einstein statistics describing the distribution of the photons in the allowed energy states impose a limitation upon the randomness. Ignoring this for the moment, we observe that the emission of a photon from some point of a radiating surface at some instant is uncorrelated with that from any other point at any other instant. That is to say, the source emits incoherent radiation. Thus the photons leave the body in an essentially random manner. When these photons are intercepted by an ideal photon or thermal detector, the fluctuations in their number or power cause the detector output to exhibit corresponding fluctuations. Even if the detector is not ideal, this may be true. For example, the detector may have a quantum efficiency less than unity and internally generated noise which is small compared to the detected photon noise. Under these circumstances the detector is photon noise limited but not ideal.

As pointed out in the chapter introduction, the photon noise that limits a detector arises from the radiating background to which the detector is exposed. This background may consist of sources of radiation external to the detector but within the field of view, or it may be the internal surfaces of the detector envelope. Frequently these internal surfaces are cooled in order that their emission be kept low. The background most frequently encountered is terrestrial. In a scanning problem, for example, radiation falls upon the detector from a background defined by the instantaneous field of view. The detector is required to determine whether a signal source is superimposed upon this background. If the detector is photon noise limited, then the random arrival of photons from the background sets a limit to the detectability of the signal.

As with other noise sources, it is possible to vary the signal-to-noise ratio from the detector by electrically filtering the output. In addition, it will be shown that the signal-to-noise ratio of a photon noise limited detector depends upon the spectral response and field of view. Thus the signal-to-noise ratio may be varied by spectrally filtering the input to the detector.

The expressions for the fluctuations in the emitted radiation power and the rate of photon emission were derived in Sections 2.5.2 and 2.5.3. We wish here to determine how the characteristics of the detector influence the detectivity to be realized under photon noise limitations. We will show that not all detectors have the same photon noise limited detectivity; photon detectors having spectral responses which extend only into the near infrared can, under certain conditions, detect a signal in the presence of photon noise more readily than detectors whose responses extend farther into the infrared. Under other conditions long wavelength detectors will outperform short wavelength ones. The purpose of this section is to derive the relationships describing the performance of photon noise limited detectors.

9.4.1 Photon Noise Limitations of Thermal Detectors

Because thermal detectors are sensitive to the absorbed radiant power intensity, whereas photon detectors respond to the rate of photon absorption, the photon noise limitations of the two types differ. The photon noise limitation of thermal detectors is imposed by fluctuations in the absorbed power, because of the quasi-random arrival of the photons, whereas photon detectors are limited by fluctuations in the rate of photon absorption. We shall consider first the photon noise limited thermal detector.

We shall assume a thermal detector that is coupled to its environment by radiation interchange alone, that is, there is no energy interchange through the mechanisms of convection or conduction. We assume that the detector is at a temperature T_1, surrounded by a uniform environment at a temperature T_2, and that it has an emissivity ε which is independent of T_1 and of wavelength. From the discussion of the noise spectrum of the power emitted by a radiating body in Section 2.5.2, we recall that dp_P described the noise spectrum in the optical frequency interval dv, where

$$dp_P = 2Ah\nu M(\nu,T) \frac{\exp{(h\nu/kT)}\,dv}{\left[\exp{(h\nu/kT)} - 1\right]}. \qquad (9.105)$$

$M(\nu,T)$ was the power per unit area per unit frequency interval emitted by a radiating body.

$$M(\nu,T) \equiv \frac{2\pi h\nu^3/c_0{}^2}{\left[\exp{(h\nu/kT)} - 1\right]}. \qquad (9.106)$$

The noise power spectrum $p_P(f)$ represented the mean square deviation from the mean of the radiant power and was given by

$$p_P(f) \equiv \int_0^\infty dp_P = \int_0^\infty 2Ah\nu M(\nu,T) \frac{\exp{(h\nu/kT)}}{\left[\exp{(h\nu/kT)} - 1\right]}\,dv. \qquad (9.107)$$

Thus $p_P(f)$ is termed the mean square noise power per unit bandwidth of the emitted radiation. The integral was found to be

$$p_P(f) = 8A\sigma k T_2{}^5, \qquad (9.108)$$

where we have added the subscript 2 to T referring to the emitting background. Note that the noise power spectrum is frequency independent or "white." The mean square noise power $\overline{p_P{}^2}$ in a bandwidth Δf is therefore

$$\overline{p_P{}^2} = 8A\sigma k T_2{}^5\,\Delta f. \qquad (9.109)$$

Consider now the interaction of this radiation noise with the photon noise limited thermal detector. The area A in this instance represents the receiving area of the detector. We assume the detector is sensitive to all possible wavelengths and therefore can detect all the incident noise. Since it has an emissivity ε, which is identical with the absorptivity, the received mean square noise power will be

$$\overline{p_P{}^2} = 8A\varepsilon\sigma kT_2{}^5\,\Delta f. \tag{9.110}$$

However, there will also be a contribution to the noise due to random fluctuations in the radiation power emitted by the detector. The detector is itself a radiating body, having an emissivity ε and a temperature T_1. The quasi-random emission of photons from it, which carry away heat, will also show up as photon noise. By the same argument as above, the mean square noise power of the emitted radiation in a bandwidth Δf will be

$$\overline{p_P{}^2} = 8A\varepsilon\sigma kT_1{}^5\,\Delta f. \tag{9.111}$$

Thus the total mean square radiation noise power in the bandwidth Δf is

$$\overline{p_P{}^2} = 8A\varepsilon\sigma k(T_1{}^5 + T_2{}^5)\,\Delta f. \tag{9.112}$$

In the particular instance in which the detector and the surroundings are in equilibrium and therefore are at the same temperature, the mean square noise power will be

$$\overline{p_P{}^2} = 16A\varepsilon\sigma kT^5\,\Delta f. \tag{9.113}$$

Remembering that $\overline{p_P{}^2}$ represents the mean square noise power, we find the rms noise power for values of $A = 1\ \mathrm{mm}^2$, $\varepsilon = 1$, $T = 300°\mathrm{K}$, and $\Delta f = 1$ cps to be

$$\left(\overline{p_P{}^2}\right)^{\frac{1}{2}} = 5.55 \times 10^{-12}\,\frac{\mathrm{watt}}{(\mathrm{cps})^{\frac{1}{2}}}. \tag{9.114}$$

We wish now to determine the photon noise limited detectivity of a photon noise limited thermal detector. The noise equivalent power, the incident radiant power required to give a signal voltage equal to the noise voltage in a specified bandwidth, will be numerically equal to $\left(\overline{p_P{}^2}\right)^{\frac{1}{2}}/\varepsilon$, since the thermal mechanism will transduce the photon noise equally as well as the radiation signal. Therefore D^*, the square root of the detector area per unit NEP in a 1 cps bandwidth, will be

$$D^* = \frac{\varepsilon}{[8\varepsilon\sigma k(T_1{}^5 + T_2{}^5)]^{\frac{1}{2}}} = \frac{4.0 \times 10^{16}\varepsilon^{\frac{1}{2}}}{(T_1{}^5 + T_2{}^5)^{\frac{1}{2}}}\,\frac{\mathrm{cm}(\mathrm{cps})^{\frac{1}{2}}}{\mathrm{watt}}. \tag{9.115}$$

Note that D^* is independent of A, as is to be expected.

In many practical instances the temperature of the background, T_2, will be room temperature, 290°K. For many detectors, such as thermopiles and bolometers, the detector temperature will also be 290°K. Figure 9.9

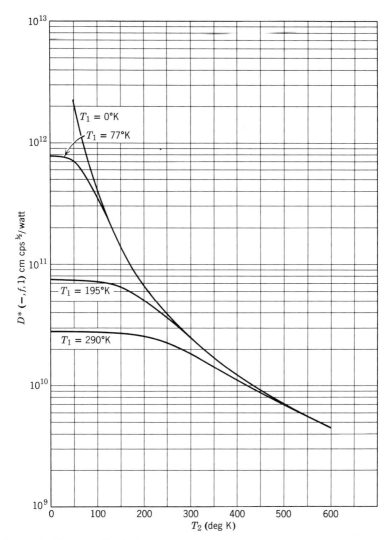

Figure 9.9. Photon noise limited D^* of thermal detectors as a function of detector temperature T_1 and background temperature T_2. Viewing angle of 2π steradians and unit quantum efficiency.

shows the photon noise limited detectivity for an ideal thermal detector having an emissivity of unity, operated at 290°K and lower, as a function of background temperature.

We see that the highest possible D^* to be expected from a thermal detector operated at room temperature and viewing a background at room temperature is 1.98×10^{10} cm cps$^{1/2}$/watt. Even if the detector or background, not both, were cooled to absolute zero, the detectivity would improve only by the square root of two. This is a basic limitation of all thermal detectors. We shall see in Section 9.4.3 that photon noise limited photon detectors have higher detectivities as a result of their limited spectral responses.

9.4.2   Temperature Noise in Thermal Detectors

Another approach to determining the photon noise limited performance of thermal detectors is through the concept of temperature noise. A thermal detector in contact with its environment by conduction and radiation exhibits random fluctuations in temperature, known as temperature noise, because of the statistical nature of the heat interchange with its surroundings. If the conduction interchange is negligible compared to the radiation interchange, we would expect temperature noise to become identified with photon noise. The following discussion demonstrates this mathematically. Consider first of all the magnitude of the temperature fluctuations of the detecting material. The material, having a heat capacity C, changes its temperature T by the incremental amount ΔT in response to the energy increment $\Delta \mathscr{E}$ according to

$$\Delta \mathscr{E} = C \, \Delta T. \tag{9.116}$$

The thermodynamic system composed of the material and surroundings possesses many degrees of freedom. Tolman[11] states that the mean square fluctuations in energy $\overline{\Delta \mathscr{E}^2}$ of a system having many degrees of freedom is given by

$$\overline{\Delta \mathscr{E}^2} = kT^2 C, \tag{9.117}$$

where k is Boltzmann's constant. Therefore

$$\overline{\Delta T^2} = \frac{kT^2}{C}. \tag{9.118}$$

Einstein[12] showed that C in this case is the harmonic mean of the heat capacities of the material and surroundings.

$$C = \frac{C_A C_B}{C_A + C_B}, \tag{9.119}$$

where C_A and C_B are the heat capacities of the body and surroundings, respectively. If the surroundings have a much greater heat capacity than the body, which is the usual situation, the mean heat capacity becomes that of the body.

Next consider the spectral content of the fluctuations. Let the detecting material be at an incremental temperature difference ΔT above its surroundings. Heat will flow from the material to the surroundings at a rate proportional to ΔT, the proportionality constant being the heat transfer coefficient K between the material and the surroundings. The heat transfer equation is

$$\frac{d(\Delta \mathscr{E})}{dt} = K \, \Delta T, \tag{9.120}$$

where $d(\Delta \mathscr{E})/dt$ is the rate of flow of heat. However, we see from Eq. 9.116 that the rate of flow of heat can also be expressed as

$$\frac{d(\Delta \mathscr{E})}{dt} = C \, \frac{d(\Delta T)}{dt}. \tag{9.121}$$

If the material is at a higher temperature than its surroundings, heat flows from the material to the surroundings and $d(\Delta \mathscr{E})/dt$ is negative. Equating Eqs. 9.120 and 9.121 we arrive at a differential equation describing the heat transfer.

$$-C \frac{d(\Delta T)}{dt} = K \, \Delta T. \tag{9.122}$$

The solution of this is

$$\Delta T = \Delta T_0 \exp\left(-t/\tau\right), \tag{9.123}$$

where ΔT_0 is the value of ΔT at $t = 0$ and $\tau \equiv C/K$. If we now allow heat to flow into the body from an external source, for example, a radiating background, the differential equation becomes

$$C \frac{d(\Delta T)}{dt} + K \, \Delta T = P(t), \tag{9.124}$$

where $P(t)$ is the fluctuation in the power from the external source. In order to solve this equation, we write

$$P(t) = P_f \exp\left(j2\pi ft\right), \tag{9.125}$$

where $j \equiv \sqrt{-1}$. Assume that the mean square value of P_f, denoted $\overline{P_f^2}$, is independent of f and can be written as

$$\overline{P_f^2} = P_0 \, \Delta f, \tag{9.126}$$

where P_0 is a constant. We then solve Eq. 9.124 and find the modulus of the solution to be

$$\overline{\Delta T_f^2} = \frac{\overline{P_f^2}}{K^2 + 4\pi^2 f^2 C^2} = \frac{P_0 \Delta f}{K^2 + 4\pi^2 f^2 C^2}. \tag{9.127}$$

The value $\overline{\Delta T_f^2}$ represents the frequency dependence of the mean square temperature fluctuations. We wish to determine the value of $\overline{\Delta T^2}$, the mean square value of the fluctuations over all frequencies. To do this we integrate Eq. 9.127 over all frequencies, assuming K and C are frequency independent.

$$\overline{\Delta T^2} = P_0 \int_0^\infty \frac{df}{K^2 + 4\pi^2 f^2 C^2} = \frac{P_0}{4KC}. \qquad (9.128)$$

In order to determine the value of P_0 we equate the expressions for $\overline{\Delta T^2}$ found in Eqs. 9.118 and 9.128. Substituting the value obtained into Eq. 9.127, we find the expression for the spectrum of the mean square fluctuations in ΔT.

$$\overline{\Delta T_f^2} = \frac{4KkT^2 \Delta f}{K^2 + 4\pi^2 f^2 C^2}. \qquad (9.129)$$

We can now determine the mean square value of the power fluctuations in the incident radiation. The mean square power fluctuations $\overline{p_P^2}$, which is simply the quantity $\overline{P(t)^2}$, is determined from Eqs. 9.127 and 9.129 to be

$$\overline{p_P^2} = 4kT^2 K \Delta f. \qquad (9.130)$$

We see that the thermal capacity C does not enter into the expression for the power fluctuations. This is of fundamental importance, since it indicates that the photon noise limited performance of a thermal detector is independent of detector material and volume.

In order to determine the value for the heat transfer coefficient K, we recall that K is defined as the proportionality constant between the rate of flow of energy and the temperature increment. (See Eq. 9.120.) For a gray body this becomes

$$K = 4\varepsilon\sigma T^3 A. \qquad (9.131)$$

Substituting this into Eq. 9.130 we determine the fluctuations in the power flow between the detecting material and source to be given by

$$\overline{p_P^2} = 16\, A\varepsilon\sigma kT^5 \Delta f. \qquad (9.132)$$

Note that this is the same expression which was deduced for the photon noise in thermal detectors given by Eq. 9.113. Therefore we have shown that temperature noise in thermal detectors in thermal equilibrium with their surroundings, in which radiation exchange with their surroundings predominates, is identified with photon noise.

9.4.3   Photon Noise Limitations of Photon Detectors

The previous two sections developed the photon noise limitations of thermal detectors. We shall now consider photon noise in photon detectors. Our derivation of the photon noise limited detectivity will be modified in two ways. First, because photon detectors respond to the number of photons rather than to the total radiant power, the analysis must consider the fluctuations in the rate of photon emission rather than the fluctuations in the radiant power. Second, we must introduce the spectral response of the detector by considering the fluctuations only in the spectral band to which the detector is sensitive. We will assume our detector has a quantum efficiency, the number of excited carriers produced per quantum incident upon the detector, less than or equal to unity from zero wavelength to the cutoff wavelength, and zero at longer wavelengths. Whereas thermal detectors, responding uniformly to all wavelengths, have a spectral $D_\lambda{}^*$ equal to their black body D^*, this, of course, is not true of photon detectors. We shall discuss initially the photon noise limited spectral $D_\lambda{}^*$, and later the photon noise limited black body D^*.

9.4.3.1 Spectral photon noise limited detectivity of photon detectors. To determine the value of $D_\lambda{}^*$ for photon detectors, we must equate the root mean square fluctuations in the rate of photon arrival from the background in the spectral range of the detector to the rate of arrival of photons from a source in the spectral range of the detector. The signal power is then determined knowing the spectral distribution of the signal. We divide this power into the square root of the detector area to determine the value of $D_\lambda{}^*$.

The determination is straightforward, assuming the signal radiation to be monochromatic and the background to be that of a black body. The average rate of arrival at the detector of area A of photons of frequency ν from the background at temperature T_2 is given by

$$\bar{N}A = \frac{M(\nu, T_2)}{h\nu} A \, d\nu, \tag{9.133}$$

where $\bar{N}$ is the average rate of arrival per unit area. The mean square deviation $\overline{N^2}$ in the rate of arrival per unit area, of photons which are governed by Bose–Einstein statistics, is given by

$$\overline{N^2} = \bar{N} \frac{\exp(h\nu/kT_2)}{\exp(h\nu/kT_2) - 1}. \tag{9.134}$$

Recalling the assumption of quantum efficiency less than or equal to unity at wavelengths less than the cutoff wavelength λ_0, and zero quantum efficiency at longer wavelengths, we determine $p_N(f)$, the frequency dependence of the mean square fluctuations in the rate of generation of current carriers due to the arrival of photons in the spectral range of the detector.

$$p_N(f) = A \int_{\nu_0}^{\infty} \eta(\nu)\overline{N^2} = \frac{A}{h} \int_{\nu_0}^{\infty} \eta(\nu) \frac{M(\nu, T_2)}{\nu} \frac{\exp(h\nu/kT_2)}{\exp(h\nu/kT_2) - 1} \, d\nu, \tag{9.135}$$

where $\eta(\nu)$ is the quantum efficiency and $\nu_0 \equiv hc_0/\lambda_0$.

Introducing the value of $M(\nu, T_2)$, Eq. 9.106, we find

$$p_N(f) = \frac{2\pi A}{c_0^{\ 2}} \int_{\nu_0}^{\infty} \eta(\nu) \frac{\nu^2 \exp (h\nu/kT_2)}{[\exp (h\nu/kT_2) - 1]^2} \, d\nu. \qquad (9.136)$$

Equation 9.136 states the frequency dependence of the mean square fluctuations per unit bandwidth in the rate of generation of current carriers due to the arrival of photons from the radiating background in the spectral interval from ν_0 to infinity. This is similar to Eq. 2.114, except that here we consider only those fluctuations in one hemisphere in the spectral interval from ν_0 to infinity. The next step is to determine the intensity of monochromatic radiation from the signal source required to give an average rate of carrier generation equal to the root mean square fluctuations in the rate of arrival of photons from the background in a bandwidth Δf. The rate of generation n_s of carriers excited by photons from the monochromatic source of power P_s incident on the detector is given by

$$n_s = \frac{\eta(\nu_s)P_s}{h\nu_s}, \qquad (9.137)$$

where $\eta(\nu_s)$ is the quantum efficiency for photons of frequency ν_s.

Fink[13] shows that the root mean square fluctuations in the bandwidth Δf is given by the square root of the quantity $2\Delta f$ times the frequency dependence of the mean square fluctuations. Assuming that ν_s lies in the spectral range of the detector, we equate the root mean square fluctuations in the bandwidth Δf to the rate of carrier excitation.

$$n_s = [p_N(f) \cdot 2 \, \Delta f]^{1/2}. \qquad (9.138)$$

Combining Eqs. 9.136, 9.137, and 9.138 we see

$$P_{s\,min} = \frac{h\nu_s(4\pi A \, \Delta f)^{1/2}}{c_0 \eta(\nu_s)} \left[\int_{\nu_0}^{\infty} \eta(\nu) \frac{\nu^2 \exp (h\nu/kT_2) \, d\nu}{[\exp (h\nu/kT_2) - 1]^2} \right]^{1/2}. \qquad (9.139)$$

Equation 9.139 gives the monochromatic radiation power $P_{s\,min}$ from the signal source needed to give rise to a signal from the detector equal to the background photon noise in the spectral interval to which the detector responds and in the electrical bandwidth Δf.

The value of D_λ^* is obtained by dividing $P_{s\,min}$ for unit bandwidth into the square root of the area. Thus

$$D_\lambda^* = \frac{c_0 \eta(\nu_s)}{2h\nu_s \pi^{1/2} \left[\int_{\nu_0}^{\infty} \frac{\eta(\nu)\nu^2 \exp (h\nu/kT_2) \, d\nu}{[\exp (h\nu/kT_2) - 1]^2} \right]^{1/2}}. \qquad (9.140)$$

Let us now determine the value of D_λ^* at the peak sensitivity, occurring in our ideal detector at $v_s = v_0$. Assuming that $\eta(v)$ is independent of wavelength, Schuldt[14] has shown that Eq 9.140 is given by

$$D_\lambda^* = \frac{c_0[\eta(v_0)]^{\frac{1}{2}}}{2\pi^{\frac{1}{2}}h^{\frac{1}{2}}v_0^2(kT_2)^{\frac{1}{2}}} \left\{ \sum_{m=1}^{\infty} \exp\left(-\frac{mhv_0}{kT_2}\right)\left[1 + \frac{2kT_2}{mhv_0} + 2\left(\frac{kT_2}{mhv_0}\right)^2\right]\right\}^{-\frac{1}{2}}.$$
(9.141)

An approximate solution, obtained by retaining only the first term in the series expansion of Eq. 9.141, is

$$D_\lambda^* \approx \frac{c_0[\eta(v_0)]^{\frac{1}{2}}\exp(hv_0/2kT_2)}{2\pi^{\frac{1}{2}}h^{\frac{1}{2}}v_0^2(kT_2)^{\frac{1}{2}}\left[1 + \frac{2kT_2}{hv_0} + 2\left(\frac{kT_2}{hv_0}\right)^2\right]^{\frac{1}{2}}}.$$
(9.142)

The approximation is valid for $hv_0 \gg kT_2$. Assuming $T_2 = 290°K$, representative of a terrestrial background, the approximation holds for $\lambda_0 \ll 50\ \mu$.

Figure 9.10 shows the spectral D_λ^* determined from Eq. 9.141 plotted as a function of wavelength for selected values of background temperature assuming unit quantum efficiency. Note that D_λ^* exhibits a minimum. For detectors viewing a terrestrial background, $290°K$, the minimum is at $14\ \mu$. Detectors having short cutoff wavelengths see less of the photon noise from the background than those having cutoff wavelengths near the minimum. This accounts for the rapid increase in the photon noise limited values of D_λ^* as the wavelength decreases in the near infrared. We shall see in Chapter 10 that the D_λ^* values of photon detectors of various types follow this envelope. As the wavelength is increased past the minimum, the increase in the amount of photon noise received is more than compensated by the increase in the number of photons per watt of signal power. Thus the value of D_λ^* increases with wavelength in this region.

In our determination of the photon noise limitations of thermal detectors, we considered not only those noise contributions arising from photons emitted by the background and received by the detector but also those arising from photons emitted by the detector to the background. In photon detectors we must distinguish between the various photon detecting mechanisms. In photoemissive, photovoltaic, or photoelectromagnetic detectors, only background fluctuations contribute to the noise. Thus Eqs. 9.140, 9.141, and 9.142 are valid for these detectors. On the other hand, photoconductive detectors depend upon the change in concentration of charge carriers upon irradiation.

In the dark the concentration of free carriers is determined by thermal excitation, the lattice vibrations causing electrons to acquire sufficient energy to break electron bonds and become free. (See the discussion in Chapter 6.) The concentration of free carriers is determined by both the

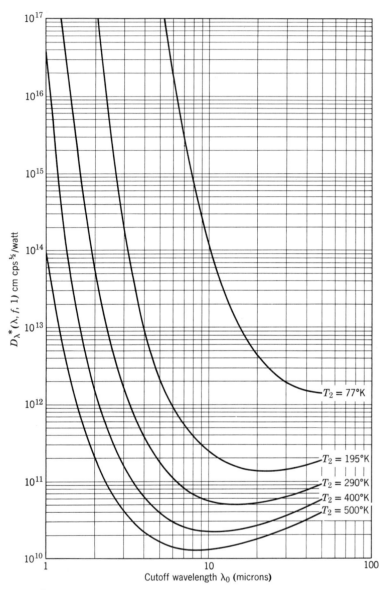

Figure 9.10. Photon noise limited D^* at peak wavelength, assumed to be cutoff wavelength, for background temperatures of 77°K, 195°K, 290°K, 400°K, and 500°K. Viewing angle of 2π steradians and unit quantum efficiency.

generation and the recombination rates. Fluctuations in these rates are termed generation-recombination, or gr noise. (See Chapter 7.) With radiation falling on the detector, the carriers are undergoing continuous photoexcitation and recombination, both spontaneous and induced. Van Vliet[15] showed that at equilibrium in a photoconductor the total noise power, photon and gr, can be no less than twice the photon noise power alone. Since the value of D_λ^* depends upon the square root of the noise power, the photon noise limit for photoconductors is the square root of two times poorer than that of photoemissive detectors. That is to say, the values of D_λ^* given by Eqs. 9.140, 9.141, and 9.142 and Fig. 9.10 must be divided by the square root of two when applied to photoconductors.

In addition, another factor tends to degrade the photon noise limit of photoconductors. Our derivation of D_λ^* considered only photons coming from the forward hemisphere. For photoemissive, photovoltaic, or PEM detectors only those photons arriving at the front surface contribute to the noise. In a photoemissive detector, which we assume here has an opaque photoemissive surface, only photons arriving from the forward hemisphere will free electrons. In a diffused junction photovoltaic cell photons arriving from the forward hemisphere will reach the junction and produce photoexcitations, whereas those from the rear will be absorbed before they reach the junction. In a PEM cell only those absorbed at the front surface will contribute to the diffusion current which gives rise to the signal. However, for a photoconductor, photons arriving from any direction which are absorbed will produce free carriers. Thus, for an uncooled photoconductor the total photon noise power must again be doubled. This causes Eqs. 9.140, 9.141, and 9.142 to be divided again by the square root of two when applied to an uncooled photoconductor. However, the back hemisphere of a cooled photoconductor will contribute relatively few excitations. Thus the second factor of the square root of two can be omitted from a cooled photoconductor.

To sum it up, Eqs. 9.140, 9.141, and 9.142 and Fig. 9.10 are valid for photoemissive, photovoltaic, and PEM detectors, whether cooled or uncooled. For a photoconductor at room temperature D_λ^* must be divided by two; for a cooled photoconductor it must be divided by the square root of two.

9.4.3.2 Black body photon noise limited detectivity of photon detectors.

Having determined the value of D_λ^*, we turn to the black body photon noise limited detectivity, D^*. The expression we determined for the mean square fluctuations in the arrival of background photons, Eq. 9.136, is, of course, still valid, but we now consider the signal as arising from a black body rather than being monochromatic. We shall still assume our detector has a quantum efficiency less than or equal to unity for wavelengths shorter than the cutoff and zero quantum efficiency for longer wave-lengths.

Equation 9.137 related the rate of arrival of incident photons from a monochromatic source to the source power. We wish to determine the rate of absorption of photons emitted from a black body at a temperature T_3, having wavelengths in the spectral sensitive range of the detector. This is given by

$$n_s' = \int_{\nu_0}^{\infty} \eta(\nu) \frac{M(\nu, T_3)}{h\nu} \, d\nu. \tag{9.143}$$

In terms of the total radiation P_s' from the black body, we express the signal photon absorption rate as

$$n_s' = \frac{\dfrac{1}{h} \displaystyle\int_{\nu_0}^{\infty} \eta(\nu) \dfrac{M(\nu, T_3)}{\nu} \, d\nu}{\sigma T_3^4} P_s'. \tag{9.144}$$

This equation, relating the rate of absorption of signal photons to the black body radiation power, is analogous to Eq. 9.137 for monochromatic radiation. Expressing n_s' in terms of n_s, we find

$$n_s' = G n_s \frac{P_s'}{P_s}, \tag{9.145}$$

where

$$G \equiv \frac{\nu_s}{\sigma T_3^4 \eta(\nu_s)} \int_{\nu_0}^{\infty} \frac{\eta(\nu) M(\nu, T_3)}{\nu} \, d\nu. \tag{9.146}$$

In order to evaluate G we shall make the assumptions, as before, that $\eta(\nu)$ is independent of wavelength and $\nu_s = \nu_0$. The exact expression for G has been shown by Schuldt[14] to be

$$G = \frac{2\pi}{c_0^2} \frac{\nu_0}{\sigma T_3} \frac{k^3}{h^2} \sum_{m=1}^{\infty} \exp\left(\frac{-mh\nu_0}{kT_3}\right) \left[\frac{1}{m}\left(\frac{h\nu_0}{kT_3}\right)^2 + \frac{2}{m^2} \frac{h\nu_0}{kT_3} + \frac{2}{m^3}\right]. \tag{9.147}$$

Assuming $h\nu_0 \gg kT_3$ as before, Eq. 9.147 reduces to

$$G \approx \frac{2\pi\nu_0^3 k}{\sigma T_3^3 c_0^2} \exp\left(-\frac{h\nu_0}{kT_3}\right)\left[1 + \frac{2kT_3}{h\nu_0} + 2\left(\frac{kT_3}{h\nu_0}\right)^2\right]. \tag{9.148}$$

Note that G depends only on ν_0 and the source temperature T_3 and is independent of background temperature T_2.

Proceeding as we did before, we equate the rate of absorption of signal photons to the rms value of the photon noise in the bandwidth Δf. The value of $P_{s\,\min}'$ is then

$$P_{s\,\min}' = \frac{P_{s\,\min}}{G}, \tag{9.149}$$

and the value of D^* is

$$D^* = D_\lambda^* G. \qquad (9.150)$$

Figure 9.11 is a graph of G as a function of λ_0 assuming the target is either a 290°K, 400°K, or 500°K black body. Since G is less than unity for all wavelengths, the value of D^*, the detectivity for black body radiation, must be always less than D_λ^*, the detectivity for monochromatic radiation. Figure 9.12 shows the photon noise limited D^* determined from Eq. 9.150 assuming unit quantum efficiency, for a 290°K (terrestrial) background, and either a 290°K or 500°K source. The application of the photon noise limited D^* (290°K, f, 1) to the problem of terrestrial mapping is found in Section 9.4.3.4.

Examining Fig. 9.12 we note two somewhat surprising facts. First,

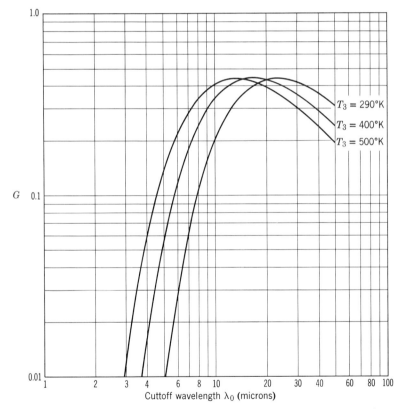

Figure 9.11. Dependence of the function G upon detector cutoff wavelength λ_0 for black body source temperatures of 290°K, 400°K, and 500°K and viewing angle of 2π steradians.

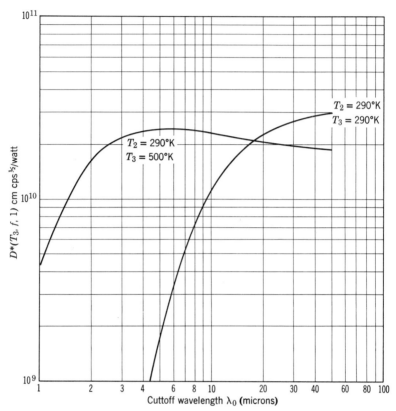

Figure 9.12. Photon noise limited D^* for 290°K and 500°K source temperatures as a function of detector cutoff wavelength λ_0 for a background temperature of 290°K, viewing angle of 2π steradians, and unit quantum efficiency.

although the photon noise limited D_λ^* was seen to have a strong dependence upon cutoff wavelength in the region from 3 to 7 μ, the 500°K black body D^* shows little dependence. It is apparent that the reduction in photon noise from the 290°K background as the value of λ_0 is reduced is accompanied by a corresponding decrease in the number of photons from the 500°K source. In other words, the detector is less susceptible to disturbance from the background radiation, but is also less able to "see" the photons from the signal source, since as λ_0 decreases it responds to a smaller and smaller fraction of them. At wavelengths shorter than 3 μ, we note that the reduction in photons from the source more than equals the reduction in the photon noise from the background and the value of D^* drops as λ_0 decreases.

The second surprising fact is the relationship between the two curves of

Fig. 9.12 at long wavelengths. Photon noise limited detectors having cutoff wavelengths beyond approximately 18 μ can detect a 290°K source against the photon noise from a 290°K background more readily than a 500°K source against the photon noise from a 290°K background. Since the limiting noise is the same in both cases, the explanation lies in the relative number of photons which give rise to the signal per watt of source power for the two sources. There are a greater number of photons emitted per second having wavelengths shorter than, say, 30 μ, per watt of power from a 290°K body than from a 500°K body. For wavelengths shorter than approximately 18 μ the inverse is true.

9.4.3.3 Photon noise limit with filtering and shielding. If we are looking at a monochromatic source, we can improve the performance of a photon noise limited detector by providing the detector with a narrow band optical filter transmitting the source wavelength. In so doing, we shall be making use of the monochromatic value of detectivity, which we have seen is always greater than the detectivity for a 500°K source. Of course, this will do no good if the filter is at the same temperature as the background since it will radiate in the spectral regions where it is opaque just as does the background. Using a cooled filter will improve the photon noise limited performance of a cooled detector looking at a monochromatic source. We see from Fig. 9.10 that the improvement at wavelengths shorter than 3 μ may be several orders of magnitude.

The use of a cooled filter can also improve the performance of a cooled photon noise limited detector viewing a grey or black source. If the spectral response of the detector extends to wavelengths much greater than those which the source radiates, then the detector may see an undesirable amount of photon noise from the background. By using a cooled filter it is possible to reduce effectively the spectral response of the detector to those wavelengths which the source radiates and thereby improve the detectivity.

It is instructive to draw an analogy between photon noise and electrical noise. If the signal we are detecting is confined to a narrow electrical bandwidth, we must reduce the bandwidth of a receiver to the signal bandwidth in order to cut out unwanted electrical noise and increase the signal-to-noise ratio. So too, if the source radiation is confined to a small range of optical frequencies, we must reduce the optical bandwidth (spectral response) of a photon noise limited detector in order to cut out unwanted photon noise and increase the signal-to-noise ratio. The ultimate performance of photon noise limited detectors is obtained by getting a proper match between the spectral emission from the source and the spectral response of the detector for any given background temperature.

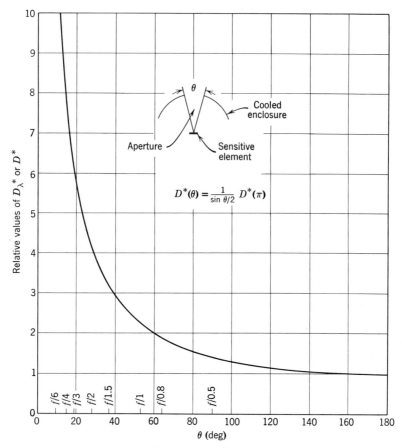

Figure 9.13. Relative increase in photon noise limited D_λ^* and D^* achieved by using cooled aperture in front of Lambertian detector.

Finally, we wish to consider the dependence of the photon noise limited detectivity upon the viewing angle. In order to reduce the amount of photon noise from the background, the detector should view as little of the background as possible. If the detector is cooled, this is obtained by mounting the sensitive element in a cooled enclosure having an aperture in it through which the detector can view the target. The detector used in an optical system should have an aperture just large enough to admit the cone of radiation impinging upon it from the optical element in front of it. In many detectors, called Lambertian, the sensitive element acts as a perfectly diffuse surface. In this case the intensity of the received radiation, and therefore the signal, is proportional to the cosine of the angle which the radiation makes with the normal to the element. Since the amount of radiation reaching a Lambertian detector is proportional to the square of the sine of the half angle of the cone extending from the center of the sensitive element to the aperture, the photon noise arriving at the detector is proportional to the sine of the half angle. The relative increase in D_λ^* or D^* obtained by shielding is shown as a function of angle in Fig. 9.13. The f/numbers of several apertures are also indicated.

As an example of the improvement attained through spectral filtering and limiting the viewing angle, the value of D^* for a lead selenide photoconductor operated at $77°K$ is shown in Fig. 9.14. The improvement obtained through use of a cooled filter and aperture is marked.

Jones[16] has proposed the figure of merit D^{**} ("dee double star") to remove the degeneracy in D^* due to the field of view considerations. He defines D^{**} to be

$$D^{**} = \left(\frac{\Omega}{\pi}\right)^{\!1/2} D^*. \qquad (9.151)$$

where Ω is the solid angle which the detector views. For the special case in which D^* is constant within a cone of half angle $\theta/2$, we have

$$\Omega = \pi \sin^2\left(\frac{\theta}{2}\right), \qquad (9.152)$$

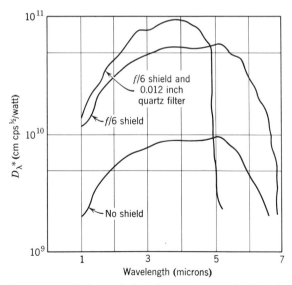

Figure 9.14. Improvement in lead selenide photoconductor obtained through use of cooled filter. (Courtesy Santa Barbara Research Center and *Missiles and Rockets*.)

and therefore

$$D^{**} = D^* \sin\left(\frac{\theta}{2}\right). \tag{9.153}$$

For a Lambertian detector, and for $\theta/2 = 90°$, Eq. 9.153 reduces to

$$D^{**} = D^*. \tag{9.154}$$

The units of D^{**} are $\dfrac{\text{cm cps}^{\frac{1}{2}} \text{ steradian}^{\frac{1}{2}}}{\text{watt}}$.

9.4.3.4 Photon noise limit when viewing earth through atmosphere. A problem which arises in thermal mapping concerns detecting small changes in temperature in a terrestrial background seen through the atmosphere. Radiation from the earth, assumed to be at an average temperature of 290°K, peaks at 10 μ. It is necessary to be able to resolve very small variations in this average temperature from point to point on the terrain. These variations may be of the order of a fraction of a degree Kelvin. Thus the individual radiating elements of the terrain have effectively the same radiant spectral distribution but vary slightly in the power density emitted from point to point. The performance of the thermal mapping system is determined, therefore, by the ability of the detector to discriminate a target at a temperature slightly greater or less than 290°K against a 290°K background. The ultimate case, in which the detector is photon noise limited, then revolves about the ability of the detector to detect a signal having the spectral distribution of a 290°K source against the photon noise arising from a 290°K background. The problem has been considered by Morton et al.,[17] who derive the relationship between the photon noise limited D^* of the detector having a cutoff wavelength λ_0 under two conditions. The first condition, that of no atmospheric absorption, has already been discussed here. The appropriate expression is that for D^* (290°K, f, 1) for a background temperature of 290°K. This is plotted in Fig. 9.12.

In the second condition, it is assumed that the earth is viewed through an atmospheric path in which absorption is appreciable. Figure 5.1a shows that an atmospheric window exists extending from roughly 8 to 14 μ. Most of the energy from a 290°K source which is transmitted by the atmosphere will be found in this window. For our calculations we shall assume that the atmosphere is completely opaque in all regions except the 8 to 14 μ window, in which it is completely transparent. In the regions in which the atmosphere is opaque it will be absorbing energy and therefore emitting it. This will give rise to photon noise that the detector sees. In the region in which the atmosphere is transparent, the signal and also photon noise from the earth background will reach the detector. For simplicity in our derivation, we shall assume the effective radiation temperature of the atmosphere to be also 290°K.

Thus for atmospheric absorption the signal is limited to that from a 290°K body lying between 8 and 14 μ. The noise comes from the radiating background over all wavelengths below the detector cutoff wavelength. At wavelengths shorter than 8 μ there will be noise but no signal and D^* will be zero. In the 8 to 14 μ region D^* will have a value which varies with cutoff wavelength. As the wavelength increases beyond 14 μ the signal will remain constant at the value obtained at 14 μ, but the noise will increase. Thus the signal-to-noise ratio, and therefore the value of D^*, will exhibit a maximum in the 8 to 14 μ region.

Consider now the rate of absorption of signal photons, n_s''. In an extension of Eq. 9.143, we have for cutoff wavelengths less than 8 μ

$$n_{s1}'' = 0, \qquad (\lambda_0 < 8 \ \mu). \tag{9.155}$$

For wavelengths in the 8 to 14 μ interval

$$n_{s2}'' = \int_{\nu_1}^{\nu_2} \eta(\nu) \frac{M(\nu, T_2)}{h\nu} \, d\nu, \qquad (8\mu < \lambda_0 < 14\mu), \tag{9.156}$$

where
$$\nu_1 = \frac{hc_0}{\lambda_{01}}, \qquad 8 \ \mu < \lambda_{01} < 14 \ \mu$$

and
$$\nu_2 = \frac{hc_0}{\lambda_{02}}, \qquad \lambda_{02} = 8 \ \mu.$$

For wavelengths greater than 14 μ

$$n_{s3}'' = \int_{\nu_3}^{\nu_2} \eta(\nu) \frac{M(\nu, T_2)}{h\nu} \, d\nu, \tag{9.157}$$

where
$$\nu_3 = \frac{hc_0}{\lambda_{03}}, \qquad \lambda_{03} = 14 \ \mu.$$

In analogy to Eq. 9.145 we write

$$n_s'' = G'' n_s \frac{P_s''}{P_s}, \tag{9.158}$$

where P_s'' is the total radiant power emitted by a body at 290°K and G'' varies according to the spectral interval as

$$G_1'' = 0, \qquad \text{for } \lambda_0 < 8 \ \mu; \tag{8.159}$$

$$G_2'' = \frac{\nu_s}{\sigma T_2^4 \eta(\nu_s)} \int_{\nu_1}^{\nu_2} \eta(\nu) \frac{M(\nu, T_2)}{\nu} \, d\nu, \qquad \text{for } 8 \ \mu < \lambda_{01} < 14 \ \mu$$

$$\text{and } \lambda_{02} = 8 \ \mu; \tag{9.160}$$

$$G_3'' = \frac{\nu_s}{\sigma T_2^4 \eta(\nu_s)} \int_{\nu_3}^{\nu_2} \eta(\nu) \frac{M(\nu, T_2)}{\nu} \, d\nu, \qquad \text{for } \lambda_{03} = 14 \ \mu. \tag{9.161}$$

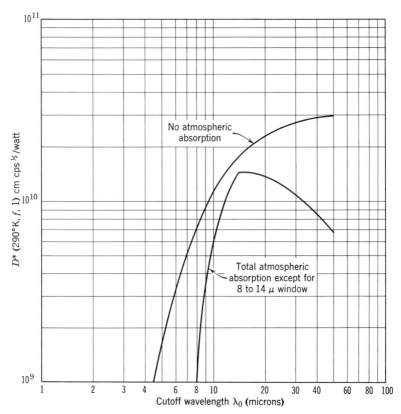

Figure 9.15. Photon noise limited $D^*(290°\text{K}, f, 1)$ for 290°K background temperature as a function of detector cutoff wavelength λ_0. Illustrating the effects of atmospheric absorption. Viewing angle of 2π steradians and unit quantum efficiency.

These G factors are evaluated similarly to the series expansion and approximation given by Eqs. 9.147 and 9.148. Proceeding as we did in Section 9.4.3.2 we find the value of $D^*(290°\text{K}, f, 1)$ in the three spectral intervals to be

$$D^*(290°\text{K}, f, 1) = D_\lambda^*(\lambda, f, 1)G_1'' = 0, \qquad \text{for } \lambda_0 < 8\,\mu; \quad (9.162)$$

$$D^*(290°\text{K}, f, 1) = D_\lambda^*(\lambda, f, 1)G_2'', \qquad \text{for } 8\,\mu < \lambda_0 < 14\,\mu; \quad (9.163)$$

$$\text{and} \quad D^*(290°\text{K}, f, 1) = D_\lambda^*(\lambda, f, 1)G_3'', \qquad \text{for } \lambda_0 > 14\,\mu. \quad (9.164)$$

Figure 9.15 illustrates $D_\lambda^*(290°\text{K}, f, 1)$ as a function of λ_0 assuming no atmospheric absorption, computed from Eq. 9.150, and also assuming total absorption except for the 8 to 14 μ window, computed from Eqs. 9.162, 9.163, and 9.164.

Note that in the absence of atmospheric absorption, as might be encountered when mapping a terrestrial background at a very short distance, the value of $D^*(290°K, f, 1)$ improves as the detector cutoff wavelength increases. When atmospheric absorption is appreciable, such as when mapping the earth from an airplane, then the optimum long wave limit is seen to be 14 μ. Detectors having long wavelength limits less than or greater than this value will have more severe photon noise limits and therefore will not perform as well. A cooled filter having a 14 μ cutoff may also be used to alter the long wavelength limit of a cooled photon noise limited detector having a response extending beyond 14 μ.

9.4.4 Photon Noise Limit of a Narrow Band Quantum Counter

Having considered the photon noise limits of thermal and photon detectors, we now turn to the narrow band quantum counter. A description of the operation of this type of detector appears in Section 8.3.9. Although it might appear that because of its limited spectral response, the narrow band quantum counter would be superior to the photon detector, Gelinas[18] indicates that such is not the case under most operating conditions. The analysis which follows will demonstrate this.

The derivation of the photon noise limit of a narrow band quantum counter follows a pattern similar to that employed in Section 9.4.3.1 for the photon detector. The narrow band quantum counter responds with a quantum efficiency $\eta(v_0)$ only to a narrow band of radiation centered at frequency v_0 and of width Δv_0, where $\Delta v_0 \ll v_0$. Outside this interval, the quantum efficiency is zero. The average rate of arrival per unit detector area of photons in the band Δv from the background having temperature T_2 is given by

$$\bar{N} = \frac{M(v_0, T_2)}{h v_0} \Delta v = \frac{2\pi v_0^2 \Delta v_0}{c_0^2 [\exp(h v_0/k T_2) - 1]}. \tag{9.165}$$

The mean square deviation $\overline{N^2}$ in the rate of arrival is

$$\overline{N^2} = \bar{N} \frac{\exp(h v_0/k T_2)}{[\exp(h v_0/k T_2) - 1]}. \tag{9.166}$$

The frequency dependence $p_N(f)$ of the mean square fluctuations in the rate of generation of current carriers due to the absorption of photons from the background is

$$p_N(f) = A\eta(v_0)\overline{N^2} \Delta v_0 = \frac{2\pi A v_0^2 \exp(h v_0/k T_2)\eta(v_0) \Delta v_0}{c_0^2 [\exp(h v_0/k T_2) - 1]^2}. \tag{9.167}$$

Photons of frequency v_0 from a monochromatic source of power P_s are absorbed by the detector. The rate of generation n_s of current carriers by the source is

$$n_s = \eta(\nu_0) \frac{P_s}{h\nu_0} . \tag{9.168}$$

Setting the rate of generation of current carriers by the source equal to the root mean square fluctuations in the generation of carriers by the background gives

$$n_s = [p_N(f) \cdot 2\,\Delta f]^{1/2}, \tag{9.169}$$

or

$$P_{s\,\min} = \frac{2\pi^{1/2} h\nu_0^2 (A\,\Delta f)^{1/2} \exp\,(h\nu_0/2kT_2)(\Delta\nu_0)^{1/2}}{c_0[\eta(\nu_0)]^{1/2}[\exp\,(h\nu_0/kT_2) - 1]} , \tag{9.170}$$

where $P_{s\,\min}$ is the power required to give rise to a signal from the detector equal to the photon noise from the background. The value of D_λ^* is

$$D_\lambda^* = \frac{(A\,\Delta f)^{1/2}}{P_{s\,\min}} = \frac{c_0[\eta(\nu_0)]^{1/2}[\exp\,(h\nu_0/kT_2) - 1]}{2\pi^{1/2} h\nu_0^2 \exp\,(h\nu_0/2kT_2)(\Delta\nu_0)^{1/2}} . \tag{9.171}$$

In terms of operating wavelength λ_0, D_λ^* is

$$D_\lambda^* = \frac{[\eta(\lambda_0)]^{1/2} \lambda_0^3 [\exp\,(hc_0/\lambda_0 kT_2) - 1]}{2\pi^{1/2} hc_0^{3/2} \exp\,(hc_0/2\lambda_0 kT_2)(\Delta\lambda_0)^{1/2}} . \tag{9.172}$$

At short wavelengths, where $h\nu_0/kT \gg 1$, Eq. 9.172 reduces to

$$D_\lambda^* = \frac{[\eta(\lambda_0)]^{1/2} \lambda_0^3 \exp\,(hc_0/2\lambda_0 kT_2)}{2\pi^{1/2} hc_0^{3/2}(\Delta\lambda_0)^{1/2}} . \tag{9.173}$$

For ease in specifying the characteristics of a narrow band quantum counter we define a quality factor Q.

$$Q \equiv \frac{\lambda_0}{\Delta\lambda_0} . \tag{9.174}$$

Thus, from Eq. 9.172

$$D_\lambda^* = \frac{[\eta(\lambda_0)]^{1/2} \lambda_0^{5/2} [\exp\,(hc_0/\lambda_0 kT_2) - 1]Q^{1/2}}{2\pi^{1/2} hc_0^{3/2} \exp\,(hc_0/2\lambda_0 kT_2)} . \tag{9.175}$$

Figure 9.16 illustrates the dependence of D_λ^* upon λ_0, according to Eq. 9.175, assuming a background temperature of 290°K, quantum efficiency of unity, and selected values of Q. The general shape of the curves is similar to that for the photon noise limited photon detector. (See Fig. 9.10.) That is to say, the value of D_λ^* exhibits a minimum, rises slowly at

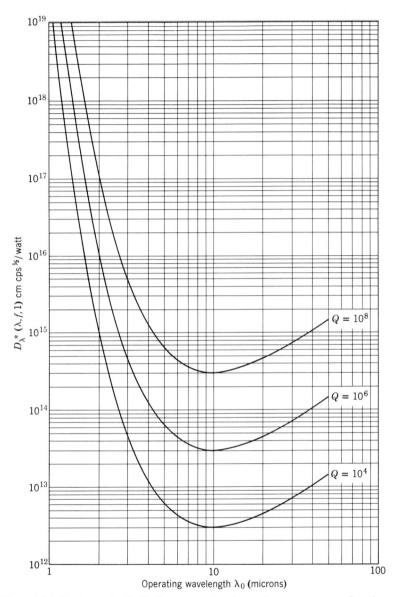

Figure 9.16. Photon noise limit of a narrow band quantum counter as a function of operating wavelength λ_0 for a 290°K background, viewing angle of 2π steradians and unit quantum efficiency.

wavelengths greater than that of the minimum, and rises sharply at wavelengths less than that of the minimum. Note that $D_\lambda{}^*$ improves as Q is increased. This is to be expected, since increasing Q reduces the photon noise from the background. Comparing the curve of Fig. 9.10 for $T_2 = 290°\text{K}$ to that for $Q = 10^4$, shown in Fig. 9.16, we note the narrow band quantum counter is much superior to the photon detector. Even for $Q = 10^2$, which would reduce the value of $D_\lambda{}^*$ by a factor of ten at each wavelength, the photon noise limited narrow band quantum counter is superior in $D_\lambda{}^*$ to the photon noise limited photon detector. Of course this is due to better background rejection. Thus narrow band quantum counters, in general, are better suited to detecting monochromatic sources emitting at their operating wavelength than photon detectors.

Proceeding in a manner similar to that of Section 9.4.3.2, we now determine the black body photon noise limited D^* of a narrow band quantum counter. By analogy to Eq. 9.146, the value of G, the ratio of D^* to $D_\lambda{}^*$, is

$$G = \frac{v_0}{\sigma T_3^4} \frac{M(v_0, T_3)}{v_0} \Delta v_0 = \frac{2\pi h v_0^3 \Delta v_0}{\sigma T_3^4 c_0^2 [\exp(h v_0 / k T_3) - 1]}, \qquad (9.176)$$

where T_3 is the temperature of the radiant signal source. Expressing G in terms of λ_0 and multiplying it by $D_\lambda{}^*$, Eq. 9.172, we find

$$D^* = \frac{\pi^{1/2} c_0^{1/2} [\eta(\lambda_0)]^{1/2} [\exp(h c_0 / \lambda_0 k T_2) - 1](\Delta \lambda)^{1/2}}{\sigma T_3^4 \lambda_0^2 \exp(h c_0 / 2\lambda_0 k T_2)[\exp(h c_0 / \lambda_0 k T_3) - 1]}, \qquad (9.177)$$

or

$$D^* = \frac{\pi^{1/2} c_0^{1/2} [\eta(\lambda_0)]^{1/2} [\exp(h c_0 / \lambda_0 k T_2) - 1]}{\sigma T_3^4 \lambda_0^{3/2} Q^{1/2} \exp(h c_0 / 2\lambda_0 k T_2)[\exp(h c_0 / \lambda_0 k T_3) - 1]}. \qquad (9.178)$$

The value of D^* as a function of wavelength for selected values of Q, assuming unit quantum efficiency, a background temperature T_2 of $290°\text{K}$, and a source temperature T_3 of $500°\text{K}$, is plotted in Fig. 9.17. Again, the general shape of the curve for D^* is similar to that for the photon noise limited photon detector (see Fig. 9.12) in that each exhibits a maximum. Note that increasing Q reduces D^*. As the operating bandwidth is narrowed, photon noise from the background is rejected, but signal photons are rejected even more. Reducing the operating bandwidth reduces the photon noise by the square root of the bandwidth, but reduces the signal linearly with the bandwidth. Comparing the curve for $T_2 = 290°\text{K}$, $T_3 = 500°\text{K}$, from Fig. 9.12 to that for $Q = 10^4$ in Fig. 9.17, we see that the photon noise limited photon detector is much superior to the narrow band quantum counter for the conditions indicated. In general, for viewing continuous sources against a terrestrial background the photon noise limited photon detector is superior.

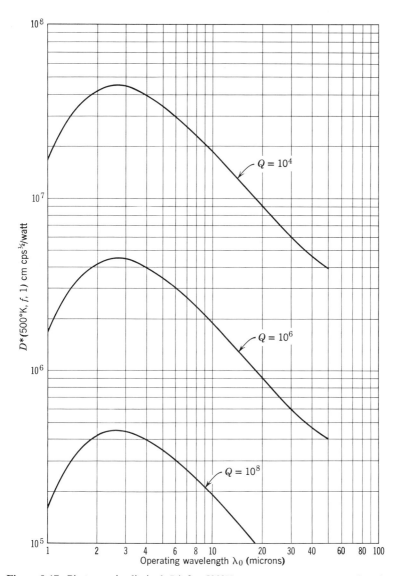

Figure 9.17. Photon noise limited D^* for $500°K$ source temperature as a function of operating wavelength λ_0 of a narrow band quantum counter. Background temperature is $290°K$. Viewing angle is 2π steradians. Unit quantum efficiency.

9.4.5   Signal Fluctuation Limit

As was pointed out in Section 9.1, under certain conditions detectors may be limited by signal fluctuations rather than photon noise from a radiating background. For example, if the background could be at absolute zero, and the detector were ideal, then the detectability of a signal against the background would be dependent upon the statistical nature of the arrival of signal photons. This problem has been treated by Gelinas and Genoud.[19]

Consider the emission of photons to be a random process. This is not strictly true, as we have seen in Section 9.4, but no serious error is introduced by such an assumption. During the interval in which the detector is viewing the signal source, at least one photon from the source must be absorbed by the detector. Since the photons arrive in a random manner, only the probability of detecting one within the viewing time can be specified. We assume that the probability of detecting a photon is proportional to the observation time. If the average number of signal photons absorbed in a given time interval t is $\bar{N}$, then the probability $P(S, \bar{N})$ of detecting S photons in an equal time interval is given by the Poisson distribution function.[20]

$$P(S, \bar{N}) = \frac{\exp(-\bar{N})\bar{N}^S}{S!}.$$
(9.179)

We require that it must be, say, 99% probable, that at least one photon will be detected in the time interval t. This is equivalent to stating that there is 1% probability that no photons will be detected. Thus we can write

$$P(0, \bar{N}) = 0.01 = \exp(-\bar{N}),$$
(9.180)

or

$$\bar{N} = \ln 100 = 4.61.$$
(9.181)

Therefore, in order that it is 99% probable that at least one photon is detected during the time interval, it is required that on the average 4.61 photons be detected during an equal time interval.

Equation 9.181 can be restated in terms of minimum detectable power. The minimum detectable power $P_{S\,\text{min}}$ required is

$$P_{S\,\text{min}} = \frac{4.61}{t} \frac{hc_0}{\lambda},$$
(9.182)

where λ is the wavelength of the photons.

As stated in Section 9.4.1, a bandwidth Δf is associated with an observation interval $(2t)^{-1}$. Therefore, Eq. 9.182 may be written as

$$P_{S\,\text{min}} = 9.22 \frac{hc_0}{\lambda} \Delta f.$$
(9.183)

As Gelinas and Genoud[19] point out, the minimum detectable power $P_{S\,min}$ is independent of area and depends linearly on bandwidth. Thus it is meaningless to calculate D^*, since D^* presupposes the minimum detectable power to be proportional to the square root of detector area and square root of bandwidth. In order to determine whether signal fluctuations may limit a given detector in a given application, it is necessary to substitute wavelength and bandwidth in Eq. 9.183 and compare with the value determined for the minimum detectable monochromatic power for photon noise from a radiating background, Eq. 9.141. The minimum detectable monochromatic power for photon detectors limited by photon noise from a 290°K background decreases very rapidly as the wavelength decreases in the near infrared and visible. (See Fig. 9.10.) Gelinas and Genoud[19] point out that it is in this region of the spectrum that detectors may be limited by signal fluctuations. Moreover, short observation times (large bandwidths) tend to cause signal fluctuations to dominate over photon noise from a 290°K background.

9.5 MEANS OF ACHIEVING PHOTON NOISE LIMITED PERFORMANCE

The discussion in Section 9.4.3 states the performance to be expected from ideal photon detectors. However, it does not indicate how to attain ideal behavior, which is treated in this section. Because embodiment of the approach varies with the operating requirements for each detector, only the broad outlines of the approach are given.

Achievement of ideal behavior in thermal detectors has already been considered. It was shown in Section 9.4.2 that photon noise limited performance in thermal detectors is obtained by causing radiation interchange to be the dominant means of heat transfer between the detector and its surroundings. The temperature noise limited performance so obtained was shown to be identical to the photon noise limited performance. Of course, the absorptivity, and therefore the emissivity, of the surface of the detector should be made as close to unity as possible. (See Eq. 9.115.)

The approach to photon noise limited performance in photon detectors is through reduction of all internal noise sources. The quantum efficiency should also be made to approach unity. Antireflection films can be used to enhance the absorptance, especially in materials having a high index of refraction.[21] As pointed out in Section 9.2, internal noise sources include current, thermal, and generation-recombination noise. The first two can usually be made to be small. Current noise, discussed in Section 7.4.1, occurs at potential barriers at the surface or within the detecting element. The noise voltage was shown to depend upon the volume of the noise source, the electrical frequency, and the current through the detecting element. Elimination of current noise is a problem in technology involving, for example, methods of preparing ohmic contacts to crystals. Because current noise is not a fundamental limitation to detectors, it will be assumed here that it will always be possible to minimize current noise with respect to other noise sources.

The other two types of internally generated noise, both fundamental in nature, are related to the thermal environment of the detector. Thermal noise, discussed in Section 7.2, arises from the interchange of energy between the detecting material and its surroundings. Generation-recombination noise, discussed in Section 7.4.2, is due to fluctuations in the thermally induced carrier excitation and recombination rates. The approach to a photon noise limited detector is through (1), suppression of thermal noise below gr noise, followed by (2), suppression of gr noise below photon noise. Consider the first step.

The thermal noise voltage $v_{N,th}$, given by Eq. 7.8, is

$$v_{N,th} = (4kTR\Delta f)^{\frac{1}{2}}, \qquad (9.184)$$

where R is the detector resistance. The gr noise voltage $v_{N,gr}$, given by Eq. 7.77 is

$$v_{N,gr} = i_N R = \frac{2I\tau^{\frac{1}{2}}}{N^{\frac{1}{2}}} R(\Delta f)^{\frac{1}{2}}, \qquad (9.185)$$

where I is the bias current, N is the total number of carriers in the semiconductor, and τ is the lifetime. We would like to minimize thermal noise with respect to gr noise. The ratio of the two is

$$\frac{v_{N,th}}{v_{N,gr}} = \frac{(qkT)^{\frac{1}{2}}nA\mu^{\frac{1}{2}}}{I\tau^{\frac{1}{2}}}, \qquad (9.186)$$

where n is the carrier concentration, A is the cross-sectional area, and μ is the mobility. Therefore, to cause gr noise to dominate over thermal noise we require low operating temperatures, low carrier concentration, low mobility, and small cross-sectional area. In a given semiconductor the carrier concentration and mobility requirements indicate material which is slightly p-type. Generation-recombination noise can also be made to be greater than thermal noise by application of a sufficiently high bias current. However, the amount of current is limited by Joulean heating of the sample and by the onset of current noise.

Consider now the second step. The manner in which gr noise can be suppressed below photon noise has been discussed by Petritz.[22] Two methods are possible: by reducing the lattice temperature so that few carriers are thermally excited, and by increasing the lifetimes associated with lattice excitation until they exceed that associated with photon excitation. Since, as was pointed out in Section 6.8, lifetimes add reciprocally, the shortest dominates. If the lifetimes associated with all lattice excitation processes can be made longer than that associated with photon excitation, then gr noise will be less than photon noise, and the detector will therefore be photon noise limited. This is the key to photon noise limited performance.

Three mechanisms exist by which carriers may recombine in a semiconductor. A short discussion is found in Section 6.8. Blakemore[23] and Bube[24] discuss each in detail. These mechanisms include radiative recombination, Auger (pronounced "oh-zhā") recombination, and recombination at lattice imperfections. Each has a lifetime associated with it. The radiative recombination process is the inverse of generation of free carriers by photons. Photon noise limited performance requires the radiative lifetime be shorter than the lattice imperfection or the Auger lifetimes.

The lifetime due to recombination at imperfections is difficult to predict for any material. Since it depends upon the perfection of the lattice, that is, such things as dislocation density, impurity atom density, and density of vacancies, it is determined in large part by the ability and experience of the persons who prepare the material. In general, given sufficient time, it should be possible to produce most semiconductor materials sufficiently pure and free from defects to cause the lifetime due to imperfection recombination to exceed both the Auger and radiative lifetimes.

Assuming that it is possible to do so, the remaining requirement is that the Auger lifetime be longer than the radiative lifetime. Each is fundamental to the semiconductor material, depending upon several semiconductor parameters including energy gap, effective masses, dielectric constant, and temperature. See Blakemore[23] for the appropriate analytic expressions. It is possible to vary each by varying the density of free carriers through doping the crystal. Careful study of the dependence of each upon temperature and purity for a given material will indicate how much cooling will be required in order to cause the Auger lifetime to exceed the radiative lifetime. In such a manner it is possible to predict theoretically the upper limit of operating temperature and the associated lifetime in order to achieve photon noise limited performance in a given semiconductor.

The following example illustrates the approach. Suppose that it is required to construct an intrinsic detector operating in the 8 to 14 μ atmospheric window, having, say, a 14.5 μ cutoff wavelength. The detector must have a response time of roughly 1 μsec and operate with a minimum amount of cooling.

The energy gap required for 14.5 μ response is 0.086 ev. It is necessary to assume values of the important parameters for a semiconductor having this gap. By analogy with other narrow gap semiconductors, it may be expected that a material with a gap of 0.086 ev would have an electron effective mass approximately 0.05 electronic mass units, a hole effective mass of approximately 0.20 electronic mass units, and a dielectric constant of 20. Using these values, the temperature dependences of the Auger and

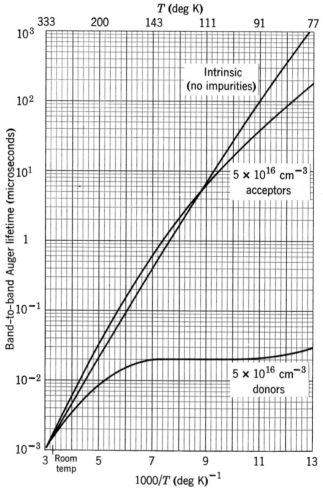

Figure 9.18. Dependence of Auger lifetime upon temperature for assumed semiconductor. (After Blakemore[25])

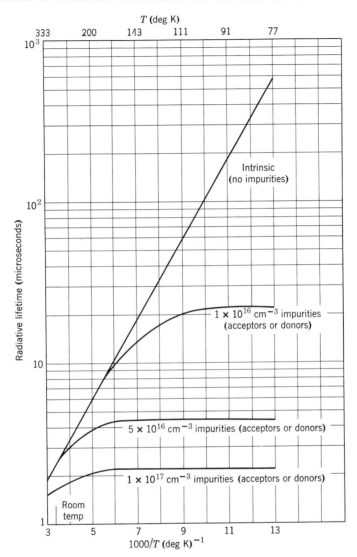

Figure 9.19. Dependence of radiative lifetime upon temperature for assumed semiconductor. (After Blakemore.[25])

radiative lifetimes due to Blakemore[25] for several material purities are shown in Figs. 9.18 and 9.19.

From Figs. 9.18 and 9.19 it can be seen that for intrinsic material the room temperature radiative lifetime is roughly 2 μsec whereas the room temperature Auger lifetime is 0.001 μsec. Thus the Auger lifetime, shorter than the radiative, dominates at room temperature in intrinsic material. As the intrinsic material is cooled, both Auger and radiative lifetimes increase, with the Auger increasing more rapidly. At 77°K the Auger lifetime is approximately 1000 μsec, whereas the radiative is approximately 600 μsec. Thus at 77°K in intrinsic material the radiative lifetime, being shorter, dominates. A detector made from intrinsic material operating at 77°K would be photon noise limited.

However, the lifetime of the material is roughly 600 μsec. For many applications this is much too great. Also, intrinsic material must be of very high purity, which is difficult to prepare. Consider the curves for material doped with 5×10^{16} acceptors/cm^3 (p-type material). Below approximately 110°K the Auger lifetime is longer than the radiative lifetime. At this temperature the radiative lifetime is approximately 6 μsec. Thus cooling material doped with 5×10^{16} acceptors/cm^3 to below 110°K would result in a photon noise limited detector with a 6 μsec response time.

The example above illustrates the general approach to the problem of achieving photon noise limited performance for a given detector requirement and illustrates the interrelationship between maximum operating temperature and detector response time.

In a practical situation it would be necessary to select those materials known to have the proper energy gap and perform the calculations based upon known values of their effective masses, indices of refraction, and dielectric constants. Evaluating each material and comparing the results obtained would indicate which would best satisfy the operational requirements.

REFERENCES

1. E. Burstein and G. S. Picus, IRIS meeting, Feb. 3, 1958.
2. W. E. Spicer, *J. Appl. Phys.* **30**, 1381 (1959).
3. T. S. Moss, *Optical Properties of Semiconductors*, Butterworths, London (1959); *Photoconductivity in the Elements*, Butterworths, London (1952).
4. Smith, Jones, and Chasmar, *The Detection and Measurement of Infrared Radiation*, Oxford University Press, London (1957).
5. C. Hilsum and O. Simpson, *Proc. Inst. Elect. Engrs.* **106**, Part B, Suppl., 15 (1959).
6. See, for example, F. E. Terman, *Radio Engineering*, 2nd Ed., McGraw-Hill Book Company, New York (1937), p. 95.
7. J. N. Shive, *J. Appl. Phys.* **18**, 398 (1947).
8. R. C. Jones, *J. Opt. Soc. Amer.* **43**, 1 (1953).
9. Chasmar, Mitchell, and Rennie, *J. Opt. Soc. Amer.* **46**, 469 (1956).
10. D. J. Lovell, *Thermistor Infrared Detectors*, NAVORD Report 5495 Part 1, AD 160106.
11. R. C. Tolman, *The Principles of Statistical Mechanics*, Oxford University Press, London (1938), p. 632.
12. A. Einstein, *Ann. Physik* **14**, 354 (1904).

13. D. J. Fink, *Principles of Television Engineering*, McGraw-Hill Book Co., New York (1940), p. 186.
14. S. B. Schuldt, private communication.
15. K. M. van Vliet, *Proc. Inst. Radio Engrs.* **46**, 1004 (1958).
16. R. C. Jones, *Proc. IRIS* **5**, No. 4, 35 (1959) (Paper unclassified); *J. Opt. Soc. Amer.* **50**, 1058 (1960).
17. Morton, Schultz, and Harty, *RCA Rev.* **20**, 599 (1959).
18. R. W. Gelinas, *Infrared Detection by Ideal Irasers and Narrow Band Counters*, RAND Report P-1844 (Oct. 26, 1959).
19. R. W. Gelinas and R. H. Genoud, *A Broad Look at the Performance of Infrared Detectors*, RAND Report P-1697 (May 11, 1959).
20. See, for example, E. U. Condon and H. Odishaw, eds., *Handbook of Physics*, McGraw-Hill Book Company, New York (1958), p. 1–155.
21. Farber, Kruse, and Saur, *J. Opt. Soc. Amer.* **51**, 115 (1961).
22. R. L. Petritz, *Photoconductivity Conference*, eds. Breckenridge, Russell, and Hahn, John Wiley and Sons, New York, 1956, p. 49; *Proc. Inst. Radio Engrs.* **47**, 1458 (1959).
23. J. S. Blakemore, *Semiconductor Statistics*, Pergamon Press (1961), Chapter VI.
24. R. H. Bube, *Photoconductivity of Solids*, John Wiley and Sons, New York, 1960, p. 303.
25. J. S. Blakemore, private communication.
26. R. J. Havens, *J. Opt. Soc. Amer.* **36**, 355(A) (1946); *Proc. IRIS* **2**, No. 1, 5 (1957) (Paper unclassified).

Appendices

I. DETERMINATION OF $D_\lambda^*(\lambda, f, \Delta f)$ FROM $D^*(T, f, \Delta f)$

The spectral D_λ^* of an infrared photon detector may be determined by allowing a measured amount of radiation in a known narrow spectral interval to fall upon the detector and measuring the rms signal-to-noise voltage ratio in unit bandwidth and at a specified chopping frequency. Difficulties are encountered in this method because of the uncertainty in measuring the intensity of the monochromatic radiation falling upon the detector.

An alternate, and more widely used, method exists for determining D_λ^*. The black body D^* is first determined as described in Section 8.2. Since the irradiance at the detector can be calculated from the temperature of the black body, no measurement of the irradiance is required. Knowledge of the relative spectral response R_λ is required, obtained as described in Section 8.2 by measuring the relative signal from the detector for equal amounts of radiant energy per unit spectral interval. The value of D^* is then related to the value of D_λ^* by

$$D^*(T, f, \Delta f) = \frac{\int_0^\infty \mathscr{H}_\lambda(T) D_\lambda^*(\lambda, f, \Delta f)\, d\lambda}{\int_0^\infty \mathscr{H}_\lambda(T)\, d\lambda}. \tag{A}$$

If the integral in the numerator is replaced by a summation, the value of $D^*(T, f, \Delta f)$ is given approximately by

$$D^*(T, f, \Delta f) \approx \frac{\sum_i \mathscr{H}_i D^*(\lambda_i, f, \Delta f)}{\sum_i \mathscr{H}_i}$$

$$\approx D_\lambda^*(\lambda, f, \Delta f) \frac{\sum_i g_i R_i}{R_\lambda}, \tag{B}$$

where $\mathscr{H}_i$ is the irradiance in the ith spectral interval centered at λ_i, g_i is the ratio of the irradiance in the ith interval to that over all wavelengths, R_i is the relative response averaged over the ith interval, and R_λ is the relative response at the wavelength λ. In practice, the numerator is evaluated by numerical integration techniques. The spectral radiance for a black body

384

at temperature T is plotted as a function of wavelength, normalized so the total radiance over all wavelengths is unity. The relative response of the detector is also plotted as a function of wavelength. The product of the average values of the functions in a small spectral interval, of the order of 0.2 μ, is formed, and the products in all the small spectral intervals over the entire spectral range of the detector are summed. The value of $D_\lambda^*(\lambda, f, \Delta f)$ is then determined from Eq. B.

A useful relationship is the value of the ratio of $D_\lambda^*(\lambda, f, \Delta f)$ evaluated at the peak of the spectral response curve to $D^*(500°K, f, \Delta f)$. Knowing the value of the 500°K black body D^* for any photon detector, the peak value of D_λ^* can readily be determined. This ratio is computed as described above, where we now take λ to be the wavelength of peak response. However, if we assume the detector to be an ideal photon counter, the ratio is simply the reciprocal of the function G, found in Eq. 9.147. This can be seen by examining Eq. 9.150. The dependence of G upon wavelength is shown in Fig. 9.11. We see that for detectors with short peak wavelengths, below about 4 μ, the ratio of D_λ^* at the peak to D^* for a 500°K black body may be one or more orders of magnitude. For longer wavelengths, the ratio is greater than unity but less than ten. The ratio is evaluated for several detectors in Appendix I of Chapter 10.

II. HAVENS' LIMIT

A thermodynamic approach to the ultimate performance of thermal detectors has been derived by R. J. Havens.[26] Although Havens' limit is an energy rather than a power limit, and is in no way connected with the photon noise limit, we include it here since it is occasionally mentioned. Havens analyzed the performance of any thermal detector as if it were a heat engine with an efficiency of 10%. He found the minimum energy q per square millimeter needed to give rise to a signal equal to the noise (thermal noise in electrical detectors, Brownian noise in mechanical detectors) to be 2.5×10^{-12} joule for operation at 300°K. For response times shorter than 0.04 sec the minimum energy was found to be independent of time, whereas for longer times it became a function of time, dependent upon the square root of the time constant for bolometers and thermopiles. As an example, for a response time of 0.04 sec, the Havens' limit value of D^* is

$$D^* = \frac{\sqrt{A\,\Delta f}}{NEP} = \frac{\sqrt{A}}{\dfrac{q}{t}\sqrt{2t}} = 5.65 \times 10^9 \text{ cm cps}^{1/2}/\text{watt}.$$

This is approximately one-fourth the photon noise limit.

10
Comparative performance
of elemental detectors

10.1 INTRODUCTION

In Chapters 8 and 9 we discussed the various detection mechanisms
upon which infrared detectors are based and studied methods of rating
the performance of detectors. With this background, we can now consider
the measured performance of available elemental detectors. The aim of
this chapter is to give representative values of the parameters of importance
so that the user may have a basis for selecting the best detector for his
application. The data pertaining to each detector at each operating
temperature are presented in four places. Table 10.1 lists numerical values
of several parameters of importance pertaining to each detector including
black body D^*, spectral D_λ^*, response time, resistance, and peak and
cutoff wavelengths. The value of D_λ^* as a function of wavelength is shown
in Figs. 10.11a through 10.14 for detectors operated at room temperature,
195°K, 77°K, and below 77°K. The next series of graphs, Figs. 10.15
through 10.18, depicts the frequency response of detectors operated at the
above temperatures. The last series, Figs. 10.19 through 10.22, shows the
noise spectra, again according to operating temperature.

Although most of the detectors listed respond in the infrared, data on
several detectors responding only in the visible are presented. The
information has been included for the benefit of the user who must select
a detector for applications involving the visible spectrum, such as star
tracking. With the exception of the photomultiplier, all the detectors for
the visible spectrum use either the photoconductive or the photovoltaic
effect.

The data have been obtained from various journal publications and manufacturer's literature. The sources, referred to in Table 10.1, are listed following the bibliography. Information obtained from different sources concerning the same type of detector will frequently vary. Since it would be undesirable to mix data on a given type of detector obtained from different laboratories, wherever possible all measurements have been made on one detector at one laboratory. In several cases, however, it has been necessary to mix data, and in others data are lacking. Whereas a great amount of unclassified information on D_λ^* exists for various detectors, in several cases information on frequency response and noise spectra cannot be obtained. Even less information is available on the important parameter, responsivity. Because of this lack it has been omitted from the data presented. Finally, where much data are available, those pertaining to representative detectors, rather than to the best ever made, are given.

In addition to the included information, the user may need to know for a particular detector such items as the sensitivity contour of the detecting element, the variation in signal and noise with applied bias, or the variation in signal with intensity of radiation. Since such information is unavailable for most of the detectors, we will show here general forms of behavior to be expected, using as an example lead selenide. The sensitivity contour of a lead selenide photoconductor is shown in Fig. 10.1, taken from Potter, Pernett and Naugle.[1]

The contour is obtained by exploring the sensitive surface with a light spot which is small compared to the surface area. Typical surfaces exhibit a pronounced irregularity.[2] In scanning applications, in which the image may be small compared to the size of the detector element, such irregularities appear as false changes in signal, termed "scanning noise."

Figure 10.2, also from Potter, Pernett and Naugle,[1] shows the signal, noise, and signal-to-noise ratio found in a PbSe photoconductor as a function of bias voltage.

It can be seen that an optimum bias exists that maximizes the signal-to-noise ratio. The broad plateau in the value of the ratio indicates that in the neighborhood of its optimum value the bias may be varied considerably without much influence on the signal-to-noise ratio.

Figure 10.3 depicts the dependence of responsivity upon irradiance found in many detectors.

The response of most detectors is linearly dependent upon radiation intensity over a wide range of intensities. At extremely high levels the signal may approach a square root dependence as indicated. Note that this applies to the detector only. Even at low levels, at which the signal is proportional to the intensity, the voltage appearing across a load

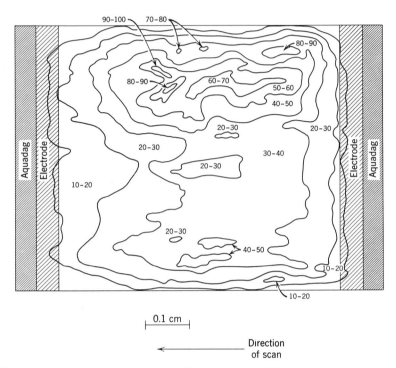

Figure 10.1. Sensitivity contour of a PbSe detector. (From Potter, Pernett, and Naugle.[1])

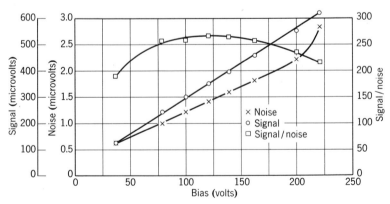

Figure 10.2. Signal, noise, and signal-to-noise ratio as a function of bias for a PbSe detector. (From Potter, Pernett, and Naugle.[1])

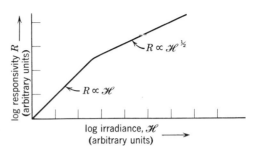

Figure 10.3. Dependence of responsivity upon irradiance.

resistor in a constant voltage circuit may not be linear with intensity because of a change in relative impedance between the detector and load as the radiation intensity changes.

10.2 SOME GENERAL REMARKS CONCERNING ELEMENTAL PHOTON DETECTORS

10.2.1 Comparison of Thin Film and Single Crystal Detectors

Because photon detectors find much wider use than thermal detectors and are made from a much greater variety of materials, we shall discuss certain generalizations applicable to them. These remarks apply to all the detectors considered in the later sections with the exception of the photomultiplier, bolometer, thermocouple, and Golay cell.

The three basic effects of use in photon infrared detectors are the photoconductive, the photovoltaic, and the photoelectromagnetic. Others, such as the photoemissive and the Dember effects, have found little if any application in elemental infrared detectors. Photoconductivity is most widely used, followed in order by the photovoltaic and PEM modes.

Photoconductive detectors are prepared either in the form of thin polycrystalline films or as slices of single crystals. The thin film detectors, confined to the lead salts and thallous sulfide, are produced by chemically reacting materials in aqueous solution so as to precipitate a photosensitive film on a substrate, or by vacuum sublimation of the desired material onto a substrate. The end result is a film about 1 μ thick composed of many small crystallites. On the other hand, the single crystal detectors are prepared by growing large single crystals and using transistor techniques to slice them into small elements, prepare the surfaces, and mount them upon substrates. The thickness of a slice is generally much greater than that of a thin film, of the order of 1 mil or more. Certain generalizations can be made concerning the merits of each. The thin polycrystalline film detectors exhibit characteristics less desirable in general than those of the single crystal detectors. As we have already seen exemplified by PbSe, a deposited polycrystalline layer exhibits nonuniformity in the point-to-point response obtained when a small diameter light spot is scanned over the surface of it. Although single crystal detectors may also exhibit nonuniformity due to a nonuniform distribution of impurities or a nonuniform surface characteristic, it is in general of lesser extent. The reproducibility of thin film detectors is poorer than for single crystal ones. That is, two thin film detectors of the same type are less likely to exhibit identical performance characteristics than two single crystal detectors of the same type. Again, all polycrystalline film detectors are limited by current noise at low frequencies, whereas some of the single crystal ones are not. Since the thin films are much thinner than the single crystals, they are more susceptible to damage by exposure to direct sunlight, ultraviolet radiation, and high ambient temperature. Certain of the thin film detectors also exhibit double time constants which are generally undesirable. We shall consider this in more detail in the discussion of the lead salts.

10.2.2 Comparison of Extrinsic and Intrinsic Materials

Detectors may be either intrinsic or extrinsic, that is the incident photons may produce either a free electron-free hole pair by direct excitation across the forbidden gap, or they may produce a free electron-bound hole, or bound electron-free hole pair by excitation of an impurity level. The intrinsic type of detector requires a material having the proper forbidden gap width for the spectral response required, whereas the extrinsic detector achieves its spectral response through use of a doping element having the proper excitation energy. The intrinsic detector generally exhibits fewer detrimental qualities than the extrinsic one. Proper doping is of prime importance to the extrinsic detector where very close control must be maintained over the concentration of one or more impurity elements. Although intrinsic detectors must sometimes be doped to optimize their performance through manipulation of the resistance, the control of the doping is usually not as severe. In addition, a major disadvantage of the extrinsic detector lies in its generally lower responsivity and "washed out" optical absorption. The amount of impurity that can be introduced in a semiconductor is limited by the solubility and by impurity band conduction. In the latter instance, if the impurity concentration is too high, conduction can take place from impurity atom to impurity atom without making use of the conduction or valence bands. Thus the concentration of centers available for extrinsic photoexcitation is generally several orders of magnitude below the concentration of the lattice sites available for intrinsic excitation. The result is that the responsivity of extrinsic detectors is usually less than that of intrinsic detectors. The optical absorption coefficient is much less also, requiring a relatively large length over which the absorption takes place. Whereas absorption in intrinsic detectors takes place in a length of the order of a micron, the absorption length in extrinsic detectors may be several centimeters. This means that extrinsic detectors must be long in the direction of the incident radiation to absorb the radiation. Since the absorption takes place in a large volume rather than a small one, the problem of imaging the incident radiation on the detector is more severe. On the other hand, the number of narrow energy band semiconductors is somewhat restricted, whereas a wide variety of doped materials having small excitation energies exist. For this reason, response at wavelengths greater than about 7 μ is generally obtained through use of extrinsic materials.

10.2.3 Detector Impedance

One of the points of interest in comparing detectors is the cell impedance, which in almost all detectors is purely resistive. Low resistance detectors are considered to be those having values of resistance up to 100 ohms. Medium resistance detectors have resistances between 100 ohms and 1 megohm, whereas high resistance detectors have values greater than 1 megohm. The resistance is of importance for four reasons. First, the detector resistance determines the amplifier requirements. Medium resistance detectors offer use of the most simple amplifier techniques. Second, most forms of detector noise depend upon the resistance. For example, detectors limited by thermal noise have a noise voltage proportional to the square root of the detector resistance. Third, detectors having extremely high resistances have associated with them problems in electrical pickup and microphonic noise not associated with low resistance detectors, whereas very low resistance detectors require transformer coupling to the amplifier, with its attendant problem of magnetic pickup. Finally, very high resistance detectors may have RC time constants considerably greater than the material lifetime, resulting in reduced frequency response.

The resistance of photoconductive and PEM detectors is controlled by the material resistivity (the magnetoresistivity in the PEM case) and the detector geometry, whereas the resistance of photovoltaic detectors is determined by the junction resistance. Since the resistivity of a semiconductor increases with decreasing temperature over most of the temperature range of interest, cooled photoconductors have much greater resistance than uncooled photoconductors using the same material. In some instances the detector geometry can be varied to influence the resistance. The resistance of a square sensitive area of a photoconductor or PEM detector is independent of the size of the square, since the resistance is proportional to the length-to-width ratio, unity for a square. For this

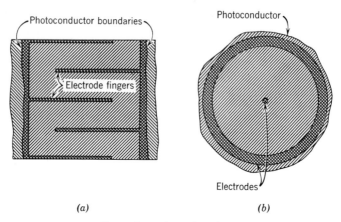

(a)   *(b)*

Figure 10.4. Electrode configurations for low resistance.

reason resistance is frequently quoted in ohms per square. By making the length-to-width ratio greater or less than unity, the resistance can be varied. However, the shape in many instances is dictated by the optical requirements. Most requirements are for square aspects, although infrared spectroscopy requires a long narrow element on which to image the exit slits. In some cases the electrodes may be formed in an interleaving or comb structure to reduce the resistance, with the photoconductor overlaying all (see Fig. 10.4a.)

Since that portion of the radiation falling on the electrodes is lost, the electrodes are made very narrow. Another geometry which will reduce the detector resistance consists of a center circular electrode surrounded by an annular ring electrode, with the photoconductor between the two, shown in Fig. 10.4b. A dead spot occurs in the center of this construction. The resistance of photovoltaic detectors, depending as it does upon the characteristics of the junction, is not open to much variation. Since the choice of line or area junctions is dictated mainly by the intended use, the user must put up with the resistance typical of the type employed.

10.2.4   Size and Shape of Sensitive Area

The size and shape of the sensitive area are virtually unlimited in photoconductive detectors, save for problems in packaging. Standard sizes are about 1 mm square, although element sizes ranging from 0.1 mm on edge to 1 cm on edge are available in many materials. This is not true for all photovoltaic detectors or for PEM cells. We have already seen in Chapter 8 that grown junction photovoltaic cells have long, narrow, sensitive areas. Diffused junction detectors, however, may have their sensitive regions shaped to the optical requirements. PEM detectors may use small square areas, up to about 2 mm on a side, or rectangles up to 2 mm by 2 cm. The requirement of obtaining a high magnetic field strength in the gap between the pole pieces of necessity keeps the gap width no greater than about 2 mm. The length of the pole pieces, however, which determines the maximum value of the element length, may be up to 2 cm.

10.2.5 Temperature of Operation

It is convenient to group the detectors according to their temperature of operation. Since cooling requirements play a large part in the over-all system complexity and reliability, the user will attempt to get by with the least complex cooling system. However, since detectors operating at lower temperatures usually surpass in performance those operating at higher temperatures, in many applications he will require a cooled detector. Thus he weighs the disadvantages of cooling complexity against the advantages of increased detector performance. The comparison of performance of detectors is facilitated by grouping the detectors according to the temperature of operation. If the detector is cooled, the operating temperature is most often a stable point associated with the coexistence of two states of some material. The four most important operating temperatures are:

1. Room temperature, taken here to be 295°K.
2. Dry ice point, the sublimation temperature of solid carbon dioxide (CO_2), 194.6°K.
3. Liquid nitrogen point, the boiling point of nitrogen, 77.3°K.
4. Liquid helium point, the boiling point of helium, 4.2°K.

Other stable points are the liquid neon point, 27.2°K, and the liquid hydrogen point, 20.4°K.

10.2.6 Photocell Housings

A large variety of detector geometries and envelope constructions exist dependent upon the mode and temperature of operation. Photoconductivity offers an advantage in simplicity of construction. In this mode the semiconductor is employed simply as a polycrystalline thin film or single crystal slab with electrical leads attached. The construction of uncooled film type photoconductive detectors is shown in Fig. 10.5.

Figure 10.5a shows the simplest of all constructions, the sensitive element lying between sublimed electrodes and usually overcoated with a protective infrared transmitting lacquer. Figure 10.5b depicts the film sublimed in an evacuated enclosure. Figure 10.5c shows the construction of a sublimed film detector of the "side-looking" variety.

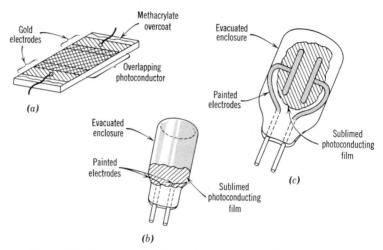

Figure 10.5. Construction of uncooled photoconductive detectors.

The design of cooled photoconductive and photovoltaic detectors is shown in Fig. 10.6.

A double-walled vacuum enclosure, termed a "dewar," serves first the purpose of maintaining a coolant, such as liquid nitrogen or dry ice, in the central well without allowing frost to collect over the exterior wall of the cell or the coolant to evaporate rapidly. In addition, some detectors require a vacuum construction to prevent performance deterioration due to the effects of the atmosphere on the surface of the detector. Obviously a vacuum construction requires an infrared transmitting window in the optical path. Early detector designs used a bubble window, consisting of a very thin glass membrane made by forming a hemispherical depression in the cell wall. Such a window does not transmit well at long wavelengths and is fragile. Modern detectors make use of infrared transmitting windows joined by glass-sealing techniques to the main detector body.

Thin film detectors, shown in Fig. 10.6a, have the sensitive surface prepared either on a glass substrate block, called a "rectangle," which is glued to the inner surface of the well, or by sublimation directly on the inner surface of the well. The film is in contact with evaporated electrodes, usually of gold, which in turn contact painted silver paste or Aquadag leads. The use of wire electrodes is to be avoided because of the microphonic problems they introduce. Although the electrodes serve to define to some extent the sensitive area in the direct sublimation method, the sensitive regions surrounding the electrodes also contribute to the sensitivity. Thus a "masking factor" is quoted, which, when divided into the over-all detectivity, specifies the performance of the area between the electrodes. "Side-looking" versions of thin film detectors have the advantage of accepting a horizontal beam of radiation while remaining vertical to prevent spilling of the coolant.

Extrinsic type photoconductors require cooling to preclude thermal excitation of the impurity levels. As mentioned earlier, they also require a long optical path. This is frequently achieved by orienting a rod of the photoconductor along the axis of the detector. Figure 10.6b shows such a construction. A cooled reflective enclosure acts as an integrating chamber, causing radiation which passes through the semiconductor to be reflected back upon it to increase the amount absorbed. The cold walls of the enclosure also lower the background photon noise which would otherwise limit certain of the detectors. The photoconductor in rod form is frequently soldered to a Kovar cup at the end of the well, which forms one electrical connection and seals to the glass well. The other connection is

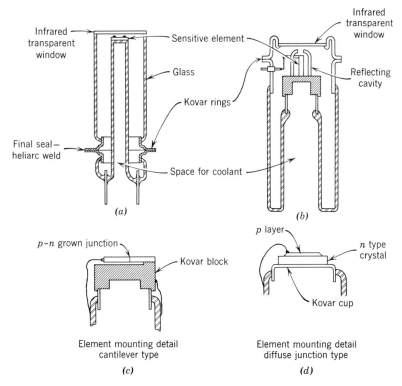

Figure 10.6. Construction of cooled photoconductive and photovoltaic detectors.

made by a lead soldered near the opposite end of the rod. Such a detector can be made side-looking also.

Neither the intrinsic photoconductor nor the photovoltaic detector requires a long optical path. The intrinsic photoconductor may be simply mounted on the end of the well in the manner depicted in Fig. 10.6a. Construction of a grown junction photovoltaic cell is shown in Fig. 10.6c. The rod containing the junction is supported in a cantilever manner normal to the cell axis. A Kovar cup at the support end forms one electrical contact, whereas the other is by a wire soldered to the opposite end. Some versions support the other end by an insulated pin to give mechanical rigidity. Diffused junction photovoltaic elements may be mounted directly on Kovar cups as shown in Fig. 10.6d, a wire electrode making contact to the diffused region at the surface.

Construction details of detectors designed to operate at temperatures close to absolute zero are shown in Fig. 10.7. A convenient operating point is the temperature of liquid helium, 4.2°K.

To prevent excessive evaporation of the expensive helium, a double dewar is used. The helium is contained in an inner well surrounded by liquid nitrogen. Vacuum insulates the helium from the nitrogen and the nitrogen from the atmosphere. The sensitive element looks out an infrared transmitting window on the bottom of the dewar through an aperture in a

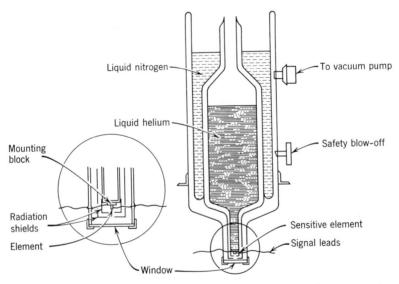

Figure 10.7. Construction of detectors incorporating double dewars. (Courtesy Hofman Laboratories, Inc.)

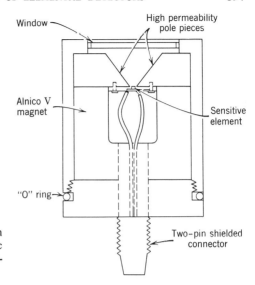

Figure 10.8. Construction of an uncooled photoelectromagnetic detector. (Courtesy of Minneapolis-Honeywell.)

radiation shield which serves to reduce the radiant heat leak-ins to the helium and the amount of photon noise coming from the detector housing.

A front surface mirror oriented at 45 degrees to the axis may be used to reflect radiation from a horizontal path through the window. Such double dewars are bulky and expensive but necessary for economical operation in the liquid helium temperature range.

Figure 10.8 shows the design of a photoelectromagnetic detector. The detector incorporates a permanent magnet, with shaped, high permeability pole pieces to direct the flux parallel to the surface of the sensitive element. Electrical contact is made by soldered or plated connections to those sample ends which are at right angles to the plane of the radiation and the magnetic field. The infrared transmitting window is sealed to a brass housing by epoxy resin. Although a sealed housing is not strictly required, it is of advantage in protecting the fragile photosensitive element and keeping metal filings from being picked up by the magnet. Because of the large mass of the magnet, which would require a large cooling capacity, PEM detectors are operated either at room temperature or cooled only to the vicinity of the ice point, 0°C.

Construction of a thermistor bolometer is shown in Fig. 10.9. The "active" thermistor flake is mounted in an enclosure together with a temperature compensating flake. Radiation passing through an infrared transmitting window falls only on the active flake. The front surface of the flake is covered with a heat absorbing film such as black paint to make more effective use of the radiation falling on it. The flake is mounted on

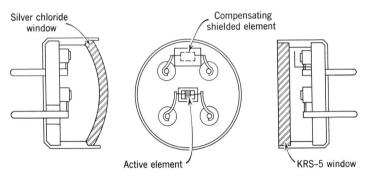

Figure 10.9. Construction of a thermistor bolometer. (Courtesy Barnes Engineering Company.)

a dielectric base with a very thin film of glue providing thermal contact to the base which may be of sapphire or glass. The base in turn is mounted on a metal block incorporated into the housing. A second flake, shielded from the radiation, is used in a bridge circuit to compensate for excursions in the ambient temperature.

10.2.7 Infrared Transmitting Windows and Immersion Optics

The optical properties of selected infrared transmitting materials and the requirements which detector windows must meet have been discussed in Chapter 4. It is sufficient to say at this point that the materials in common use are infrared grade quartz for detectors having optimum response in the near infrared at wavelengths shorter than 3 μ, sapphire for the region between 3 and 6 μ, arsenic trisulfide, magnesium oxide, and calcium fluoride for the region between 6 and 10 μ, and barium fluoride for the 1 to 12 μ interval. High resistivity n-type silicon and germanium also find use at wavelengths from 1.0 to 15 and 1.7 to 25 μ, respectively,

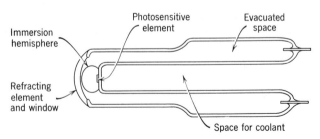

Figure 10.10. Cooled detector with immersed optics. (Courtesy Santa Barbara Research Center.)

although they should be overcoated with antireflecting films, discussed also in Chapter 4, to minimize reflection losses.

A means of increasing the effective D^* of a detector is achieved by immersion optics. In this technique a material of high index of refraction, such as strontium titanate ($n = 2.2$) or germanium ($n = 4$), is ground to a solid hemisphere or hyperhemisphere having a surface polished to optical tolerances. The sensitive element is glued, or "immersed," on the flat side of the hemisphere with a thin film of an infrared transparent adhesive. Rays refracted by the hemisphere are focused to an image which is reduced in area by n^2 below the one obtained without the hemisphere. This permits use of an element of area reduced by n^2 over the requirement without the hemisphere. The apparent area of the detector, that is, the area "seen" by an optical system looking back at the detector, remains the same. The signal-to-noise ratio, and thereby the detectivity, is improved by n. Figure 10.10 depicts the use of immersion optics in a cooled detector. An infrared transmitting lens, part of an Abbé condensing system, appears in the place normally occupied by the window.

10.3 THE LEAD SALT PHOTOCONDUCTORS

Having discussed a number of generalizations applicable to detectors, we shall now consider the performance of selected types. The lead salt photoconductors, comprising lead sulfide, lead selenide, and lead telluride, find very wide use in infrared systems. All are thin film detectors of impedance in the 0.5 to 100 megohm range, and are prepared by either vacuum sublimation or chemical deposition from solution. All have been investigated thoroughly, and all are readily available commercially. We shall begin our discussion by considering lead sulfide. The reader is referred to the works of Moss,[3] Smith,[4] and Cashman,[5] for extensive discussion of the lead salts.

10.3.1 Lead Sulfide (PbS)

Lead sulfide is a material used very extensively in infrared detectors, finding widespread application in guidance systems for air-to-air missiles. With a spectral response extending to about 3 μ, it is capable of detecting the 2.7 μ H_2O and CO_2 band in the plume of heated gas from a hydrocarbon burning jet engine. It exhibits excellent performance at room temperature, and even better performance when cooled to solid carbon dioxide or liquid nitrogen temperature. Although photovoltaic and PEM responses have been reported, photoconductivity is the only mechanism widely used.

Lead sulfide was first investigated extensively in Germany before and during World War II. Since then reports concerning it have appeared in the published literature of many countries. Two principal methods of preparation exist: by chemical deposition from solution and by sublimation in vacuum. In both methods PbS is prepared in a thin polycrystalline film deposited on a glass or quartz substrate between electrodes, usually made of gold. The chemical deposition method has been described by Kicynski.[6] The substrate is placed at the bottom of a beaker containing an aqueous solution of lead acetate and thiourea, a sulfurbearing compound. These are stirred continuously while a solution of sodium hydroxide is added. A mirror-like coating of PbS precipitates over the surface of the substrate and the walls of the vessel. The substrate is removed and allowed to dry. Baking the surface in air or oxygen is required for the surface to show optimum photosensitivity.

The sublimation method[7] utilizes a crucible containing PbS which is heated in an evacuated chamber under a small pressure of oxygen. Sublimation of the PbS occurs, a film condensing upon a substrate maintained at a lower temperature. Some techniques entail sublimation upon a wall of the vessel and resublimation to the substrate.

In either method of preparation the effect of oxygen, introduced either during the air bake or during the sublimation, is critical. Unoxidized films have been shown to be n-type. The addition of the oxidant serves to change the film to p-type, a resistance maximum being reached when a critical amount of oxygen is added to the film. The optimum photoconductive response has been found to occur in films oxidized to near maximum resistance.

The chemical deposition method is much more widely used than the sublimation one. The former method results in a layer having a more uniform sensitivity over the surface, shown by scanning a small spot of light across the surface. Fluctuations in the signal from point to point may vary by 50% or more for sublimed films. Chemical deposition also allows a closer control of the amount of oxidation received by the film, this being determined by the amount of oxidant introduced into the solution. In addition, chemical deposition allows preparation of a larger number of detectors with a smaller amount of equipment than does the sublimation method.

The layer deposited by either method is a polycrystalline film. Electron microscopy has shown the film to be composed of crystallites having dimensions of the order of 0.1 μ packed closely together. Mahlman[8] shows that the addition of the oxidant to the chemical solution causes a reduction in crystallite size from about 0.5 μ to about 0.1 μ.

The physical construction of the uncooled detector varies in the two methods of preparation. The chemical deposition method results in a thin film on a glass or quartz substrate, similar to that shown in Fig. 10.5a, which can be protected from atmospheric effects by overcoating with a thin film of butyl methacrylate. Leads are attached to gold electrodes in contact with the film. The detector may be used as is, or may be mounted in an enclosure which provides more security from physical damage. The sublimed detectors, however, consist of thin films between painted graphite or evaporated gold contacts in a vacuum enclosure of the types shown in Figs. 10.5b and 10.5c. Envelopes for both types suitable for operation at low temperatures are shown in Fig. 10.6a. Immersion optics have been used successfully on the uncooled detector. Cells constructed in both manners are subject to deterioration in performance at high ambient temperatures. The upper limit for prolonged storage is about 100°C. Performance will also deteriorate under prolonged ultraviolet irradiation. A decrease in performance with time has been associated with migration of sodium ions from the glass substrate into the lead sulfide film. This has been corrected through use of quartz or other substrate materials not containing sodium. Some films have been found to polarize, that is, exhibit a resistance dependent upon direction of current flow, under prolonged electrical bias.

The mechanism of the photoconductive response in PbS, also applicable to the other lead salts, deserves special mention because of two competing theories postulated to account for the photoconductive behavior. On the one hand, the usual theory of photoconductivity states that photons falling on a crystal liberate free carriers, increasing the conductivity. On the other hand, the so-called barrier modulation theory, based upon the polycrystalline nature of the film, states that potential barriers exist throughout the material at the intergranular boundaries. Photons falling on the film produce carriers which are trapped at the barriers. The space charge due to the trapped carriers reduces the potential barriers, causing the conductivity to increase through an increase in the effective carrier mobility. This theory has been developed by Slater,[9] among others.

It now appears that the theory applicable to the lead salt photoconductors is the former, that is, the conductivity increase upon irradiation is due to the increase in the number of free carriers. The experiments of Woods[10] showed the relative change in conductivity upon irradiation equalled the relative change in Hall coefficient over a wide range of values. Since the relative change in Hall coefficient was equal to the relative change in carrier concentration, the conductivity change was shown to be due to the added carriers.

The D^* of PbS detectors operated at 295°K, 195°K, and 77°K is given in Table 10.1 and plotted in Figs. 10.11a, 10.12, and 10.13. The values given are representative of cells available from commercial sources. It should be pointed out that detectors optimized to operate at a given temperature, whether made from PbS or any other material, will usually not exhibit optimum performance at other temperatures. Note that in the vicinity of 2 μ, PbS is far superior to any other detector. The spectral responses of the lead salt photoconductors shift to longer wavelengths as the detectors are cooled. Thus lead sulfide has a cutoff wavelength, defined here as the wavelength at which D_λ^* is 50% of the maximum value, of 2.5 μ at 295°K, 3.0 μ at 195°K, and 3.3 μ at 77°K. Even longer cutoff wavelengths at 77°K are available in special PbS types. These detectors, however, exhibit values of D_λ^* at the peak less than that found in the normal type. Figures 10.15, 10.16, and 10.17 depict the frequency response of the detectors at the three operating temperatures, the 3 db points being 640 cps, 350 cps, and 350 cps, respectively. The time constants shown, 250 μsec, 455 μsec, and 455 μsec at the three temperatures, are typical, although detectors are available having time constants ranging from a few microseconds to several milliseconds. Frequently the response at 77°K will be slower than that at 195°K. The cooled detectors may have double time constants which cause the responsivity to decrease less rapidly than the normal drop of 6 db per octave at high frequencies. The detectivity of PbS is in general proportional to the time constant, obeying the so-called McAllister relation.[11] Long time constant cells therefore exhibit superior detectivity compared to short time constant ones. Lead sulfide detectors should be selected having the longest time constant consistent with the frequency response requirements of the system. Electrical compensation techniques can be used to extend the useful frequency response of the slower detectors.

The noise spectra of lead sulfide at the three operating temperatures are indicated in Figs. 10.19, 10.20, and 10.21. It can be seen that PbS is definitely current noise limited; however, it follows a power law that differs from a simple inverse frequency dependence. Noise spectra for different PbS detectors will differ, but all will exhibit approximately an inverse frequency dependence.

10.3.2 Lead Selenide (PbSe)

The second member of the lead salt family, PbSe, is a thin film intrinsic photoconductor similar to PbS but exhibiting a response to longer wavelengths.[12] As with PbS, it may be operated at ambient, cooled to 195°K, or cooled to 77°K. The detectivity of PbSe at 295°K, 195°K, and 77°K is listed in Table 10.1 and shown in Figs. 10.11a, 10.12, and 10.13. It is used chiefly as an uncooled photoconductor in the 3 to 4 μ range or as a 77°K photoconductor in the 4 to 6 μ range. The cooled detector is used in air-to-air missile guidance and airborne ground mapping systems, where its long cutoff wavelength and short response time are used to advantage. As was true for PbS, the spectral response of PbSe extends to longer wavelengths upon cooling.

The method of preparation of PbSe detectors is similar to that of PbS, both sublimation and chemical deposition techniques being used. In both cases a polycrystalline film of the order of 1 μ thick is prepared. Oxygen is again the most important impurity, converting the film from n- to p-type.

The chemical deposition method is the more widely used. The manner of preparation is similar to that of PbS, except that selenourea replaces the thiourea. The mirror-like coating formed on the substrate is sensitized by baking in air. The substrate may be mounted in a protective enclosure for ambient operation or a cooled enclosure for low temperature operation. The protective enclosure helps to improve the stability of the PbSe layer, which tends to deteriorate with age. Immersion optics utilizing a strontium titanate hemisphere have been successfully applied to the cooled detector, as shown in Fig. 10.10.

The frequency response at the three temperatures is shown in Figs. 10.15, 10.16, and 10.17. Note that PbSe is much faster than PbS, having time constants of 4 μsec, 125 μsec, and 48 μsec at 295°K, 195°K, and 77°K, respectively. Whereas PbS can be obtained in a wide range of time constants, this is not true for PbSe.

The noise spectra of PbSe operated at 295°K, 195°K, and 77°K are depicted in Figs. 10.19, 10.20, and 10.21. They are seen to follow approximately a $1/f$ power law, indicating current noise limitation.

10.3.3   Lead Telluride (PbTe)

The third and last member of the lead salt family is PbTe, an intrinsic photoconductor. The preparation and study of it have not received as much attention as have those of PbS and PbSe. The construction and performance of PbTe photoconductive detectors are discussed by Beyen et al.[13] and by McGlauchlin.[14] In contrast to PbS and PbSe, PbTe is not very sensitive at ambient temperature or at 195°K, but must be cooled to 77°K. The performance of detectors operated at this temperature is listed in Table 10.1.

Lead telluride at 77°K has roughly the same spectral response and time constant as PbSe at that temperature, but its detectivity is much poorer. This may be because production-type PbTe detectors have not been made in the quantities that PbSe detectors have. It has also a much higher resistance, and should be gridded in the low resistance configuration shown in Fig. 10.4 to ease the amplifier impedance requirements. It finds application as does PbSe in missile guidance and airborne mapping systems.

Lead telluride is prepared almost always by vacuum sublimation, the chemical deposition method offering no advantage. As was true for the other lead salts, the effect of oxygen upon the photoconductivity is critical. If the layer is overoxidized, a peak in the spectral response occurs in the vicinity of 1.4 μ, which can have a much higher value of D_λ^* than that at the normal 4 μ peak. This 1.4 μ peak has associated with it a time constant much longer than the normal 25 μsec. Lead telluride detectors properly prepared do not exhibit this peak.

The spectral response of PbTe at 77°K is shown in Fig. 10.13. In the region in which it exhibits its best performance, many detectors are seen to surpass it. Figures 10.17 and 10.21 depict the frequency response and noise spectrum. The noise spectrum is of interest because of the exhibition of an almost exact $1/f$ power law dependence.

10.4 GERMANIUM (Ge)

We shall now consider detectors based upon impurity excitation in germanium and germanium-silicon alloys. Neither Ge nor Si is used as an intrinsic infrared detector because their wide band gaps, 0.67 ev and 1.12 ev, indicate cutoff wavelengths of about 1.85 μ and 1.10 μ, respectively. Since operation in this region is admirably accomplished by PbS, intrinsic Ge and Si infrared detectors are not used. However, Ge and Si photo-diodes have found application as detectors for visible radiation. Extrinsic Ge detectors operating at 77°K and below are excellent detectors, having responses extending much farther into the infrared. Although extrinsic Si detectors have not appeared, Ge-Si alloys have been investigated to some extent.

We have seen in Table 6.2 that many impurities soluble in Ge have shallow donor or acceptor levels. Of particular interest in Ge are the gold levels. Dunlap[15] investigated extensively the impurity states due to Au in Ge and found one donor level 0.05 ev above the valence band, and two acceptor levels, one 0.15 ev above the valence band and one 0.20 ev below the conduction band. Woodbury and Tyler[16] reported the existence of a third acceptor level lying 0.04 ev below the conduction band. Morton, Hahn, and Schultz[17] have investigated the effects of compensating the gold with arsenic. Arsenic contributes electrons to the lattice which fill the Au levels and change the crystal from p-type to n-type, the electrons arising from As donors compensating first of all the 0.15 ev Au level, then the 0.20 ev Au level, and finally the 0.04 ev Au level. Johnson and Levinstein[18] discuss in detail the optical properties of Au in Ge.

10.4.1 Gold-Doped Germanium (Ge:Au), p-Type

Photoconductive detectors made from gold-doped germanium, uncom-pensated or compensated with antimony, have been reported by Lasser,

Cholet, and Wurst,[19] by Beyen et al.,[13] and by Levinstein.[20] In order to obtain the long wavelength response associated with the Au levels, it is necessary to cool the detectors to 77°K. In the uncompensated Ge:Au detector, photons of wavelength less than or equal to about 9 μ free holes from the 0.15 ev Au level. The photoconductivity is therefore p-type. Levinstein[20] states the solubility of Au in Ge to be less than 10^{16} atoms/cm^3, causing less than 10% of the incident radiation falling on a detector element of normal dimensions to be absorbed. Gold-doped germanium detectors use elements in the form of long thin rods mounted in the integrating chamber type of enclosure shown in Fig. 10.6b.

The spectral response of Ge:Au at 77°K is listed in Table 10.1 and shown in Fig. 10.13. At about 1.5 μ the detectivity reaches a peak due to intrinsic excitation of Ge. In many applications this is undesirable and must be avoided by using a Ge filter in front of the detector. A secondary broad maximum having a D_λ^* value of 1.75 × 10^{10} cm cps$^{1/2}$/watt occurs at about 5 μ due to photoexcitation of the Au levels. The value of D_λ^* diminishes slowly with increasing wavelength to beyond 8 μ.

Gold-doped germanium detectors can be made to be background limited by cooling to about 65°K, the temperature obtained by using a vacuum forepump to reduce the vapor pressure above liquid nitrogen. The spectral response at this temperature is shown in Fig. 10.14. The maximum value of D_λ^* is thus increased to more than twice that obtained at 77°K.

The frequency responses of Ge:Au at 77°K and 65°K are shown in Figs. 10.17 and 10.18. The response is flat at both temperatures to beyond 20 kc. Levinstein[20] states the time constant, which is inversely proportional to the number of singly charged Au ions, is of the order of 0.1 μsec. The noise spectra are shown in Figs. 10.21 and 10.22. Current noise dominates below about 40 cps, going over to gr noise at higher frequencies.

10.4.2 Gold, Antimony-Doped Germanium, (Ge:Au,Sb) n-Type

By adding antimony and gold in the proper proportions to germanium, it is possible to produce an n-type detector having a shorter cutoff wavelength, a longer lifetime, and a better detectivity at wavelengths less than 3.5 μ than that found in Ge:Au. The addition of two Sb atoms for every Au atom, producing a donor level sufficiently shallow to be thermally ionized at 77°K, compensates (fills with electrons) the 0.15 ev and 0.20 ev Au acceptor levels. Electrons may then be excited from the 0.20 ev level to the conduction band. Therefore, n-type photoconductivity found in Ge:Au,Sb requires photons of energy greater than 0.20 ev or wavelength shorter than about 6 μ.

The requirements for proper compensation make preparation of Ge:Au,

Sb elements more difficult than simple Ge:Au. On the one hand, if insufficient Sb is added, the number of compensated atoms at the 0.20 ev level will be reduced, with a corresponding reduction in detectivity. On the other hand, if too much Sb is added, excess electrons will be donated to the lattice and the accompanying decrease in resistivity will reduce the detectivity.

The value of D_λ* for Ge:Au,Sb detectors operated at 77°K is listed in Table 10.1 and shown in Fig. 10.13 to be 2.5 × 10^{10} cm cps$^{1/2}$/watt at 3 μ, slightly less than twice that of Ge:Au at the same wavelength. Lasser, Cholet, and Wurst[19] indicate the increased detectivity to be due to two factors: a reduction in the number of thermally excited carriers due to the requirement for 0.20 ev excitation rather than 0.15 ev excitation, and an increase in the lifetime of the excited carriers. The latter factor is the result of the decrease in the number of available recombination sites. In Ge:Au, the photoexcited holes recombine at singly charged Au sites. In Ge:Au,Sb the addition of antimony reduces the number of singly charged Au sites, thereby increasing the time constant. Although this is about 110 μsec, Levinstein[20] states that overcompensation can produce detectors with time constants in the millisecond range. The situation is complicated by the existence of a time constant of less than 1 μsec for excitation in the intrinsic region.

The frequency response and noise spectrum of Ge: Au,Sb at 77°K are shown in Figs. 10.17 and 10.21. Because of its long time constant, the frequency response is inferior to that of Ge: Au. The noise spectrum indicates that current noise is dominant in the frequency range indicated. Lasser, Cholet, and Wurst[19] indicate that by special treatment of the contacts and surface, the noise power at low frequencies can be reduced.

10.4.3   Zinc-Doped Germanium (Ge:Zn) "Zip"

Burstein et al.[21] have discussed the photoconductive response at 4.2°K of germanium doped with various donor and acceptor impurities. Of particular interest is zinc-doped germanium, referred to as "Zip" (zinc impurity photoconductor). Addition of Zn to Ge gives rise to an acceptor level 0.033 ev above the valence band. Photons of energy greater than 0.033 ev, that is, wavelength shorter than about 40 μ, produce free holes. In order to prevent thermal excitation, the Ge must be maintained at liquid helium temperature in a dewar enclosure of the type shown in Fig. 10.7. This does not lend itself to a very compact arrangement.

The response of Ge:Zn is listed in Table 10.1 and depicted in Fig. 10.14. At the wavelength of peak response, 36 μ, the value of D_λ* is 1 × 10^{10} cm cps$^{1/2}$/watt. The value in the near infrared beyond the intrinsic absorption edge is seen to be an order of magnitude less.

The frequency response and noise spectrum of Ge: Zn are shown in Figs. 10.18 and 10.22. The time constant has been found to be less than 0.01 μsec, indicating a response independent of frequency in the region shown. The noise spectrum shows the detector to be limited by current noise.

10.4.4 Zinc, Antimony-Doped Germanium (Ge:Zn,Sb)

As with gold-doped germanium, zinc-doped germanium compensated with an n-type impurity has a shorter cutoff wavelength than the uncompensated detector. The addition of Sb to Ge:Zn gives rise to a detector having peak response in the region of the atmospheric window between 8 and 14 μ. Whereas zinc-doped germanium owes its response to ionization of a Zn level close to the valence band, the addition of Sb compensates this level, leaving a higher one available for excitation to the valence band. To prevent thermal ionization of the impurity levels, the Ge:Zn,Sb detector must be operated at 50°K or lower.

The value of D_λ^* of zinc, antimony-doped germanium is given in Table 10.1 and depicted in Fig. 10.14. The detectivity at the 12 μ peak wavelength is seen to be 3×10^9 cm cps$^{1/2}$/watt. No published information is available on the frequency response or noise spectrum.

10.4.5 Copper-Doped Germanium (Ge:Cu)

A copper-doped germanium photoconductor operating at or below the temperature of liquid hydrogen has been reported.[22,23] This detector operates on the lowest Cu level, located 0.041 ev above the valence band. (See Table 6.2.) Thus it has a long wavelength limit of roughly 30 μ, with a peak at approximately 20 μ. The detector incorporates a double dewar construction similar to that shown in Fig. 10.7, the outer dewar containing liquid nitrogen and the inner either liquid hydrogen or liquid helium. Since the detector becomes background limited several degrees below the hydrogen point, it operates best with liquid helium. The performance of the detector is listed in Table 10.1, with the spectral D_λ^* value shown in Fig. 10.14. The peak D_λ^* value at 20 μ is seen to be 2.5×10^{10} cm cps$^{1/2}$/ watt with a 60° field of view. The frequency response and noise spectrum are shown in Figs. 10.18 and 10.22. The detector appears current noise limited below 1 kc and gr noise limited above.

10.4.6 Cadmium-Doped Germanium (Ge:Cd)

A cadmium-doped germanium photoconductor has been developed which has roughly the same performance as copper-doped germanium but becomes background limited a few degrees above the hydrogen point.[22] This represents an advantage in that a two-stage nitrogen-hydrogen mini-cooler may be employed rather than requiring liquid helium. The preparation of the Ge:Cd detector, however, is more difficult than that of the Ge:Cu detector owing to the difficulty of incorporating Cd into the Ge lattice. The characteristics of the Ge:Cd detector are listed in Table 10.1 and shown in Figs. 10.14, 10.18, and 10.22. Two peaks in D_λ^* occur, at 16 μ and 20 μ. The value of D_λ^* at 16 μ is 1.8 × 10^{10} cm cps$^{1/2}$/watt, using a 60° field of view. The detector appears to be current noise limited below 500 cps and *gr* noise limited above.

10.4.7   Germanium-Silicon Alloys (Ge-Si:Au; Ge-Si:Zn,Sb)

Johnson and Christian[24] have shown that it is possible to obtain semi-conductors with a range of forbidden band energies by preparing alloys of Ge and Si of varying composition. The band gaps formed lie between those of Ge and Si, the values being dependent upon the molar composition. At low Si concentrations, below about 15%, the band gap increases rapidly with Si concentration. Above this value the band gap increases less rapidly with the Si concentration. The ionization energies of impurities in the Ge-Si alloy also increase as the Si concentration is increased.

Two types of extrinsic photoconductive detectors made from Ge-Si alloys have been described by Morton, Schultz, and Harty:[25] the gold-doped alloy and the zinc, antimony-doped alloy. The gold-doped p-type detector uses an alloy of 89.1% Ge and 10.9% Si. As with Au in Ge, Au in the alloy has several excitation levels. The incorporation of Si with Ge serves to increase the ionization energies of the levels. Although the lowest Au level in Ge has a long wavelength limit of approximately 25 μ, the same level in the alloy containing 10.9% Si has a long wavelength limit of about 14 μ. Thus the gold-doped detector is useful for operation in the 8 to 14 μ atmospheric window.

The spectral response of the gold-doped alloy photoconductor is listed in Table 10.1 and shown in Fig. 10.14. The peak D_λ^* value at 7.3 μ is seen to be 7 × 10^9 cm cps$^{1/2}$/watt, with a broad response extending to about 15 μ. The normal operating temperature is 50°K, obtained by pumping over a liquid oxygen-liquid nitrogen mixture. At lower temperatures the D_λ^* value improves. The response time of the detector is less than 1 μsec. Because of its high resistance, 10 megohms per square, the detector may be limited by its RC time constant rather than the material lifetime. Since the detector capacitance is of the order of 10 pf, the RC time constant is about 100 μsec, a value considerably greater than the photoconductive lifetime. However, use of a load resistor of 1 megohm will reduce the RC time constant to about 10 μsec. The frequency response shown in Fig. 10.18 assumes this has been done. The noise mechanism is generation-recombination, having the flat spectrum indicated in Fig. 10.22.

The second type of Ge-Si alloy photoconductor makes use of a Zn acceptor level. Zinc has two acceptor levels in the alloy. By compensating the first (lowest) level with Sb, the second becomes available for excitation. As with the gold-doped alloy, so also the zinc, antimony-doped alloy has its maximum response in the 8 to 14 μ atmospheric window. The long wavelength limit corresponding to the second Zn level in the alloy composed of 95.5% Ge and 4.5% Si is also about 14 μ. However, the peak response occurs at 10 μ rather than at 7.3. (See Table 10.1 and Fig. 10.14.) In addition, the maximum D_λ^* value of 1×10^{10} cm cps$^{1/2}$/watt is roughly one and one-half times that of the gold-doped alloy. The response time of the Zn, Sb detector is also less than 1 μsec. The detector resistance at the normal operating temperature of 50°K is about 20 megohms per square. Therefore, the frequency response of the zinc, antimony-doped detector will also be limited by the RC time constant. The frequency response characteristic shown in Fig. 10.18 assumes the detector is used with a load resistance small enough to minimize this effect. The noise mechanism, as with the gold-doped alloy, is generation-recombination noise, having the white spectrum shown in Fig. 10.22.

10.5 INDIUM ANTIMONIDE (InSb)

Indium antimonide is a compound semiconductor formed by melting together stoichiometric amounts of indium and antimony. It is one of a group of semiconductors known as the intermetallics, the galvanomagnetic properties of which have been reviewed by Welker and Weiss,[26] and the optical properties by Frederikse and Blunt.[27] The width of the forbidden band in InSb, listed in Table 6.1, is 0.18 ev at room temperature, corresponding to a cutoff wavelength of about 7 μ. On cooling InSb to 77°K, the band width increases to 0.23 ev, giving rise to a cutoff wavelength of about 5.5 μ. This is accompanied by an increase in the already large electron mobility from a value of 60,000 cm^2/volt sec at 295°K to 300,000 cm^2/volt sec at 77°K.

Detectors have been made from InSb based upon the photoconductive effect,[28] the photovoltaic effect,[29] and the photoelectromagnetic effect.[30] Rieke, DeVaux, and Tuzzolino[31] have discussed detectors based upon all three modes.

10.5.1 Photoconductive Mode

Indium antimonide photoconductive detectors[28] have been made for operation at 295°K, 195°K, and 77°K. The room temperature detector, in the form of a long thin strip of thickness about 10 μ, finds its greatest application in infrared spectroscopy. Although operated in the photoconductive mode, the limiting noise has been found by Moss[32] to be thermal noise. Since the detector resistance is of the order of 20 ohms per square, special low noise amplifier techniques are required involving transformer coupling to the detector. These difficulties are decreased if the detector is operated at 195°K or 77°K, the resistance at the latter temperature being 10 kohms.

The detectivities of the uncooled and cooled detectors are given in Table 10.1 and Figs. 10.11a, 10.12, and 10.13. The room temperature D_λ* is seen to be poor, about 4×10^7 cm cps$^{1/2}$/watt at 6.5 μ. However, the peak detectivity improves by nearly two orders of magnitude upon cooling to 195°K, where the peak wavelength shifts to 5.0 μ. Upon cooling to 77°K, D_λ* becomes even better, attaining a value of 6×10^{10} cm cps$^{1/2}$/ watt at 5.0 μ for a 60° field of view.

The frequency response of uncooled and cooled photoconductive InSb detectors are depicted in Figs. 10.15, 10.16, and 10.17. Since the lifetime is less than 2 μsec when cooled and is about 0.2 μsec at 295°K, the response is frequency independent over most of the region shown. The noise spectra are shown in Figs. 10.19, 10.20, and 10.21. Although thermal noise limits the detector at room temperature, the detector becomes current noise limited at frequencies below 400 cps at 195°K, and is current noise limited up to at least 10 kc at 77°K.

10.5.2 Photovoltaic Mode

Indium antimonide photovoltaic detectors,[29] operating at 77°K, achieve excellent detectivity in the 4 to 5.5 μ region. Because of the small energy gap of InSb, no photovoltaic effect is observed at 295°K. No information is available concerning 195°K photovoltaic operation. Detectors operating at 77°K have been made from grown junctions, which, as seen in Chapter 8, have narrow sensitive areas, and from diffused junctions, which have broad areas. The latter have been found to be more useful in most infrared applications, with the exception of spectroscopy. Early difficulties were encountered in achieving a uniform sensitivity over the junction, the highest sensitivity arising at the edges of the diffused region. However, this has largely been overcome. The detectors may be operated in either the unbiased photovoltaic mode or the reverse biased photodiode mode. The detector resistance of about 1 kohm at 77°K is well suited for transistor amplifiers.

The value of D_λ* for the photovoltaic detector operating at 77°K is shown in Fig. 10.13 and listed in Table 10.1. The peak detectivity of 4.3×10^{10} cm cps$^{1/2}$/watt occurs at 5.3 μ. The time constant is less than 1 μsec, indicating a flat frequency response as shown in Fig. 10.17. The noise spectrum, shown in Fig. 10.21, indicates current noise limitations below about 100 cps, going over to *gr* noise at higher frequencies.

10.5.3 Photoelectromagnetic Mode

Indium antimonide operated in the PEM mode[30] at room temperature offers excellent performance in the 5 to 7 μ interval. Difficulties associated with cooling the rather large thermal mass associated with the detector magnet militate against operating the detector at 195°K or 77°K, although cooling a few degrees below ambient temperature is possible. The construction of the detector is as shown in Fig. 10.8. Although it is rather bulky, having a weight of about 100 grams, its better detectivity and unbiased operation offer advantages over the uncooled photoconductive mode. As with the uncooled photoconductor, it has the disadvantage of low resistance and requires transformer coupling to the amplifier for optimum performance.

The value of D_λ^* is listed in Table 10.1 and shown in Fig. 10.11. The peak value of 3×10^8 cm cps$^{1/2}$/watt occurs at 6.2 μ. This relatively long wavelength response provides utility as a passive detector of radiation from ambient sources, making it of use, for example, in radiometers. The frequency response, shown in Fig. 10.15, is flat over the region indicated, the response time having been found to be no greater than 0.2 μsec. The limiting noise is thermal noise, having the frequency independent spectrum shown in Fig. 10.19. Thus both signal and noise are flat to far beyond 20 kc, an attractive feature in wide band applications.

10.6 INDIUM ARSENIDE (InAs)

Indium arsenide is another intermetallic compound semiconductor, the properties of which bear some resemblance to those of InSb. Unlike InSb, however, reaction of the elements to form the compound is not a simple matter. To keep the arsenic from disappearing because of its high vapor pressure near the melting point, it is necessary to react the constituents in a sealed quartz or Vycor vessel, within which the pressure rises to many atmospheres. Purification of InAs is also more difficult than InSb.

The width of the forbidden gap in InAs is about 0.35 ev, giving rise to a cutoff wavelength of about 3.6 μ, showing some dependence upon the mode of operation. Photocells have been made from InAs to operate in the photoconductive, photovoltaic, and PEM modes. Hilsum[33] describes operation at 295°K in the photoconductive and photovoltaic modes. Talley and Enright[34] describe operation in the photovoltaic mode at 77°K. Cholet[35] discusses operation of a sapphire-immersed photovoltaic detector operating at 295°K.

The spectral detectivity of InAs operated at 295°K in all three modes is shown in Fig. 10.11a. Although the PEM and photoconductive modes appear similar in magnitude, the photovoltaic detector appears to be more than an order of magnitude better. The information concerning the photovoltaic mode, which came from a different source than that for the other two, represents a detector at a more advanced stage of development.

The photovoltaic InAs detector is being developed as a possible successor to PbS. The discussion in Section 10.2.1 pointed out the advantages a single crystal detector (InAs) has over a film-type detector (PbS). Although the value of D_λ^* at the peak is less for InAs than for PbS, the spectral response of InAs extends somewhat farther into the infrared, making it more useful than PbS in some applications. InAs, competing with uncooled PbSe also, appears superior to it. The response time of InAs is less than 2 μsec, giving rise to the flat frequency response shown in Fig. 10.15. The noise spectrum, shown in Fig. 10.19, is white.

10.7 TELLURIUM (Te)

Tellurium is an elemental semiconductor which has found use as an intrinsic photoconductive detector. The band gap of 0.33 ev enables it to respond to radiation of wavelength less than about 3.8 μ. Since its response is quite sharply peaked in this region, it has the advantage of operating in the minimum of the atmospheric radiant background arising from the sun and earth.

The construction and performance of tellurium photoconductive detectors made from crystals prepared from the vapor phase have been described by Suits and Rice.[36] Detectors based upon crystals grown by the Czochralski method have been made by Butter and McGlauchlin.[37] Both types require cooling to 77°K for best response. Typical detector resistance at this temperature is 2000 ohms, making it ideal for direct coupling to transistor amplifiers. The spectral response of the pulled crystal type is given in Fig. 10.13, the peak D_λ^* value of 6 × 10^{10} cm cps$^{1/2}$/ watt occurring at 3.5 μ. The frequency response and noise spectrum are shown in Figs. 10.17 and 10.21. The response time of 60 μsec causes its signal response to be flat to about 1 kc. The noise spectrum shows the detector to be current noise limited.

10.8 THALLOUS SULFIDE (Tl$_2$S)

Thallous sulfide was the first infrared photoconductor to receive serious attention. Case[38] in 1917 first discovered its infrared-detecting properties, referring to the detector as the "thalofide" cell. Serious development of it as a sensitive infrared detector began with the work of Cashman[39] in the United States during World War II. It has been applied principally as a detector in infrared communication systems, using modulated high intensity discharge tube sources. Since some discharge sources emit considerable radiation in the vicinity of 1 μ, where the detector exhibits an extremely high detectivity, they form an efficient communications link. The lack of response of Tl_2S beyond about 1.3 μ renders it of little use in passive detection systems. It is not readily available.

Thallous sulfide is prepared in a manner similar to the vacuum deposition method used for the lead salt compounds. Performance of the deposited film, composed of an array of crystallites, is highly dependent upon the amount of oxygen incorporated. The detector is used in the uncooled photoconductive mode. The spectral response, shown in Fig. 10.11a, exhibits a peak D_λ^* value of 2.2 × 10^{12} cm cps$^{1/2}$/watt at 0.9 μ, considerably higher than that of any other infrared detector. The time constant of 530 μsec gives rise to a frequency response flat below approximately 100 cps as shown in Fig. 10.15. The noise spectrum has not been reported in detail, but Cashman[39] indicates the detector is current noise limited.

10.9   MERCURY TELLURIDE-CADMIUM TELLURIDE (HgTe-CdTe)

Mixed crystals of the compounds mercury telluride and cadmium telluride are semiconductors having forbidden gap widths varying according to composition of the crystals.[40] Pure HgTe has a gap width less than 0.03 ev, indicating a long wavelength limit of greater than 40 μ, whereas pure CdTe has a gap width of 1.5 ev corresponding to a long wavelength limit of 0.8 μ. By forming the two compounds from the elements, then melting them together in a sealed vessel, mixed crystals can be formed of composition varying from pure CdTe at one extreme to pure HgTe at the other. Thus it is possible to form a series of intrinsic semiconductors of long wavelength limits lying at values between 0.8 and 40 μ. For example, detectors made from a crystal composed of 90% HgTe and 10% CdTe have a 13 μ long wavelength limit. Infrared detectors made from such material would have the advantages over detectors made from extrinsic material indicated in Section 10.2.2.

Lawson et al.[40] reported that photoconductive, photovoltaic, and PEM effects were explored in the mixed crystals. A photoconductive detector made from 86% HgTe and 14% CdTe had a D_λ^* value of 1.5 × 10^7 cm cps$^{1/2}$/watt at its 6 μ spectral peak. Such poor performance is indicative of the early stages of development. Because of the lack of information, the spectral response, frequency response, and noise spectrum are not shown. Some data are listed in Table 10.1.

10.10   THERMISTOR BOLOMETER

We have discussed in Section 8.4.1 some of the properties of metal and thermistor bolometers. Since metal bolometers have found little application in infrared systems, we shall discuss here only the thermistor bolometer. The latter has found wide application in the detection of infrared from low temperature objects, two of the principal applications being in radiometers and in thermal imaging systems. A more recent application is found in horizon scanning to establish vertical reference systems in satellites. Because of its rather long time constant, the elemental bolometer is of little use in missile seekers or trackers, which require fast response.

The construction and properties of thermistor bolometers have been discussed by Wormser,[41] and by DeWaard and Wormser.[42] Detectors can be constructed with a wide selection of sensitive areas and a variety of response times in the millisecond range. Since the detectivity is proportional to the square root of the response time, the user should select the detector having the longest response time consistent with his frequency response requirements to obtain the highest possible detectivity. In wide band applications, however, the user will frequently select a photon detector with the proper spectral response and generally faster speed of response. We have seen that many photon detectors have peak detectivities superior to that of the bolometer.

Although the bolometer is usually mounted in a simple enclosure utilizing an infrared transmitting window as shown in Fig. 10.9, thermistor bolometers have been made exhibiting increased performance through use of immersion optics. Mosaic-type detectors have also been made which find application in thermal imaging systems. DeWaard and Wormser[42] describe construction of selective wavelength detectors which respond to only a small spectral interval by utilizing overcoating materials exhibiting selective absorption.

Table 10.1 lists data pertaining to the thermistor bolometer. The spectral response of a bolometer with a 1.5 msec response time is depicted in Fig. 10.11a. The D^* value of 1.95×10^8 cm cps$^{1/2}$/watt is poorer than that obtained by other uncooled photon detectors at wavelengths shorter than 7μ but superior at longer wavelengths. The frequency response corresponding to a 1.5 msec response time is shown in Fig. 10.15, whereas the noise spectrum is found in Fig. 10.19. Since most bolometers are limited by thermal noise, the spectrum is frequency independent.

10.11   SUPERCONDUCTING BOLOMETER

The superconducting bolometer, described in Section 8.4.2, is a laboratory instrument not available commercially. Although bolometers made from a variety of superconductors have been developed, none has appeared sufficiently reproducible, reliable, and simple to operate to warrant widespread application. In addition, many problems arise because of the precise temperature control needed to maintain the bolometer at the superconducting transition temperature. Since it would have to compete with photon detectors in other regions of the spectrum, the superconducting bolometer would appear to have its greatest utility as a detector in the 50 μ to 1 mm range. For the purpose of comparison, we have included in Table 10.1 and Figs. 10.14 and 10.18 data on the Andrews[43] niobium nitride superconducting bolometer operated at 15°K. The value of D_λ^* is 4.8 × 10^9 cm cps$^{1/2}$/watt. The very low resistance, less than 1 ohm, introduces the requirement for transformer coupling to the amplifier, with the transformer also cooled to reduce the thermal noise contribution from its windings. The response time, 500 μsec, is short for thermal detectors, but much longer than that of most photon detectors. The noise mechanism appears to be that of some unknown effect associated with superconductivity.

10.12 CARBON BOLOMETER

A bolometer made from a carbon composition resistor, operating at 2.1°K, has been described by Boyle and Rodgers.[44] Although such a bolometer is not available commercially, we include it because of its unusually good performance. The spectral response, frequency response, and noise spectrum are shown in Figs. 10.14, 10.18, and 10.22, respectively.

The bolometer was constructed by slicing the core of a carbon composition resistor into a slab 48 μ in thickness and having 19 mm^2 surface area. It was mounted in a double dewar enclosure having a liquid air jacket surrounding a liquid helium bath. The temperature was reduced below the normal boiling point of helium by pumping on the helium with a vacuum pump. The bottom of the dewar had three aligned infrared transmitting windows, the outer one being quartz at ambient temperature, the second being paraffin at liquid air temperature, and the third being quartz at liquid helium temperature. These served as a filter which removed all radiation of wavelengths less than 40 μ.

The bolometer showed a very high responsivity, of the order of 10^4 v/watt, and a response time of 10 msec. Current noise, coming from either the carbon or the contacts, limited the detector. The detectivity obtained, 4.25 × 10^{10} cm cps$^{1/2}$/watt, exceeds the photon noise value for a cooled thermal detector looking through 2π steradians at a room temperature ambient. However, the field of view was limited to a narrow cone in the forward direction, and the filtering action of the paraffin and quartz cut out most of the photon noise from sources at ambient temperature.

10.13 RADIATION THERMOCOUPLE

The radiation thermocouple, always operated uncooled, is a thermal detector which finds widespread commercial use. The fragile construction makes it of little use in applications where it would be subject to vibration and shock. It is the basic detector for use in infrared spectrometers, although photon detectors are finding use also in this application. Radiation thermocouples for spectroscopic applications are designed so that the dimensions of the blackened receiver adjacent to the junction match those of the exit slit of the spectrometer. These thermocouples are designed for high responsivity and have low thermal mass for fast response. The low resistance, of the order of 5 ohms, requires transformer coupling to the amplifier input so that the thermal noise of the thermocouple will become the factor limiting the detectivity of the system. Commercially available thermocouples are made of either thin wire junctions or evaporated film junctions. DeWaard and Wormser[42] discuss the properties of commercially available thermocouples. Smith, Jones, and Chasmar[45] review in detail the theory of operation.

The value of $D_\lambda{}^*$ for a well-constructed radiation thermocouple, listed in Fig. 10.11a and Table 10.1, is 1.4×10^9 cm cps$^{1/2}$/watt. Thus it is about an order of magnitude from the photon noise limit for thermal detectors. The response time of 36 msec gives rise to the response shown in Fig. 10.15, which is frequency dependent even below 10 cps. Since thermocouples are thermal noise limited, the noise spectrum is white, shown in Fig. 10.19.

10.14 GOLAY CELL

Although the Golay cell,[46] described in Section 8.4.4, is a very sensitive uncooled thermal detector which is available commercially, it has found little use in military systems because of its fragile construction and slow response. Its main application appears in infrared spectroscopy, where laboratory conditions impose less stringent requirements for ruggedness and speed of response. Data on the Golay cell have been included in Table 10.1 and Figs. 10.11a, 10.15, and 10.19 for purposes of comparison with more useful detectors. As is shown, it approaches within an order of magnitude of the photon noise limit of thermal detectors, having a $D_\lambda{}^*$ of 1.67×10^9 cm cps$^{1/2}$/watt. The very slow response time of 20 msec indicates it should be used at frequencies below 10 cps. Performance is limited by temperature noise, having the frequency independent spectrum shown in Fig. 10.19.

10.15 DETECTORS FOR THE VISIBLE SPECTRUM

In this section data are presented on six detectors which respond in the visible spectrum but show little response to infrared radiation. All six operate uncooled. Three of them, cadmium sulfide, cadmium selenide, and the 1N2175 silicon photo-duo-diode, operate in the photoconductive mode; two of them, selenium-selenium oxide and gallium arsenide, are photovoltaic detectors; and the sixth, the 1P21 photomultiplier, utilizes the photoemissive effect. Data including detectivity, frequency response, and noise spectra are shown in Table 10.1 and Figs. 10.11b, 10.15, and 10.19.

10.15.1 Cadmium Sulfide (CdS)

Cadmium sulfide is a compound semiconductor which has found widespread use as a compact and inexpensive detector of visible radiation. It exhibits the greatest change in resistance upon illumination of any known photoconductor, up to six orders of magnitude change between the dark and illuminated condition. Its dark resistance is of the order of 5×10^{11} ohms per square. In order to reduce the dark and illuminated resistances to a more useful range, a grid configuration is sometimes used. Its energy band gap is 2.4 ev, indicating a cutoff wavelength of about 0.5 μ.

Bube[47] lists the following five general methods for the preparation of photoconductive CdS: single crystals prepared by vapor deposition, photoconductive powder, sintered layers and sintered pellets, evaporated layers, and vapor deposited layers. The most widely used method is that of the sintered layer, High purity CdS powder, mixed with small amounts of copper and chlorine impurities, is spread on a ceramic substrate and fired at an elevated temperature. The resulting sintered layer, incorporating the impurities within the CdS lattice, adheres to the substrate.

The response of CdS is slow compared with that of most of the other photon detectors. This long lifetime is due to deep trapping centers within the forbidden band. The decay time in response to a radiation pulse is usually longer than the rise time. Hysteresis effects following long periods in the dark are observable.

Although the relative response and frequency response are widely reproduced, little data are available on the spectral detectivity or noise spectra. The data presented here are from an Eisenman and Cussen NOLC report.[48] These reports from Naval Ordnance Laboratories, Corona, California, are excellent sources of information on detectors. Most of them are classified. The unclassified report referred to, the data of which are reproduced in Figs. 10.11b, 10.15, and 10.19, shows a detector having a peak D_λ^* of 3.5×10^{14} cm cps$^{1/2}$/watt at 0.5 μ. The response time is 53 msec. The detector is limited by current noise.

10.15.2 Cadmium Selenide (CdSe)

Cadmium selenide is a room temperature photoconductor bearing much similarity to CdS. Because its energy gap is less than that of CdS, 1.8 ev instead of 2.4 ev, its spectral response extends to longer wavelengths, into the near infrared. The peak detectivity of 2.1×10^{11} cm cps$^{1/2}$/watt occurs at 0.7 μ. The change in resistance upon illumination is less than that found in CdS, and the response time, about 12 msec, is shorter. It is also current noise limited. The data concerning detectivity, frequency response, and noise spectra, obtained from Eisenman and Cussen,[48] are presented in Table 10.1 and Figs. 10.11b, 10.15, and 10.19.

10.15.3 Selenium-Selenium Oxide (Se-SeO)

Selenium-selenium oxide photovoltaic photocells, sometimes referred to as "barrier layer" photocells, are characterized by a large output voltage per watt of radiant power. Thus they are used to operate relays directly, or devices such as automatic diaphragm controls for cameras without amplification. When used with a visual correcting filter such as the Kodak Wratten 61, the response approximates the visibility curve, that is, the spectral response of the human eye. They are used therefore as the sensing element in photometers. Table 10.1 and Figs. 10.11b and 10.15 present data from Eisenman and Cussen.[48] The peak D_λ^* value of 1.2×10^{11} cm cps$^{1/2}$/watt at 0.55 μ is far below that of CdS. The response time is about 1 msec. The noise spectrum is unknown.

10.15.4 Gallium Arsenide (GaAs)

Gallium arsenide is a compound semiconductor which has been developed for star-tracking applications. It operates in the photovoltaic mode at room temperature. The energy band gap value of 1.45 ev indicates a spectral response extending to approximately 0.85 μ. Since GaAs decomposes at its melting point, it must be prepared in a sealed container in a manner similar to that used for InAs. Table 10.1 and Figs. 10.11b, 10.15, and 10.19 summarize data reported by Cholet, Slawek, and Repper.[49] The value of D_λ^* at the 0.8 μ peak is 4.5×10^{11} cm cps$^{1/2}$/watt. The response time is about 1 msec. The detector is current noise limited. Immersion techniques have been applied successfully.

10.15.5 1N2175 Silicon Photo-Duo-Diode

Although several germanium and silicon photodiodes are available, we shall discuss only one type. The 1N2175 Si photo-duo-diode[50] is an *n-p-n* diffused junction detector which is operated in the photoconductive mode. The detector consists of two identical junctions arranged symmetrically so that the bias can be applied in either direction. The over-all dimensions have been kept small so that many can be arranged in a small volume for applications such as reading punched cards and tapes. Table 10.1 and Figs. 10.11*b*, and 10.15 give its detectivity and frequency response. The peak D_λ^* value of 2.5×10^{10} cm cps$^{1/2}$/watt occurs at 0.95 μ.[51] The frequency response is flat to about 5 kc. The noise spectrum is unknown.

10.15.6   1P21 Photomultiplier

Although the photomultiplier is available in a variety of configurations with several types of photoemissive surface, only the model 1P21 will be considered here. It uses a cesium-antimony photoemissive surface, referred to as an S-4 surface,[52] which exhibits its maximum response at 0.4 μ. Jones[53] has reviewed the performance of several photomultipliers. Table 10.1 and Figs. 10.11*b*, 10.15, and 10.19 give data on the detectivity, frequency response, and noise spectrum. The frequency response, limited by the transit time spread of the electrons emitted by the multiplier dynodes, extends to about 100 Mc. The photomultiplier is limited by shot noise, having a white spectrum. Thus D_λ^* is frequency independent below about 100 Mc.

10.16   NOTE CONCERNING FIGURES

Where data of the measured frequency response of detectors in Fig. 10.15 through 10.19 are lacking, estimates have been made, based on known response time values. These curves are labeled "estimated."

Where data, of the measured noise spectra for detectors in Fig. 10.15 through 10.19 are lacking, assumptions have been made, based on known noise mechanisms. These curves are labeled "assumed."

Where data, sufficient to make reasonable estimates or assumptions are lacking, the curves are labeled "unknown."

TABLE 10.1. Performance of elemental detectors

Material	Photon or Thermal	Mode of Operation	Film or Single Crystal	N-type, P-type, or Intrinsic	Operating Temperature (Deg. K)	Wavelength of Peak Response $\lambda_p(\mu)$	Cutoff Wavelength (50% value) $\lambda_0(\mu)$	D^* (500° K, f, 1) cm cps$^{1/2}$/watt (Measuring Frequency Indicated)	$D_{\lambda p}^*$ (λ_p, f, 1) cm cps$^{1/2}$/watt (Measuring Frequency Indicated)	Response Time (μsec)	Calculated Optimum Chopping Frequency (cps)	Resistance per Square (ohm)	Noise Mechanism	References
1 PbS	P	PC	F	I	295	2.1	2.5	4.5×10^8 90 cps	1.0×10^{11} 90 cps	250	640	1.47 meg	Current	34
2 PbS	P	PC	F	I	195	2.5	3.0	4.0×10^9 1000 cps	1.7×10^{11} 1000 cps	455	350	4 meg	Current	35
3 PbS	P	PC	F	I	77	2.5	3.3	4.0×10^9 90 cps	8.0×10^{10} 90 cps	455	350	5 meg	Current	8, 35
4 PbSe	P	PC	F	I	295	3.4	4.2	3.0×10^7 90 cps	2.7×10^8 90 cps	4	40 kc	50 kohm	Current	8
5 PbSe	P	PC	F	I	195	4.6	5.4	7.5×10^8 900 cps	6×10^9 900 cps	125	1270	40 meg	Current below 6 kc	9
6 PbSe	P	PC	F	I	77	4.5	5.8	2.2×10^9 90 cps	1.1×10^{10} 90 cps	48	3300	5 meg	Current	8, 36
7 PbTe	P	PC	F	I	77	4.0	5.1	3.8×10^8 90 cps	2.7×10^9 90cps	25	6500	32 meg	Current	7, 8
8 Ge:Au	P	PC	SC	P	77	5.0 (excluding intrinsic peak)	7.1	7.5×10^9 900 cps	1.75×10^{10} 900 cps	<1	Frequency independent above 40 cps	1.0 meg	Current below 40 cps, gr above	2, 7
9 Ge:Au	P	PC	SC	P	65	4.7 (excluding intrinsic peak)	6.9	1.7×10^{10} 900 cps	4×10^{10} 900 cps	<1	Frequency independent above 40 cps		Current below 40 cps, gr above	7
10 Ge:Au,Sb	P	PC	SC	N	77	No clearly defined peak exists except for intrinsic excitation.		2.9×10^9 90 cps	2.5×10^{10} at 3μ 90 cps	110	1500	1.0 meg	Current	8
11 Ge:Zn (Zip)	P	PC	SC	P	4.2	36	39.5	4.0×10^9 800 cps	1.0×10^{10} 800 cps	<0.01		300 kohm	Current	8, 27

420

					50	12	15							
12 Ge:Zn,Sb	P	PC	SC					2×10^9 900 cps	3×10^9 900 cps					1
13 Ge:Cu	P	PC	SC	P	<20	20	27	1×10^{10} (60° field of view) 900 cps	2.5×10^{10} (60° field of view) 900 cps			0.1 meg	Current below 1 kc; gr above 1 kc	5
14 Ge:Cd	P	PC	SC	P	<25	16	21.5	7×10^9 (60° field of view) 500 cps	1.8×10^{10} (60° field of view) 500 cps				Current below 500 cps; gr above 500 cps	4, 5
15 Ge-Si:Au	P	PC	SC	P	50	7.3	10.1	3.1×10^9 90 cps	7.0×10^9 90 cps	0.1	Frequency independent to approximately 1 Mc	10 meg	gr	19
16 Ge-Si:Zn,Sb	P	PC	SC	P	50	10	13.3	4.0×10^9 100 cps	1.0×10^{10} 100 cps	0.1	Frequency independent below approximately 1 Mc	20 meg	gr	19
17 InSb	P	PC	SC	I	295	6.5	7.3	1.4×10^7 800 cps	4.3×10^7 800 cps	0.2	Frequency independent to 500 kc	20	Thermal	29
18 InSb	P	PC	SC	I	195	5.0	6.1	5×10^8 900 cps	2.5×10^9 900 cps	<1	Frequency independent above 500 cps	60	Current below 400 cps	9
19 InSb	P	PC	SC	P	77	5.0	5.4	1.2×10^{10} (60° field of view) 900 cps	6×10^{10} (60° field of view) 900 cps	<2		10 kohm	Current	6
20 InSb	P	PV	SC	PN	77	5.3	5.6	8.6×10^9 900 cps	4.3×10^{10} 900 cps	<1	Frequency independent above 500 cps	1 kohm	Current below 100 cps gr above	2, 3, 30
21 InSb	P	PEM	SC	I	295	6.2	7.0	1.0×10^8 400 cps	3.0×10^8 400 cps	0.2	Frequency independent below 100 kc	20	Thermal	18, 25

TABLE 10.1. (Continued)

Material	Photon or Thermal	Mode of Operation	Film or Single Crystal	N-type, P-type, or Intrinsic	Operating Temperature (Deg. K)	Wavelength of Peak Response $\lambda_p(\mu)$	Cutoff Wavelength (50% value) $\lambda_0(\mu)$	D^* (500°K, f, 1) cm cps$^{1/2}$/watt (Measuring Frequency Indicated)	$D^*\lambda_p$ (λ_p, f, 1) cm cps$^{1/2}$/watt (Measuring Frequency Indicated)	Response Time (μsec)	Calculated Optimum Chopping Frequency (cps)	Resistance per Square (ohm)	Noise Mechanism	References
[22] InAs	P	PC	SC	N	295	3.6	3.8	1.4×10^7 90 cps	1.4×10^8 90 cps	0.2	Frequency independent below 100 kc			12
[23] InAs	P	PV	SC	PN	295	3.4	3.7	2.5×10^8 90 cps	2.5×10^9 750 cps	<2		50	Assumed thermal	23, 24
[24] InAs	P	PEM	SC	N	295	2.5	3.4	1.4×10^7 90 cps	1.4×10^8 90 cps	0.2	Frequency independent below 100 kc		Assumed thermal	12
[25] Te	P	PC	SC	P	77	3.5	3.8	4.0×10^9 900 cps	6.0×10^{10} 900 cps	60	2700	2 kohm	Current	17
[26] Tl$_2$S	P	PC	F	I	295	0.9	1.1		2.2×10^{12} 90 cps	530	300	5 meg	Current	14
[27] 86% HgTe 14% CdTe	P	PC	SC	I	295	6	6.5	5×10^6	1.5×10^7			~1		13
[28] Thermistor Bolometer	T	Bolometer			295			1.95×10^8 (1.5 millisec) 10 cps	1.95×10^8 (1.5 millisec) 10 cps	1500	Frequency independent below 30 cps	2.4 meg	Thermal	21
[29] Radiation Thermocouple	T	Thermoelectric effect			295			1.4×10^9 5 cps	1.4×10^9 5 cps	3.6×10^4	<5	5	Thermal	32
[30] Golay Cell	T	Expansion of air			295			1.67×10^9 10 cps	1.67×10^9 10 cps	2×10^4	<5		Temperature	16
[31] NbN Bolometer	T	Superconducting Bolometer			15			4.8×10^9 360 cps	4.8×10^9 360 cps	500		0.2	Unknown	22

422

Detector	T				2.1			4.25×10^{10} 13 cps	4.25×10^{10} 13 cps	10^4	16	0.12 meg		
32 Carbon Bolometer		Bolometer											Current	20
33 CdS	P	PC	SC F Sintered	N	295	0.5	0.51		3.5×10^{14} 90 cps	5.3×10^4	3	5×10^{11}	Current	10
34 CdSe	P	PC	SC Sintered		295	0.7	0.72		2.1×10^{11} 90 cps	1.2×10^4	13	1.5×10^{11}	Current	10
35 Se-SeO	P	PV	SC	PN	295	0.55	0.69		1.2×10^{11} 90 cps	910	160	3 kohm Area dependent	Current	10
36 GaAs	P	PV	SC	PN	295	0.8	0.89		4.5×10^{11} 400 cps	1000	160	4.6 meg Area dependent	Current	11
37 IN 2175 Photo-duodiode	P	PC	SC	PN	295	0.95	1.07		2.5×10^{10} 400 cps	8	20 kc	4×10^9		28, 33
38 1P21 Photo-multiplier	P	PE	F		295	0.40	0.53		5×10^{14} 1000 cps	<0.01	Frequency independent to about 100 Mc		Shot	15, 31

Notes (Numbers correspond to those in column 1)

1,2,3 Detectors with time constants ranging from about 1 μsec to 10,000 μsec are available. The detectivity will vary with time constant according to the McAllister relation. The cutoff wavelength may also be shifted to greater values with a sacrifice in detectivity. Detectors operating at 77°K may exhibit double time constants.

6 May exhibit double time constant.

7 Resistance may be reduced by grid type electrodes. Detector is background limited. Performance at 90°K same as at 77°K. May have second time constant for 1.5 μ radiation.

8 Exhibits long time constant for intrinsic excitation (less than 2 μ). Detectivity improved by cooling to 65°K. See below.

9 Exhibits long time constant for intrinsic excitation (less than 2μ).

10 Detectivity at 90°K equals that at 77°K. Exhibits wavelength dependent time constant.

12 Not readily available.

15 Spectral response can be changed by varying alloy composition. Frequency response may be limited by RC time constant.

16 Spectral response can be changed by varying alloy composition. Frequency response may be limited by time RC constant.

19 Responsivity is superior to InSb PV, 77°K.

20 May be either broad area diffused junction or line type of grown junction. May be operated with or without bias voltage.

21 Maximum dimensions approx. 2 × 10 mm. Should be transformer coupled to amplifier. Sensitive to magnetic pickup from electrical mains.

22 Not readily available.

23 Detector is sapphire immersed.

24 Not readily available.

25 Peak detectivity in solar and earth background minimum.

26 Not readily available.

27 Not readily available.

28 Detectors with time constants ranging from about 1 msec to 50 msec are available.

29 Widely used in infrared spectroscopy.

30 Fragile, microphonic.

31 Not readily available. Noise appears to arise from some unknown mechanism associated with superconductivity.

32 Not readily available. Use of quartz and paraffin filters cut out response at wavelengths shorter than 40 μ.

33 Highest responsivity of any photoconductor.

34 Responds to longer wavelengths and is faster than CdS.

35 Used in exposure meters.

36 Useful for star tracking.

37 Very small over-all size.

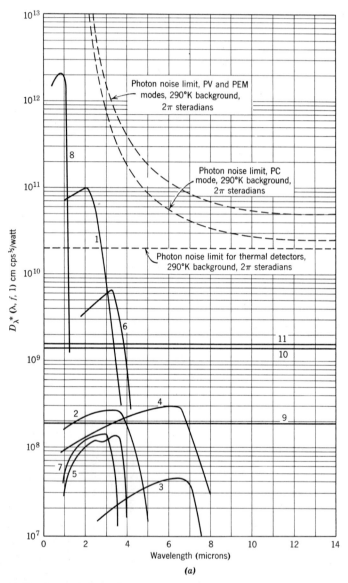

Figure 10.11a. Spectral D_λ^* of room temperature detectors. 1. PbS, PC (250 μsec) (90 cps). 2. PbSe, PC (90 cps). 3. InSb, PC (800 cps). 4. InSb, PEM (400 cps). 5. InAs, PC (90 cps). 6. InAs, PV (frequency unknown) (sapphire immersed). 7. InAs, PEM (90 cps). 8. Tl$_2$S, PC (90 cps). 9. Thermistor bolometer, (1500 μsec) (10 cps). 10. Radiation thermocouple, (36 msec) (5 cps). 11. Golay cell, (20 msec) (10 cps).

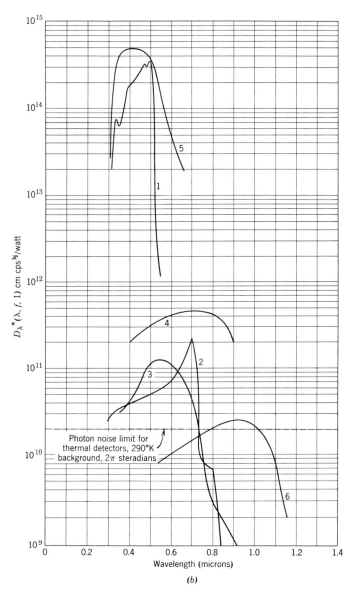

Figure 10.11b. Spectral D_λ^* of room temperature detectors responding in the visible spectrum. 1. CdS, PC (90 cps). 2. CdSe, PC (90 cps). 3. Se-SeO, PV (90 cps). 4. GaAs, PV (90 cps). 5. 1P21 photomultiplier (measuring frequency unknown). 6. 1N2175 Si photo-duo-diode, PV (400 cps).

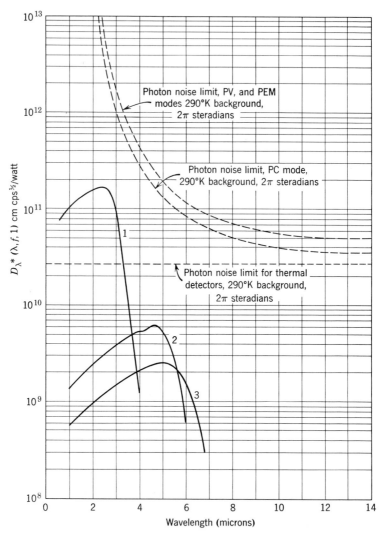

Figure 10.12. Spectral D_λ^* of detectors operating at 195°K. 1. PbS, PC (1000 cps). 2. PbSe, PC (900 cps). 3. InSb, PC (900 cps).

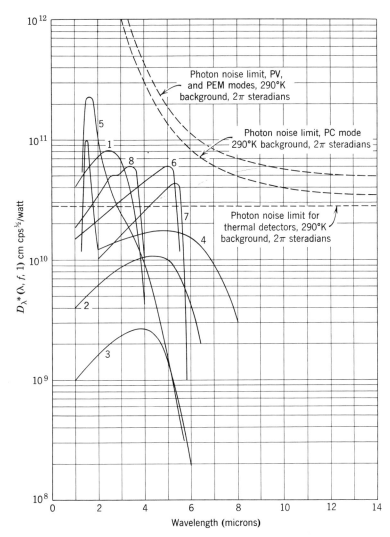

Figure 10.13. Spectral D_λ^* of detectors operating at 77°K. 1. PbS. PC (90 cps). 2. PbSe, PC (90 cps). 3. PbTe, PC (90 cps). 4. Ge:Au, PC (900 cps). 5. Ge:Au,Sb, PC (90 cps). 6. InSb, PC (900 cps) (60° field of view). 7. InSb, PV (900 cps). 8. Te, PC (900 cps).

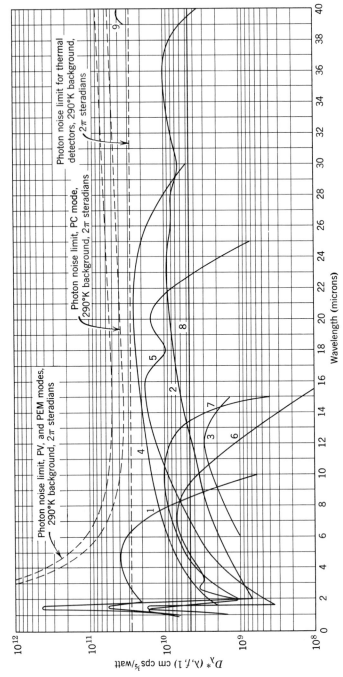

Figure 10.14. Spectral $D_\lambda{}^*$ of detectors operating at temperatures below 77°K. 1. Ge:Au, 65°K, PC (900 cps). 2. Ge:Zn, 4.2°K, PC (800 cps). 3. Ge:Zn,Sb, 50°K, PC (900 cps). 4. Ge:Cu, 4.2°K, PC (900 cps, 60° field of view). 5. Ge:Cd, 4.2°K, PC (500 cps, 60° field of view). 6. Ge-Si:Au, 50°K, PC (90 cps). 7. Ge-Si:Zn,Sb, 50°K, PC (100 cps). 8. NbN superconducting bolometer, 15°K (360 cps). 9. Carbon bolometer, 2.1°K (13 cps).

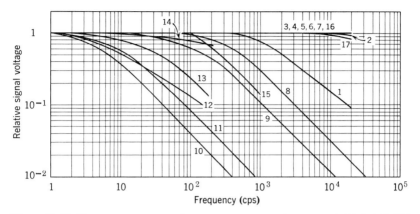

Figure 10.15. Frequency response of room temperature detectors. 1. PbS, PC (250 μsec). 2. PbSe, PC. 3. InSb, PC (estimated). 4. InSb, PEM (estimated). 5. InAs, PC (estimated). 6. InAs, PV (estimated). 7. InAs, PEM (estimated). 8. Tl$_2$S, PC (estimated). 9. Thermistor bolometer (1500 μsec) (estimated). 10. Radiation thermocouple (36 msec) (estimated). 11. Golay cell (20 msec) (estimated). 12. CdS, PC. 13. CdSe, PC. 14. Se-SeO, PV. 15. GaAs, PV. 16. 1P21 photomultiplier. 17. 1N2175 Si photo-duo-diode.

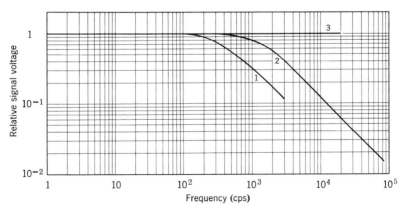

Figure 10.16. Frequency response of detectors operating at 195°K. 1. PbS, PC. 2. PbSe, PC (estimated). 3. InSb, PC (estimated).

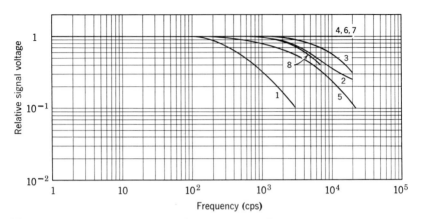

Figure 10.17. Frequency response of detectors operating at 77°K. 1. PbS, PC. 2. PbSe, PC. 3. PbTe, PC. 4. Ge:Au, PC. 5. Ge:Au,Sb, PC. 6. InSb, PC (estimated). 7. InSb, PV. 8. Te, PC.

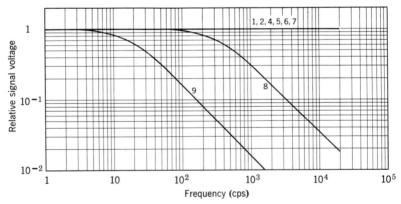

Figure 10.18. Frequency response of detectors operating below 77°K. 1. Ge:Au, 65°K, PC (estimated). 2. Ge:Zn, 4.2°K, PC (estimated). 3. Ge:Zn,Sb, 50°K, PC (unknown). 4. Ge:Cu, 4.2°K, PC (estimated). 5. Ge:Cd, 4.2°K, PC (estimated). 6. Ge-Si:Au, 50°K, PC (estimated). 7. Ge-Si:Zn,Sb, 50°K, PC (estimated). 8. NbN superconducting bolometer, 15°K (estimated). 9. Carbon bolometer, 2.1°K (estimated).

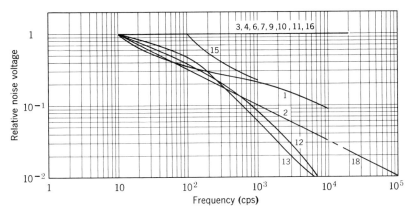

Figure 10.19. Noise spectra of room temperature detectors. 1. PbS, PC (250 μsec). 2. PbSe, PC (assumed). 3. InSb, PC (assumed). 4. InSb, PEM (assumed). 5. InAs, PC (unknown). 6. InAs, PV. 7. InAs, PEM (assumed). 8. Tl$_2$S, PC (unknown). 9. Thermistor bolometer (assumed). 10. Radiation thermocouple (assumed). 11. Golay cell (assumed). 12. CdS, PC. 13. CdSe, PC. 14. Se-SeO, PV (unknown). 15. GaAs, PV. 16. 1P21 photomultiplier. 17. 1N2175 Si photo-duo-diode, PV (unknown). 18. $1/f$ power law.

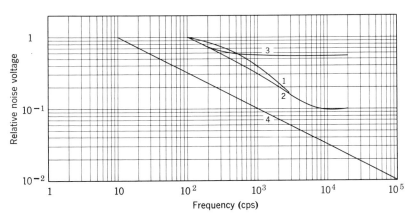

Figure 10.20. Noise spectra of detectors operating at 195°K. 1. PbS, PC. 2. PbSe, PC. 3. InSb, PC. 4. $1/f$ power law

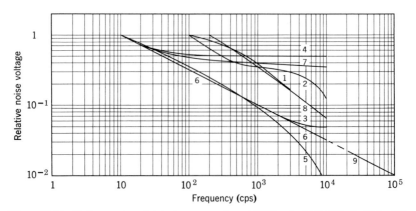

Figure 10.21. Noise spectra of detectors operating at 77°K. 1. PbS, PC. 2. PbSe, PC. 3. PbTe, PC. 4. Ge:Au, PC. 5. Ge:Au,Sb, PC. 6. InSb, PC. 7. InSb, PV. 8. Te, PC. 9. 1/f power law.

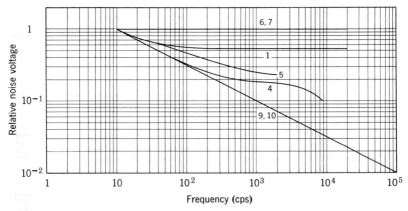

Figure 10.22. Noise spectra of detectors operating below 77°K. 1. Ge:Au, 65°K, PC. 2. Ge:Zn, 4.2°K, PC (unknown). 3. Ge:Zn,Sb, 50°K, PC (unknown). 4. Ge:Cu, 4.2°K, PC. 5. Ge:Cd, 4.2°K, PC. 6. Ge-Si:Au, 50°K, PC (assumed). 7. Ge-Si:Zn,Sb, 50°K, PC (assumed). 8. NbN superconducting bolometer, 15°K, (unknown). 9. Carbon bolometer, 2.1°K (assumed). 10. 1/f power law.

REFERENCES

1. Potter, Pernett, and Naugle, *Proc. Inst. Radio Engrs.* **47**, 1503 (1959).
2. See D. Dutton, *Photoconductivity Conference*, Breckenridge, Russell, and Hahn, eds., John Wiley and Sons, New York (1956), p. 591.
3. T. S. Moss, *Optical Properties of Semiconductors*, Butterworths Scientific Publications, London (1959); *Proc. Inst. Radio Engrs.* **43**, 1869 (1955).
4. R. A. Smith, *Advances in Physics* **2**, 321 (1953).
5. R. J. Cashman, *Proc. Inst. Radio Engrs.* **47**, 1471 (1959).
6. F. Kicynski, *Chem. and Ind.* **17**, 54 (1948).
7. Sosnowski, Starkiewicz, and Simpson, *Nature* **159**, 818 (1947).
8. G. W. Mahlman, *Phys. Rev.* **103**, 1619 (1956).
9. J. C. Slater, *Phys. Rev.* **103**, 1631 (1956).
10. J. F. Woods, *Phys. Rev.* **106**, 235 (1957).
11. E. D. McAllister, Eastman Kodak Co. report, unpublished. See also R. C. Jones, *Advances in Electronics* V, Academic Press, New York (1953), p. 1.
12. For a description of PbSe detectors see *Final Engineering Report*, April 1, 1957 to June 30, 1958, Contract AF 33(616)-5062, Santa Barbara Research Center, Goleta, Calif.
13. Beyen et al, *J. Opt. Soc. Amer.* **49**, 686 (1959).
14. L. D. McGlauchlin, private communication.
15. W. C. Dunlap, *Phys. Rev.* **91**, 1282 (1953); **97**, 614 (1955); **100**, 1629 (1956), *Photoconductivity Conference*, Breckenridge, Russell, and Hahn, eds; John Wiley and Sons, New York (1956), p. 539.
16. H. H. Woodbury and W. W. Tyler, *Phys. Rev.* **105**, 84 (1957).
17. Morton, Hahn, and Schultz, *Photoconductivity Conference*, Breckenridge, Russell, and Hahn, eds; John Wiley and Sons, New York (1956), p. 556.
18. L. Johnson and H. Levinstein, *Phys. Rev.* **117**, 1191 (1960).
19. Lasser, Cholet, and Wurst, *J. Opt. Soc. Amer.* **48**, 468 (1958).
20. H. Levinstein, *Proc. Inst. Radio Engrs.* **47**, 1478 (1959).
21. Burnstein et al, *Phys. Rev.* **93**, 65 (1954).
22. Physics Report 104-3, *Interim Report on Infrared Detectors*, Syracuse University Research Institute (August 1, 1960).
23. Talley, Bode, and Konkel, *Proc. IRIS* **5**, No. 1, 503 (1960). (Paper unclassified).
24. E. R. Johnson and S. M. Christian, *Phys. Rev.* **95**, 560 (1954).
25. Morton, Schultz, and Harty, *RCA Rev.* **20**, 599 (1959).
26. E. Welker and H. Weiss, *Solid State Physics* **3**, Academic Press, New York (1956).
27. H. P. R. Frederikse and R. F. Blunt, *Proc. Inst. Radio Engrs.* **43**, 1954 (1955).
28. D. W. Goodwin, *J. Sci. Instr.* **34**, 367 (1957); Avery, Goodwin and Rennie, *J. Sci. Instr.* **34**, 394 (1957); Nicolosi, DeVaux, and Strauss, *Electronics, Eng. Ed.* **31**, 48 (1958); L. H. DeVaux and A. J. Strauss, Electrochem. Soc. Fall Meeting, (October, 1957).
29. Avery, Goodwin, and Rennie, *op. cit.*; Nicolosi, DeVaux and Strauss, *op. cit.*; Mitchell, Goldberg and Kurnick, *Phys. Rev.* **97**, 239 (1955); Lasser, Cholet, and Wurst, *op. cit.*
30. Nicolosi, DeVaux, and Strauss, *op. cit.*; C. Hilsum and I. M. Ross, *Nature* **179**, 146 (1957); S. W. Kurnick and R. N. Zitter, *J. Appl. Phys.* **27**, 278 (1956); P. W. Kruse, *J. Appl. Phys.* **30**, 770 (1959), *Electronics* **33**, No. 13, 62 (1960).
31. Rieke, DeVaux, and Tuzzolino, *Proc. Inst. Radio Engrs.* **47**, 1475 (1959).

32. T. S. Moss, *Photoconductivity Conference*, Breckenridge, Russell, and Hahn, eds., John Wiley and Sons, New York (1956), p. 427.
33. C. Hilsum, *Proc. Phys. Soc.* **B70**, 1011 (1957).
34. R. M. Talley and D. P. Enright, *Phys. Rev.* **95**, 1092 (1954).
35. P. H. Cholet, *Proc. IRIS* **5**, No. 4, 161 (1960) (Paper unclassified); private communication.
36. G. Suits and P. Rice, Report No. 2144-240-*T*, The University of Michigan Engineering Research Institute (1958).
37. C. D. Butter and L. D. McGlauchlin, *Proc. IRIS* **5**, No. 3, 103 (1960) (Paper unclassified).
38. T. W. Case, U.S. Patent No. 1,301,227, April 22, 1919; No. 1,316,350, Sept. 16, 1919; *Phys. Rev.* **15**, 289 (1920).
39. R. J. Cashman, *J. Opt. Soc. Amer.* **36**, 356 (1946); *Proc. Inst. Radio Engrs.* **47**, 1471 (1959); U.S. Patent No. 2,448,517, Sept. 7, 1948; No. 2,448,518, Sept 7, 1948.
40. Lawson et al, *J. Phys. Chem. Solids* **9**, 325 (1959).
41. E. M. Wormser, *J. Opt. Soc. Amer.* **43**, 15 (1953).
42. R. DeWaard and E. M. Wormser, *Proc. Inst. Radio Engrs.* **47**, 1508 (1959).
43. D. H. Andrews, *Report on International Conference on Low Temperatures*, Cambridge (1946), summarized by Smith, Jones, and Chasmar, *The Detection and Measurement of Infrared Radiation*, Oxford University Press, London (1957).
44. W. S. Boyle and K. F. Rodgers, Jr., *J. Opt. Soc. Amer.* **49**, 66 (1959).
45. Smith, Jones, and Chasmar, *loc. cit.* p. 56.
46. M. J. E. Golay, *Rev. Sci. Instr.* **20**, 816 (1949).
47. R. H. Bube, *Photoconductivity of Solids*, John Wiley and Sons, New York (1960), p. 88.
48. W. L. Eisenman and A. J. Cussen, NAVORD Report 4649 (NOLC Report 404) (Unclassified).
49. Cholet, Slawek, and Repper, *A Solid State Celestial Body Sensor*, Philco Report No. 2224-2; see also Lucovsky and Cholet, *J. Opt. Soc. Amer.* **50**, 979 (1960).
50. Texas Instruments bulletin 15528, *N-P-N Diffused Silicon Photo-Duo-Diode, Type 1N2175*.
51. R. L. Aagard, private communication.
52. For the spectral characteristics of the various photoemissive surfaces, see RCA booklet CRPD-105A, *RCA Photosensitive Devices and Cathode Ray Tubes*, p. 16.
53. R. C. Jones, *Advances in Electronics* V, Academic Press, New York (1953), p. 1; **XI**, Academic Press, New York (1959), p. 87.

List of References for Data in Table 10.1 and Figs. 10.11a to 10.22

1. *Interim Report on Infrared Detectors*, Syracuse University Research Institute, Physics Report 103-6 (Sept. 1, 1958).
2. *Interim Report on Infrared Detectors*, Syracuse University Research Institute, Physics Report 103-7 (Dec. 1, 1958).
3. *Interim Report on Infrared Detectors*, Syracuse University Research Institute, Physics Report 104-1 (Feb. 1, 1960).
4. *Interim Report on Infrared Detectors*, Syracuse University Research Institute, Physics Report 104-2 (May 1, 1960).
5. *Interim Report on Infrared Detectors*, Syracuse University Research Institute, Physics Report 104-3 (Aug. 1, 1960).
6. Bratt et al, *Germanium and Indium Antimonide Infrared Detectors*, Syracuse University Research Institute, Final Report (Feb., 1960).

7. Beyen et al, *J. Opt. Soc. Amer.* **49**, 686 (1959).
8. B. Elsbach, *Properties of Detectors for Infrared Radiation*, Aerojet-General Avionics Division Technical Memorandum TM520:59-10-444 (April 7, 1959).
9. B. Elsbach and E. Banks, *Evaluation of Some Long Wavelength Infrared Detectors*, Aerojet-General Avionics Division Technical Memorandum TM527:60-10-582 (April 27, 1960).
10. W. L. Eisenman and A. J. Cussen, *Properties of Photoconductive Detectors (Photoconductive Detector Series, 31st Report)*, NAVORD Report 4649 (NOLC Report 404) (December, 1957) (Unclassified).
11. Cholet, Slawek, and Repper, *A Solid State Celestial Body Sensor*, Philco Report No. 2224-2 (May 31, 1960).
12. C. Hilsum, *Proc. Phys. Soc.* **B70**, 1011 (1957).
13. Lawson et al., *J. Phys. Chem. Solids* **9**, 325 (1959).
14. R. J. Cashman, *Proc. Inst. Radio Engrs.* **47**, 1471 (1959).
15. R. C. Jones, *Proc. Inst. Radio Engrs.* **47**, 1495 (1959).
16. M. J. E. Golay, *Rev. Sci. Instr.* **20**, 816 (1949).
17. C. D. Butter and L. D. McGlauchlin, *Proc. IRIS* **5**, No. 3, 103 (1960) (Paper unclassified); private communication.
18. P. W. Kruse, *J. Appl. Phys.* **30**, 770 (1959), *Electronics* **33**, No. 13, 62 (1960).
19. Morton, Schultz and Harty, *RCA Rev.* **20**, 599 (1959).
20. W. S. Boyle and K. F. Rodgers, Jr., *J. Opt. Soc. Amer.* **49**, 66 (1959).
21. R. DeWaard, *Proc. IRIS* **3**, No. 1, (March 1958) (Paper unclassified).
22. Smith, Jones, and Chasmar, *The Detection and Measurement of Infrared Radiation*, Oxford University Press, London (1957), p. 112.
23. P. H. Cholet, *Proc. IRIS* **5**, No. 4, 161 (1960) (Paper unclassified); private communication.
24. P. H. Cholet, private communication.
25. W. A. Farber, private communication.
26. Bulletin No. 10, April, 1959, *Golay Infra-Red Detector*, The Eppley Laboratory, Inc., Newport, R. I.
27. Data on PE 536-1 ZIP Detector from Perkin-Elmer Corp., Norwalk, Conn.
28. Bulletin 15528, *Tentative Specification, Type 1N2175 N-P-N Diffused Silicon Photo Duo-Diode*, Texas Instruments, Inc., Dallas, Texas.
29. Data on ORP10 Photoconductive Cell, Mullard Ltd.
30. Bulletin SC 725, *Indium Antimonide Diffused Junction Infrared Detector*, Texas Instruments, Inc., Dallas, Texas.
31. Form No. CRPD-105A, *RCA Photosensitive Devices and Cathode Ray Tubes*, RCA Electron Tube Division, Harrison, N. J.
32. Bulletin No. 58-A, Charles M. Reeder and Co., Detroit, Mich.
33. R. L. Aagard, private communication.
34. Naugle, Allen, and Cussen, *Properties of Photoconductive Detectors, Photoconductive Detector Series, 24th Report*, NAVORD Report 4641 (NOLC Report 396) (November, 1957) (Unclassified).
35. *Typical Characteristics of Infratron Cooled Lead Sulfide Photoconductors*, Infrared Industries, Inc., Waltham, Mass.
36. Bode et al., *Proc. IRIS* **4**, No. 3, 141 (1959) (Paper unclassified); private communication.

Appendices

Some useful engineering approximations

I. RATIO OF $D_{\lambda_p}*(\lambda_p, f, \Delta f)$ TO $D*(500°K, f, \Delta f)$ FOR SELECTED DETECTORS

In Appendix I of Chapter 9 the method for determining the ratio of the spectral $D_{\lambda_p}*$ at the wavelength of maximum response to the $D*$ value for a 500°K black body was discussed. Below are tabulated values of the ratio for certain selected materials. Values for others may be obtained by referring to Table 10.1.

Material, Mode, and Temperature	$\lambda_p(\mu)$	$\dfrac{D_{\lambda_p}*(\lambda_p, f, \Delta f)}{D*(500° \text{ K}, f, \Delta f)}$
PbS, PC, 295°K	2.2	~220
PbSe, PC, 295°K	3.4	~9
PbSe, PC, 77°K	4.5	~5
PbTe, PC, 77°K	4.0	~7
Ge:Au, PC, 77°K	5.0	~2
Ge:Zn, PC, 4.2°K	36	~2
InSb, PEM, 295°K	6.6	~3
InSb, PV, 77°K	5.3	~5
Te, PC, 77°K	3.5	~15

II. APPROXIMATE METHOD FOR CONVERTING THE VALUE OF $D^*(T_1, f, \Delta f)$ TO $D^*(T_2, f, \Delta f)$

On occasion it is required to convert the black body D^* evaluated for a particular source temperature T_1 to the value for some other temperature T_2. The exact expression can be obtained from the spectral D_λ^* by a numerical integration technique similar to that described in Appendix I of Chapter 9. An approximate solution, good if T_1 and T_2 are not widely separated, is obtained through using the concept of equivalent cutoff wavelength discussed in Section 8.2. The signal from the detector is assumed to be proportional to the amount of radiant power it receives in the spectral region to which it responds, that is, at wavelengths shorter than the equivalent cutoff wavelength λ_0, regardless of the spectral distribution of the radiation in that region. Although this assumption is, of course, not true, it does not lead to serious error if we compare the signal obtained when the detector looks at black bodies maintained at temperature no more than a few hundred degrees Centigrade apart. For a given noise level, then, the value of D^*, which is proportional to the signal per unit radiant power, is proportional to the amount of radiant power below λ_0 divided by the total power, that is, D^* is proportional to the fraction of the radiant power from a black body which lies at wavelengths less than λ_0. If we denote this fraction as $p(\lambda_0, T_2)$ for a black body at temperature T_2, we see that $D^*(T_2, f, \Delta f)$ is proportional to $p(\lambda_0, T_2)$. Note that no factor of T_2^4 enters. Therefore, knowing the value of D^* at some temperature T_1, which is usually $500°K$, we obtain that at some other temperature T_2 not far removed from T_1 by

$$D^*(T_2, f, \Delta f) = D^*(T_1, f, \Delta f) \frac{p(\lambda_0, T_2)}{p(\lambda_0, T_1)},$$

where as we have said, λ_0 is the equivalent cutoff wavelength of the detector. The value of $p(\lambda_0, T)$ can be read directly from a radiation slide rule. As an example, consider a detector having an equivalent cutoff wavelength of $7\ \mu$ and $D^*(500°K, 900, 1) = 1 \times 10^8$ cm cps$^{1/2}$/watt. We wish to determine $D^*(600°K, 900, 1)$. Reference to a radiation slide rule show us

$$p(7\mu, 500°K) = 0.38,$$

$$p(7\mu, 600°K) = 0.51.$$

Therefore

$$D^*(600°K, 900, 1) = 1 \times 10^8 \times \frac{0.51}{0.38} \text{ cm cps}^{1/2}/\text{watt}$$

$$= 1.3 \times 10^8 \text{ cm cps}^{1/2}/\text{watt}.$$

III. USEFUL FREQUENCY RANGE OF A DETECTOR

Given a detector having a response time τ, the question arises as to what is the "useful" frequency range of the detector. A somewhat common fallacy is to consider that the responsivity is essentially flat to the frequency which is the reciprocal of the response time, that is, a detector having a 1 μsec response time is flat to 1 megacycle. This is, of course, not true. As we have pointed out in Section 8.2, Eq. 8.8, at a frequency $f = 1/2\pi\tau$ the responsivity $\mathscr{R}(f)$ has already dropped to 0.71 of the value $\mathscr{R}_0$ obtained at low frequencies. For a value $f = 1/\tau$ the responsivity would be about $0.16\mathscr{R}_0$. A convenient approximation is to assume the useful range extends to $f = 1/4\tau$. At this frequency the responsivity is approximately $0.53\mathscr{R}_0$.

Index

Time constant, 274–276, 390
Transmission, of radiation through
the atmosphere, 173, 179–181,
191 ff.
of specific optical materials, 138 ff.
Transmissivity, 13, 100, 103
Transmittance, 138–139, 143
Traps, 220

Valence band, 198
Vibration-rotation spectra, 63, 68
Vidicon, 291
Visual range, 163, 179, 190–192
Vycor, 145

Water vapor, absorption of, 163, 179–
181
Wave equations, 90 ff.
Wavelength distribution of radiation,
11, 20
Welsbach mantle, 73
Wien displacement law, 29–31, 84
Windows, atmospheric, 177–181
detector, 128–129, 398–399
requirements for, 128

Xenon flash tube, 74

Zinc sulfide (Irtran-2), 140–141, 153